Museum Informatics

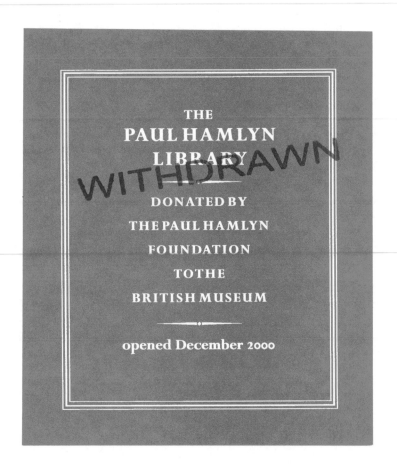

Routledge Studies in Library and Information Science

Museum Informatics

People, Information, and Technology in Museums

Paul F. Marty

and

Katherine Burton Jones

Routledge
Taylor & Francis Group

New York   London

Routledge
Taylor & Francis Group
270 Madison Avenue
New York, NY 10016

Routledge
Taylor & Francis Group
2 Park Square
Milton Park, Abingdon
Oxon OX14 4RN

© 2008 by Taylor & Francis Group, LLC
Routledge is an imprint of Taylor & Francis Group, an Informa business

Printed in the United States of America on acid-free paper
10 9 8 7 6 5 4 3 2 1

International Standard Book Number-13: 978-0-8247-2581-5 (0)

Library of Congress Cataloging-in-Publication Data

Museum informatics : people, information, and technology in museums / edited by
   Paul F. Marty and Katherine Burton Jones.
      p. cm. -- (Routledge studies in library and information science)
   Includes bibliographical references and index.
   ISBN-13: 978-0-8247-2581-5 (hardback : alk. paper)
      1. Museum techniques. 2. Interactive multimedia. 3. Museums--Data processing.
I. Marty, Paul F. II. Jones, Katherine Burton.

AM215.M87 2007
069'.1--dc22   2007010937

**Visit the Taylor & Francis Web site at
http://www.taylorandfrancis.com**

**and the Routledge Web site at
http://www.routledge.com**

Contents

Figures and Charts

Preface

Today, in the early 21st century, there is tremendous interest in how people, information, and technology can work together to enhance the museum experience for museum professionals, museum visitors, and all users of museum resources. That so many people are interested in this topic is due in no small part to the contributions of several books covering different aspects of museums and information technology, including:

- *The Wired Museum: Emerging Technology and Changing Paradigms*, edited by Katherine Jones-Garmil (American Association of Museums, 1997);
- *The Virtual and Real: Media in the Museum*, edited by Selma Thomas and Ann Mintz (American Association of Museums, 1998);
- *Digital Collections: Museums in the Information Age*, by Susanne Keene (Butterworth-Heinemann, 1998);
- *Information Management in Museums*, by Elizabeth Orna and Charles Pettitt (Gower, 1998); and
- *Theorizing Digital Cultural Heritage,* edited by Fiona Cameron and Sarah Kenderdine (MIT Press, 2007).

Despite the valuable contributions of these publications, as well as the conference proceedings and academic journals which increasingly help museum researchers and practitioners remain engaged with issues about museums and information technology, few manuscripts have provided a comprehensive look at the role of museums in the information society. In particular, there is a need for an approach that merges current theoretical and practical perspectives about people, information, and technology in museums, and does so in a way that is accessible to museum researchers and professionals worldwide. The goal of this book, therefore, is to offer a unique perspective on these topics that when combined with the works listed above will provide fresh, complementary insights for the museum community.

Museum Informatics explores the sociotechnical issues that arise when people, information, and technology interact in museums. It is designed specifically to address the many challenges faced by museums, museum professionals, and museum visitors in the information society. It examines not only applications of new technologies in museums, but how advances in information science and technology have changed the very nature of museums, both what it is to work in one, and what it is to visit one. Museum professionals face new information management challenges, from making decisions about content management software to selecting metadata standards for sharing information with other institutions. Museum visitors come with new expectations about the resources museums should provide, asking questions that rely on an unprecedented level of information access and expecting most of the answers to be available online. These issues reflect just some of the new needs, challenges, and expectations that have arisen as museums adapt to their new role in the information society.

To explore these issues, this book offers a selection of contributed chapters, written by leading museum researchers and practitioners, that cover the latest scientific and technological advances relevant to museums and information resources. The chapters are arranged into seven sections, each covering a different theme or concept fundamental to museum informatics. Each section opens with an introductory chapter that sets the stage for the contributed chapters and provides a continuous narrative about museum informatics that connects the book's chapters into one integrated whole. The contributed chapters alternate between theoretical chapters discussing the latest research on a variety of museum informatics topics, and practical chapters providing significant examples and detailed case studies useful for museum researchers and professionals. The seven sections of the book are arranged as follows:

The book opens with an introduction to museum informatics, discussing the emergence of museum informatics as an interdisciplinary field of study and examining the origin, nature, and transformation of the digital museum.

The second section explores the role of information representation in museums, analyzing the methods by which museums represent the knowledge inherent to their collections, and examining how museums have managed the information revolution in their own environments.

The third section covers information organization and management in museums, exploring how museums establish and promote information policies that connect museum resources and their users, and discussing the metadata structures that govern access to museum information.

The fourth section analyzes the growing importance and integration of interactive technologies into museum exhibits, exploring the ways in which new technologies have revolutionized the museum visitor experience, and examining the potentials for new technologies to blur the boundaries between in-house and online museum experiences.

The fifth section explores the changing needs and expectations of museum audiences, discussing the importance of understanding and evaluating the needs of museum visitors, in-house and online, and examining the potential of meeting new needs in museum education by connecting online resources to classroom environments.

The sixth section discusses the role of collaborations and consortia for meeting shared museum goals, providing examples that illustrate the challenges faced by museums when creating and sustaining museum collaborations, as well as the benefits that cooperative activities offer to participating museums.

The final section examines the future of museums and museum professionals, exploring the meaning of museum information professionals curating information resources on the cyberinfrastructure and exploring the future of museums in the information society.

In this way, *Museum Informatics* takes a fresh look at the sociotechnical interactions that occur between people, information, and technology in museums. It is our sincere hope that this book presents its information in a format accessible to multiple audiences, including academic researchers, students interested in museum careers, museum professionals, and museum visitors.

Acknowledgments

We would like to acknowledge the assistance of the students enrolled in the Museum Informatics (LIS 5590) course in the College of Information at Florida State University, and the Museums and the Web (MUSE E130) course offered through the Museum Studies program at the Harvard Extension School. The students in these courses were instrumental in testing the suitability of drafts of this manuscript for instruction in both Museum Studies and Library and Information Science programs. Thanks are also due to the teaching assistants for these two courses, Jason Springer and Adam Rozan at Harvard, and Kim Thompson at FSU. Finally, we would like to acknowledge the special assistance provided by Nancy Enterline, a recent graduate of the Florida State University College of Information, who served as the research assistant for this project.

Paul F. Marty
Katherine Burton Jones

Section 1

Introductions

1 An Introduction to Museum Informatics

Paul F. Marty

Florida State University

Museum informatics is the study of the sociotechnical interactions that take place at the intersection of people, information, and technology in museums. Researchers and professionals interested in museum informatics have spent years exploring the impact of information science and technology on the people who use museum information resources (Marty, Rayward, & Twidale, 2003). As these users become more information-savvy, and their information needs and behaviors become more complex, both museum professionals and museum visitors have had to adapt to the new role of museums in the information society.

As an example of the sociotechnical interactions that comprise the study of museum informatics, consider the following episode which took place as a university museum of world cultures moved its collections from old to new facilities (Marty, 2000b; cf. Marty, 2002). From 1998 to 2002, staff members at the University of Illinois' Spurlock Museum inventoried, packed, and moved 45,000 artifacts, while simultaneously working with architects and exhibit designers to develop and install all new exhibits and storage facilities. During this process, they stumbled across a wide variety of information management problems, most of which had lain hidden for decades in the museum's records, ledgers, and files.

One problem had been created in 1971, when two similar-looking, yet not identical, African figures were accessioned into the museum's collections, and mistakenly assigned the same accession number. For nearly thirty years these figures remained in storage, and were only removed when needed by a student, scholar, or other individual. According to the museum's records, only one figure existed, and whenever this "figure" was removed from storage, it was a gamble as to which figure would actually be found and removed. As the museum's records migrated from ledger files, to card files, to electronic databases over the years, the inconsistency of two artifacts sharing one record was neither detected nor corrected.

During this time, however, the various paper and electronic records for this "figure" began to accumulate certain oddities: official descriptions included minor yet striking contradictions; official photographs did not quite match corresponding textual descriptions; and so on. While occasionally remarked upon by visiting researchers, this problem remained unsolved until the year 2000, when "the figure" was packed into two separate boxes in preparation for the move. When the "second" figure was packed and its location entered into the museum's information systems, an error immediately registered, informing the registrar that this artifact was "already packed" in a different box. Opening both boxes, the registrar discovered that this "single artifact" was actually two slightly different objects, and that corrections had to be made in the museum's ledgers, files, and database systems.

For nearly thirty years, the museum's social systems (those that facilitated interactions between museum staff, students, and scholars) conspired with the museum's technical systems (those paper and electronic records and databases that monitored objects, information, and their use) to keep this problem (and many similar ones) hidden. By the end of the 20th century, however, these same sociotechnical interactions were taking place within a new information environment specially attuned to the unique needs of the museum, where such difficulties were less likely to remain hidden, and more likely to become a visible part of the museum's information infrastructure (Marty, 1999). This transition, along with the advances in information science and technology that made it possible, lies at the heart of museum informatics.

As this story illustrates, museums are in the midst of an information revolution, and museum professionals must work with a variety of different information resources, from the museum's collections themselves, to information about those objects, to information about the contexts in which those objects are displayed, studied, or interpreted (Knell, 2003). The past few decades have seen an important shift from the idea of museums as repositories of objects to museums as repositories of knowledge (Cannon-Brookes, 1992; Hooper-Greenhill, 1992; cf. White, 2004). The museum has truly become an information utility, where the need to provide access to information about objects, in addition to the objects themselves, is well known (Fahy, 1995; MacDonald, 1991; Washburn, 1984).

Museum professionals, naturally, have a lengthy history of working with information resources, tools, and technologies, and information management skills have always been important for museum professionals (Lord & Lord, 1997; Orna & Pettitt, 1998; Zorich, 1999). The traditional careers of museum librarian or registrar provide exemplars of museum positions that require extensive experience with information resources, and the information management skills these individuals bring to the museum are well-documented (American Association of Museums, 1994; Burkaw, 1997; Case, 1995; Danilov, 1994; Glaser & Zenetou, 1996; Koot, 2001; Reed & Sledge, 1998; Schwarzer, 2001a). Yet the challenges of museum informatics are not the same as those posed by museum librarianship or museum registration. While there is considerable overlap between these disciplines, museum informatics represents a unique field of study with its own, substantially different, required educational background and career path (Marty, 2005).

Museum informatics is an extremely interdisciplinary field of study. To meet the challenges of the museum's changing role in the information society, researchers studying museum informatics have drawn upon theories and techniques from dozens of related fields, including digital libraries, human–computer interaction, social network analysis, cognitive science, museum studies, library and information science, etc. While much early work in this area was primarily focused on questions of how information technologies should be used in museums, a number of researchers and professionals are now emphasizing the need for an underlying body of theory and methods for studying museum informatics as well as related areas such as museums and new media or digital cultural heritage (Cameron & Kenderdine, 2007; Parry, 2005).

To explore new theoretical perspectives and to develop new methodologies, researchers and professionals from around the world have joined together to form an evolving community of practice. This community is dedicated to providing guidance to museums and other cultural heritage institutions as they ask and answer important questions about museum informatics and its significance for museums. While the history of this community dates at least back to the 1960s (Ellin, 1969; Vance, 1975), the number of individuals involved with museum informatics has increased dramatically in the past decade. A widespread interest in museum informatics can now be found in a variety of arenas, and thousands of people worldwide now participate in discussions, projects, and research initiatives related to museum informatics.

Each year, an ever-increasing number of museum professionals and researchers join professional organizations and attend conferences dedicated to exploring museum informatics, including the meetings of such organizations as the Museum Computer Network, the Museum Documentation Association, the International Council of Museum's International Committee for Documentation, the International Cultural Heritage Informatics Meeting, Museums and the Web, and the Institute of Museum and Library Services' WebWise Conference. In addition, past organizations such as the Consortium for the Computer Interchange of Museum Information (CIMI) maintain archival websites of important documents for individuals interested in museum informatics issues.

A growing number of academic and professional journals publish papers about museum informatics, including past journals such as *Archives and Museum Informatics,* and current journals such as *Spectra*, a publication of the Museum Computer Network. The *Journal of the American Society for Information Science* published a special issue on museum informatics and the Web in 2000; *Curator* published a special issue on technology and museums in 2002; and the *Journal of Digital Libraries* published a special issue on digital museums in 2004. In addition, there are a number of books covering related topics, including Thomas and Mintz (1998), *The Virtual and Real: Media in the Museum*, Keene (1998), *Digital Collections: Museums in the Information Age*, Orna and Pettitt (1998), *Information Management in Museums*, Jones-Garmil (1997), *The Wired Museum: Emerging Technology and Changing Paradigms*, and Cameron and Kenderdine (2007), *Theorizing Digital Cultural Heritage*.

The researchers and professionals who attend these conferences, belong to these organizations, and publish in these journals and books, are interested in a wide variety of topics, such as integrated information management systems (Blackaby, 1997), virtual museums (Schweibenz, 1998), open source data standards (Perkins, 2001), and copyright and intellectual property (Zorich, 2000). They have explored methods of documenting and describing museum artifacts and have developed data standards that allow museum professionals to share information about museum collections between different institutions (Bearman, 1994; Gladney et al., 1998; Moen, 1998). They have studied the use of interactive, multimedia exhibits in museum galleries, the potential of online, virtual museums to expand the educational reach of the museum, and the impact of educational outreach programs designed to bring museum information resources into the classroom over the Internet (Economou, 1998; Frost, 2001; Rayward & Twidale, 2000; Teather & Wilhelm, 1999). They have studied the information needs of museum visitors and explored different methods for determining whether museum websites meet the needs of their users (Chadwick & Boverie, 1999; Cunliffe, Kritou, and Tudhope, 2001; Dyson & Moran, 2000; Kravchyna & Hastings, 2002; Thomas & Carey, 2005).

While a complete summary of museum informatics research topics is beyond the scope of this essay, even a short list of some recent projects can indicate the range of research conducted in this area and the breadth of individuals conducting this research. University researchers have studied how the ability to help museum visitors conceptualize information resources transformed the way museum professionals build relationships with their users (Cameron, 2003), and explored different methods of targeting individual user needs through personalization and pervasive computing technologies inside and outside the museum (Bowen and Filipinni-Fantoni, 2004). Museum professionals have looked at how different metadata schemas help or hinder users seeking collections data from museums (Coburn & Baca, 2004), and examined the changing expectations for online museums engaged in outreach to many different audiences (Hamma, 2004a). Researchers from IBM have studied digital imaging at the Vatican Museums (Mintzer et al., 1996), invisible watermarking at the Hermitage Museum (Mintzer et al., 2001), and pervasive computing in Egyptian national museums (Tolva, 2005). Researchers from Xerox PARC have developed new electronic guidebooks and shared listening devices for museum visitors, examining the impact of such technologies on the museum visit (Aoki & Woodruff, 2000; Woodruff et al., 2002).

Recently, there has been a growing trend to focus more on how new technologies affect the social relationships that occur inside and outside of the museum. In particular, there has been an increased focus on the information needs, seeking, and behavior of the typical users of museum information resources, both in the museum (Booth, 1998; Evans & Sterry, 1999; Galani & Chalmers, 2002; Schwarzer, 2001b) and online (Goldman & Schaller, 2004; Ockuly, 2003; Sarraf, 1999). By understanding the information needs of museum visitors, museum professionals can better serve their clientele from a variety of perspectives (Müller, 2002; Zorich, 1997). By evaluating the steps they are taking to meet these needs, museum professionals can help ensure a positive relationship between museums and museum visitors (Gillard & Cranny-Francis, 2002; Harms & Schweibenz, 2001; Hertzum, 1998; Streten, 2000).

These and other excellent research initiatives have not only dramatically increased our knowledge of museum informatics, but have also highlighted the need for an increased focus on museum informatics research in general. This need comes at a time when museums are in a state of constant upheaval with respect to their use of information technologies and the development of their sociotechnical activities. Traditional methods of information organization and access in the museum have given way to newer, more modern systems for information storage and retrieval (Doty, 1990). Accession cards and ledger files, once the primary media for storing information about museum artifacts, have been replaced by electronic databases and online public access catalogs, offering museum professionals the potential to gather more detailed information about their collections, and museum visitors greater access to the information they desire (Buck & Gilmore, 1998). Information about the objects, topics, and cultures found in the museum is now as important as the museum's collections themselves (Pearce, 1986).

These changes have not only influenced the way people think about museums; they have had a profound impact on the sociotechnical interactions that take place in museums. Museum professionals, including registrars, curators, and conservators, go to work armed with new tools for managing their unique information resources. Museum visitors, whether they are visiting the museum in person or over the Internet, have new methods of learning more about the museum and its collections. Museum users of all types, from scholars to students, have new ways of accessing and manipulating the museum's information resources. As a result, museum researchers and professionals have developed new conceptions of why museums exist and new expectations of what museums should offer. Today, it is virtually impossible to discuss museum technologies without touching in some way on how these technologies will affect all users of museum information resources, in-house and online.

When viewed from this perspective, it can be argued that the purpose of studying museum informatics is to examine the issues museum professionals and visitors face as they take advantage of advances in information science and technology while realizing that these issues exist within complex and interlocking organizational and social contexts affecting the nature of museums in general and the expectations of museum professionals and visitors in particular. The relationship between museums, museum professionals, and museum visitors is constantly evolving in response to the changing demands and problems of information organization, access, management, and use in museums. If museums are to remain relevant in the information society, museum professionals and researchers will need to embrace the growing role of museum informatics in the 21st century museum and continue to explore the sociotechnical implications of people, information, and technology interacting in museums.

2   The Transformation of the Digital Museum

Katherine Burton Jones

Harvard Divinity School

INTRODUCTION

This history of computing in museums is the history of the application of computing technologies to the work of museum professionals beginning with the work done by curators and registrars in the mid- to late 1960s. This chapter describes work that is largely about the manipulation of museum information—for the purposes of tracking and identifying objects in museums of all kinds, for cataloguing and displaying this information, for accessing information about objects, and for interpreting and telling the stories of the objects in museums. The history of museum computing is not about the development of computing machinery but is the story of how many museum professionals were able to think beyond the constraints of the day and use computing machines and emerging technologies in the service of the museum field.

This chapter discusses the catalysts for change in moving museums toward a digital age focusing primarily on developments in technology. The beginnings are covered here only briefly as the decades of development have been discussed in detail by Anderson (1999), Jones-Garmil (1997a), Light, Roberts and Stewart (1986), Sarasan and Neuner (1983), and others. The last decade, from 1994 to the present has seen a more rapid change, and more acceptance and use of the emerging technologies. This era will be the focus of later sections of this chapter.

As museums moved toward more broad and ubiquitous information sharing and the development of new audiences, they were transforming in other ways. This shift from museums providing little access to information for the public to providing much greater access—a shift toward transparency of information about collections and research—will also be emphasized.

One is tempted to break the history of museum computing into reports on each decade. This was the approach taken in *The Wired Museum* (Jones-Garmil, 1997b), but many of the transformations enabled by technologies do not fit so neatly into "decade buckets." The sections in this chapter will look at the various transformations that museum have made over the last

40 years, discussing what led to the transformation, and the opportunities or barriers in each case.

ORGANIZING OUR THOUGHTS AS WE ORGANIZED OUR INFORMATION

David Vance (1986) describes two parallel developments—one at the Smithsonian's National Museum of Natural History (NMNH) and one at the Institute for Computer Research in the Humanities (ICRH) in 1963 that led to the introduction of data processing systems in museums. At NMNH the Automatic Data Processing Committee was formed and the museum spent the next few years ramping up personnel and creating information systems, first SIIRS and then SELGEM, Self Generating Master (Creighton & Crockett, 1971). At ICRH, Professor Jack Heller created a system that came to be called GRIPHOS, General Retrieval and Information Processor for Humanities-Oriented Studies, which was initially used for and funded by the United Nations Dag Hammarskjold Library (Heller, 1973).

These two systems, GRIPHOS and SELGEM, were among the first database management systems used in museums. Vance (1986) writes: "For a decade literature on United States museums and computers gave systems more attention than the information they process" (p. 38). This was, as one might expect, due to the "new-ness" of the idea of organizing museum information electronically. Museum professionals needed some time to accommodate this somewhat radical shift or transformation in the way we worked with information.

Organizing Information

The first transformation was that of moving museum information from paper to electronic formats. In many cases our information was systematized and ready for this conversion. For years museum accession cards had collected information in a formatted or fielded manner that could translate to such an environment. In many museums a unique and logical structure for numbering objects had been adopted. One example of this is described by David Vance (1986), noting the efforts of several New York museums in 1966 (and thereafter) and the beginnings of what would become the Museum Computer Network. It is instructive to use this effort as an example of what was happening at that time. During a visit to the Metropolitan Museum of Art Jack Heller was shown the museum's card catalogue, "housed then as now in a bank of wooden file drawers one city block in length . . . Their combined information value must be unimaginable. Yet very little was accessible save by inventory number" (Vance, 1986, p. 40). Impressed by this lack of access, Heller wrote a proposal for $2.5 million to create a Fine Arts Data Bank. Vance (1986) notes that "New York museum directors (including some not of art museums) met at the Whitney Museum of American Art in February and March of 1967, to study the proposal. There they formed a 'Museum Computer Network' (MCN) to plan and seek funds" (p. 40). Funding efforts were successful.

Vance (1986) goes on to describe the process of bringing museums in New York together to discuss the creation of a trial databank. This was the beginning of one of the first union catalogs of museum data. It was not without problems even though the participating museums were using similar card systems: there was no common definition of the data categories used. The system used for this project was GRIPHOS. The structure of the system was flexible, allowing data values of any length and multiple instances of these values, using a system of tags and end of field/end of record delimiters. Vance and other participants in this "experiment" found early on that the strongest need was for standardization. Vance (1986) notes, "Our experiment could not have continued if a high degree of natural, unplanned consistency had not been found to underlie the apparent babel" (p. 41).

Use of the system outside of these pioneer museums began in the early 1970s. One of the earliest adopters of the "MCN methods and the GRIPHOS programs" (Vance, 1986, p. 41) was the Arkansas Archaeological Survey. The pioneer effort for use in the archaeological community was headed by Robert Chenhall. Chenhall and others worked with Vance to create a users guide and thesaurus for the GRIPHOS system. Vance in hindsight is somewhat critical of these early efforts but I think he underestimates the value of these efforts toward standardization. While it is true that use of GRIPHOS ceased in the early 1980s, the seeds for standardization of data fields and terminology had been sown in the art museum community and beyond.[1] SELGEM has a particularly long life in natural history museums and collections, with use documented in major institutions as recently as 1999.

Chenhall and colleagues continued the discussion through the organization of the Museum Data Bank Coordinating Committee (MDBCC), which met regularly from 1971 to 1975. The proceedings of this Committee were published and form a body of early thinking on computing in museums.[2] From the mid-1960s to the 1980s, museum professionals like Robert Chenhall, David Vance, and Everett Ellin, met to discuss how to organize information about objects in these systems. Other early efforts are well chronicled by Vance (1986) and in Light, Roberts, and Stewart (1986), Sarasan and Neuner (1983), and Jones-Garmil (1997a).

Putting Our Hands on and Minds around Our Data

The next transformation is that of computer architecture: moving from mainframe computers to minicomputers to desktop computers and then to client-server models, as museums were able to take advantage of market trends toward lower-cost computing. More importantly and equally transforming, it gave museum professionals, especially the early adopters of GRIPHOS and SELGEM, control over the development of the database systems as well as their data.

The tangential trend was that the need for museum information management sparked the development of a software market aimed at the museum community. Database engines such as dBASE, Advanced Revelation, and the like became the basis of commercial products designed solely for that museum market. Some of these products had the advantage for the museum community of being developed within a museum, by or with the advice of museum professionals. Some had the advantage of being the second generation system used by a museum, thus benefiting from awareness of data fields, data values, and standardized terminologies based on systems like *Nomenclature for Museum Cataloguing: A System for Classifying Man-Made Objects* (Chenhall, 1978) or the Art and Architecture Thesaurus (AAT).[3]

OUR TECHNICAL TRANSFORMATION

MCN was just one of several organizations that formed during the early days of museum computing; however, it has been one of the few that has remained a viable force. During the 1970s the museum community itself began to change. Through the efforts of the American Association of Museums (AAM) professional standing committees were formed for each area of museum work. For example, there are professional committees for registrars, exhibition designers, curators, and so on. It is noteworthy that it was the area of registration that first made an effort toward digitization of museum collections, and with good reason. Registrars are charged with the legal responsibility for the stewardship of objects in the museum; whether the object is on loan, purchased or acquired in some other manner, and in storage, on display, or on loan to another institution. The job of Registrar was one of the first to take advantage of the possibilities of museum computing. It was this group that began to think about standard terminology for describing museum objects. It is this group that provides the nexus for information on the objects in the collections; working with donors, shippers, exhibitors, and just about anyone who touches the object from the time it enters the museum. Registrars were among the first museum professionals to work with system developers to create the first generation of the collections management systems that we use today in museums.

As the AAM moved U.S. museums in new directions, through the reports on *Museums for a New Century* (1984) and *Excellence & Equity: Education & the Public Dimension of Museums* (1992), other museum professionals began to see the value of digital information and digitized collections in their work as well as the possibilities of using digital media as well. The efforts of the AAM, MCN, and the Getty moved museum professionals toward the digital museum as did the opportunities afforded by commercially-developed and supported systems and increasingly affordable and ubiquitous desktop computers.

DIGITAL IMAGING, MULTIMEDIA, AND THE WEB

The use of technology in museums has changed the way we work from its inception. Not only have we needed to learn about and define standards for describing objects in our care electronically, we have had to learn about the technologies. We have, by necessity, added skills like database programming, HTML coding, image capture and enhancement, and multimedia production to our list of core competencies. The need for this new skill set touches some jobs more than others. Registrars are more likely to have learned about database structures and programming, while educators have learned to create Web sites and sophisticated media productions. While some of this work has been outsourced, museum professionals need to understand technical concepts in order to direct the project.

The Impact of Technology on Jobs in the Museum

Beginning in the 1990s information and communication technologies (ICT) were introduced into the workplace. Email is the most notable of these and is the most ubiquitous with the Internet running a strong second since at least 1998. Few museums (or individuals) could operate without email. Communications to colleagues, vendors, donors, and members rely heavily on email. This technology like all others came with a need for support. Some museum professionals developed skills as system administrators; in many cases, however, museums decided to outsource the support for this service.

In *The Wired Museum* the general lack of technical expertise in museums in the areas of imaging and multimedia was noted. To remedy this it was suggested that there should be "changes in the curriculum in museum studies programs and the development of true career-track positions in the management of museum automation systems" (Jones-Garmil, 1997a, p. 60). We would add to this the need for technology training as a core competency.

To measure what was being done to train museum professionals, Margarida Loran[4] (1999) conducted a survey to answer the following questions:

> What are the current training needs in information technology in the museum sector? Are museums studies and mid-career training programs responding to these needs? If so, in what measure? If not, why not? Where are museum professionals learning what they need to know? (p. 2)

Loran's findings showed the following:

- There is a general positive climate toward providing IT training in the programs surveyed.
- Program directors rated the importance of providing IT training very high, and recognized widely the importance of the role of IT in museums/cultural institutions.
- The training provision in IT is still in an initial stage. There is a general need to better tailor this training to the needs of museums and cultural heritage institutions.
- Deterrents were noted including the lack of appropriate faculty and the lack of computing resources. Directions and assistance from the museum/cultural heritage community would also be necessary.

She found that most programs were best prepared to support the job profile of Web manager. They could also support information manager, multimedia developer, and database administrator.

If we look at the Museum Studies program[5] at the Harvard Extension School, we note that there is only one course offered on the use of technology in museums. However, the School offers a wide range of information technology courses that can be pursued for credit in the program. It seems more likely that the in-depth technology training could be supported by programs like the one at Harvard University where students can mix courses across disciplines to meet the sometimes unique needs[6] of museum professionals.

Unlike students of library science, museum professionals have not had a course of study that would fully prepare them to use technology in their work. Perhaps only with this book as a beginning synthesis can we begin the field of study called Museum Informatics.

In addition to new technical skills, museum professionals need to learn about managing intellectual property. Registrars again were among the first to realize this need and were somewhat prepared for it. Registrars[7] were traditionally responsible for managing the legal aspects of objects in the collection—from the time of donation or purchase to the use of images of objects in exhibitions and in print publications at their own institutions as well as by others. Registrars extended this knowledge into the digital world as well. They are often responsible for negotiating or at least participating in the negotiation of legal contracts for the use of digital assets of the museum.

Registrars, however, are not as likely to create content or sophisticated multimedia presentations as are educators and the technical staff of the museum. Thus, all museum professionals need to better understand laws governing copyright and fair use. Many of the media productions developed by museums, whether CD, DVD, or on the Web, involve licensing content from other organizations and individuals. We need to develop policies to protect the intellectual property of our own organization and understand the parameters of use set forth by our sister institutions.

As museums developed sophisticated media products or participated in the development of these through consortia, we realized how critical licensing of this material was. Historically the Getty Information Institute (GII) (known earlier as the Getty Art History Information Program or AHIP) played a significant role in research on what might be done to bring content holders (museums) and institutions of higher education into discussions and then partnerships in this area. In 1995 the Museum Educational Site Licensing project was launched with funding from the Getty Information Institute. In 1999, Eleanor Fink, the director of the GII, summarized the project in *The Getty Information Institute: A Retrospective*. She noted:

> In addition to ensuring a voice for the arts and humanities in planning global networks and addressing research needs to help policy makers and foundations direct support for arts and humanities computing, other challenges commanded GII's strategic planning: inadequate understanding of, and the lack of agreements related to, issues surrounding Intellectual Property Rights (IPR) were major barriers to universal access to images and art information. In response, GII launched the Museum Educational Site Licensing project (MESL), a three year initiative aimed at demystifying IPR by exploring and identifying educational uses of digital images and recommending models for site licensing. MESL participants included seven museums and seven universities. The museums provided a testbed of digital images to the universities that was made available over campus networks. Two publications summarize the findings of MESL: *Images Online: Perspectives on the Museum Educational Site Licensing Project* and *Delivering Digital Images: Cultural Heritage Resources for Education*. As a pioneering initiative designed to increase understanding of the mechanisms needed to support digital libraries or virtual databases, MESL inspired two independent business models: the Art Museum Image Consortium (AMICO) and the Museum Digital Licensing Collective (MDLC). (Establishing a Voice section, ¶8).

A number of museums participated in the AMICO[8] project for many years and now with its multiple successor projects, RLG's CAMIO and H.W. Wilson's Art Museum Image Gallery. ARTstor, a Mellon Foundation-funded project, has created an image archive product based on some AMICO participant collections and other museum collections and resources as well.

Digital Imaging

The antecedents to digital imaging in museums were the various projects that developed multimedia products using analog video disks. MOMA, the Peabody Museum at Harvard, the George Eastman House, and the Southwest Museum were among the museums that created projects with this technology. These projects recorded massive amounts of information; some disks holding as many as 54,000 images. Rus Gant worked with Dan Jones and Melissa Banta at the Peabody to transfer about 2,000 images from the slide collection covering a wide range of subject matter. Gant (personal communication, January 6, 2006) notes,

> They were put on a video disc along with their matching data record on the alternating frame. We also built a link to a database in "C" which would do a variety of searches and return either the image or the image and visual version of the data record . . . This was the first Vision Machine and featured multiple monitors which could display both the data record and the video image. We built a number of demos which showed how the data could be linked to maps and other graphics. We also took some of the objects and did 360 degree rotations which we then gave the user control over to rotate the object at will.

Once digital imaging became affordable for museums, this technology was abandoned and, unfortunately, in most cases, the museums were not able to repurpose the images.

Industry advances in the late 1980s and in the 1990s made it possible for museums to begin to incorporate digital images into collections management systems. This was based on scanning technologies that reduced the amount of heat and light generated per scan making it safer for light sensitive works of art on paper, for example, to be reformatted. Changes in the capability of databases to accommodate the storage of digital images in or associated with the text record describing the object moved museum collections management systems from text-only to images and then other types of digital media (audio, video, etc.). As in the other transformations, museum professionals needed new skills to help in understanding digital imaging and standards for the creation, documentation, and preservation of these images and other media. The Getty Information Institute, the Museum Computer Network, and Archives and Museum Informatics each had a role in promoting and developing standards, teaching museum professionals, and supporting collaborations, product development, and generally, furthering the understanding of the potential of digital imaging in museums. Digital imaging was an essential piece in moving from visually starved collections databases to those that allowed for a full range of media, as well as shifting from the use of those databases solely to store tracking information to databases that could supply richer information that could be used as a resource for the development of new media products.

Multimedia

Building on databases and the use of digital images, museums soon moved to creating multimedia products based on visual and textual content. Early efforts in this area were carried out by the National Museum of American

Art, the Barnes Collection (with support from Microsoft), the National Gallery of Art, and many others. An early and forward thinking project was developed by the Michael C. Carlos Museum at Emory University under the direction of Maxwell Anderson, funded by a Lila Wallace–*Reader's Digest* Museum Collections Accessibility Initiative grant. The museum worked with Georgia Tech's Interactive Media Technology Center (IMTC) to develop interactive programs presented on kiosks in the museum. The goal was to present in-depth information about several objects in the museum including a bat flute and a jaguar vessel, both from Mesoamerica. Visitors to the museum could navigate through the kiosk screens through menus or maps.

One segment of the presentation shows the bat flute in the hands of Mexican ethnomusicologist Antonio Zepeda who played the flute under controlled circumstances. The museum recorded a video sequence that shows the conservator taking the flute from storage, preparing it to be played, and then cleaning it and returning it to storage. This allowed the museum to explain the work of the conservator as well as allowing the viewer to hear the flute as it was played. This project was one of the first to use QuickTime VR (QTVR).

The project was managed by Elizabeth Hornor, Director of Education at the Carlos Museum. She and the consultants from IMTC planned and implemented a Web version of the kiosk and developed a database for the 15,000 works of art in the museum. Efforts by the museum in bringing works of art to life continue through the Odyssey On-Line project (http://carlos.emory.edu/ODYSSEY/).

Pioneering the Internet

Museum professionals began to get a taste of what was possible electronically in the late 1980s and early 1990s and became hungry for more. The Internet that we use today had its beginnings in the U.S. Department of Defense's Defense Advanced Research and Projects area (DARPA) in the late 1960s. In the late 1980s the network that was known as ARPA-NET was split into two segments: a private military-only network and another to cover the rapidly expanding use for research, business,and personal traffic that was appearing on ARPA-NET. The Internet was further augmented by the National Science Foundation network in 1986. (Wallace & Jones-Garmil, 1994).

Significant legislation was passed by the U.S. Congress in 1991 in the form of the High Performance Computing and Communications Act to further support the Internet. Subsequent legislation in this area and following the 1991 Act has furthered the development of the Internet. The Institute for Museum and Library Services (IMLS) has been a primary source of federal funding to allow museums to develop educational projects that make heavy use of technology and of the Web. The American Association of Museums (AAM) and the Coalition for Networked Information (CNI) among others function as legislative advocates for the museum and library communities. The AAM distributes information on relevant legislation to its members. During the early 1990s the Museum Computer Network served in an advisory role to the AAM on issues of emerging technologies. All of these organizations distributed information on relevant legislation to constituent communities. AAM and CNI continue their advocacy roles.

Museum professionals began to see the promise of electronic mail early on and this form of communication had been used by universities even earlier. In fact it was due to the use of email at university and national museums in the United States that museum professionals began to use email. By 1994 email and other Internet communications had been embraced by MCN and MCN worked with AAM and the Getty Information Institute to further the knowledge of these tools throughout the museum community in the United States. What seems like a small thing now was truly transformational.

The 1994 *Museum News* article by Wallace and Jones-Garmil was among the first to introduce the museum community to the Internet and the accompanying tools. The article included a glossary that contained definitions of the following: cyberspace, Gopher,[9] Internet, listserv, MOSAIC, packet, telnet, and World Wide Web. These terms are now either commonplace or obsolete. Gopher services were replaced by the capabilities of Web sites and other tools like discussion boards on the World Wide Web. Web browsers like MOSAIC were subsumed by the more fully featured browsers offered by Netscape and Microsoft.

Articles and conference papers about the Web began to appear in MCN's journal *SPECTRA*, *Archives and Museum Informatics* and in *Museum News*. Indeed, most articles of the summer and fall 1995 issues of *SPECTRA* focused on the Web. A few examples are:

- "Museums and the Internet: Weaving towards the Web" (Guralnick, 1995)
- "Sights on the Web" (Hermann, 1995)
- "Museum Collections and the Information Superhighway" (Economou, 1996)
- "So You Want to Build a Web Page" (Johnston & Jones-Garmil, 1995)[10]

In 1991 John Chadwick established the first listserv (and later newsgroup) for the U.S. museum community entitled museum-l.[11] The number of subscribers grew quickly and soon had international membership. Museum-l is still in operation. MCN used museum-l until it established its own listserv mcn-l in 1995. MCN-l is also still in use.

In 1994, David Bridge of the Smithsonian set up a gopher service for MCN. This service allowed the first Internet-based distribution of articles and other information for and by MCN members. As noted above it was easier for museum professionals who worked at university museums or U.S. national museums to get access to technology and provide platforms for others to use. The following example, posted by MCN *SPECTRA* editor Suzanne Quigley to CIDOC-L in 1995, gives an idea of the content that could be posted.

> Date: Tue, 14 Mar 1995 20:48:32-0500
> From: squigle@cms.cc.wayne.edu
> To: "CIDOC Distribution List" <cidoc-l@nrm.se>
> Subject: new on the MCN gopher
> Errors-to: <postmaster@nrm.se>
> Reply-to: cidoc-l@nrm.se
> Sender: maiser@nrm.se
> X-listname: <cidoc-l@nrm.se>
> Content-Type: text/plain; charset="us-ascii"
> X-Mailer: Mercury MTS v1.21
> Colleagues—
> Should you be interested, a listing of articles that have appeared in Spectra (Vol 18:1/2 through Vol 22:3 (Winter/Spring 1990/91— Winter 1994–95)) is now available via gopher at world.std.com. Choose Professional and Membership Associations from the main menu, then Museum Computer Network from the second. The next level gives you MCN files and directories.
> Spectra is, of course, not to be confused with Spectrum! BTW, the latest issue over Spectra contains a glowing review of Spectrum!
> Suzanne Quigley
> Editor, Spectra
> squigle@cms.cc.wayne.edu

In 1995 the Museum Computer Network co-sponsored a booth at the AAM Museum Expo in Philadelphia in conjunction with the AAM conference. Jones-Garmil (1995) summarized this event for MCN *SPECTRA*:

> The Museum Computer Network made its presence known at the 90th meeting of the American Association of Museums by bringing Internet Services to the MuseumExpo '95 exhibit hall. MCN and its booth partner, LibertyNet arranged of a T-1 connection and workstations to demonstrate Internet browsers, gopher, and to let conference attendees catch up with their email . . . In addition, MCN worked with the AAM Media and Technology Committee to organize a number of sessions on technology. They included Museums On-Line: Responsibilities, Issues and Solutions, Impact of the Information Superhighway on a Museum's Programs and Internal Practice, Electronic Media Issues: Copyright, Policy and Contract Considerations, and Evaluation of Interactive Multimedia (p. 6).

While providing such demonstrations, a group of MCN members (Johnston, Hermann, and Bridge primarily) worked together at the booth to develop the first MCN Web site, partly to document the technology events being held at the conference. The Museum Computer Network sponsored the AAM Directors Forum: Information Superhighway as Focus. This forum featured Brian Kahin, then Director of the Information Infrastructure Project, John F. Kennedy School of Government, Harvard University. His keynote address was followed by a panel discussion with the following participants: Dr. Oliver Strimpel, then Director of The Computer Museum (then in Boston); Ms. Christine Steiner, then General Counsel for the Getty Trust; and Dr. Maxwell Anderson, then Director of the Michael C. Carlos Museum at Emory University.

These technology showcases continued to be featured at the AAM MuseumExpo for the next two years with the sponsored partnership of MCN, AAM, and the Getty Information Institute. They were important in providing a first look for many museum directors at the Web site that had been created for their museums. Many had never seen the Web sites representing their own institutions. These Web sites as we know provide the first impression for many members of the rapidly expanding online audience.

In this space as well as in the area of imaging and educational site licensing, the Getty Information Institute took a leadership role. Eleanor Fink (1999) describes two important initiatives that were sponsored by the GII:

> Two additional GII strategic initiatives grew from the interest generated by the Profile. The first was a collaborative effort with the ACLS and CNI to form a broad coalition of arts, humanities, and social science organizations. The resulting National Initiative for a Networked Cultural Heritage (NINCH) was designed to serve as an on-going voice in the planning and development of the National Information Infrastructure, the much-publicized plan for a national telecommunications system. Today, NINCH consists of nearly seventy member organizations and serves as a pivotal resource for information on issues and policies across its constituency.
>
> The second initiative, *The Research Agenda for Networked Cultural Heritage*, was undertaken by GII to offer public policy makers and private foundations the information they needed to direct support for arts and humanities computing. The project was designed to (1) achieve consensus among technology and information experts on the research needs for arts and humanities computing and (2) to articulate and publish a research agenda. Eight papers were commissioned to outline research needs for specific domains. These papers were made available on the Internet and electronic discussions ensued. The final report, containing the papers and a summary of these exchanges, was published in 1996 (Establishing a Voice section, ¶6–7).

NINCH conducted important work in several areas but the most important to museums were the efforts toward an understanding of copyright and fair use in the digital world and in the creation of *The NINCH Guide to Good Practice in the Digital Representation and Management of Cultural Heritage Materials.*

David Green (1998) described the project:

> Our Guide would be designed as a primer on technical and information standards, metadata and best practices for a broad audience that would assume no prior knowledge. The primer would highlight the context and re-usability of material, intellectual property and longevity issues. It would also be addressed to funders, aiming to provide them with a set of key criteria for assessing the fundability of digital projects (p. 2).

NINCH formed a working group for this project that included individuals from the museum, library, archives, and humanities communities. The project was contracted to Humanities Advanced Technology and Information Institute (HATII) of the University of Glasgow. The project included interviews with organizations such as the Art Institute of Chicago, the Harvard University Libraries, and other who had conducted large-scale digitization projects. The Guide is a resource for archives, libraries, and museums, whether large or small, experienced in this area or not, in the development of digital materials.

Sadly the closure of the Getty Information Institute and the scaling back of NINCH took away two valuable resources from the museum community. The NINCH Guide is available online and still serves the community even though it has not been updated since its publication.

How Museums Responded to the Use of Internet Technologies

Being online for museums might be compared to the California Gold Rush. In the early days (1994–1998) there were a few pioneers who were willing to take a first step toward have an online presence. The Peabody Museum of Archaeology and Ethnology was one of the first museums to create Web sites. During these early days the Peabody created and hosted Web sites for the other Harvard museums with content from a publication titled *Treasures of Art and Science at Harvard University* (Chandler, Andrews, & Rossi-Wilcox, 1994). These first Web sites like the other pioneering efforts were nothing more than short informational flyers. The Peabody Museum is a university museum and resources were available to support this.

One of the first more fully formed Web sites was created by Robert Guralnick for the Museum of Paleontology at the University of California, Berkeley. This site featured information on the museum and an online exhibition on dinosaurs. Guralnick wrote several articles for MCN that helped others understand the potential of the Web for museums. Others followed with due haste. Early adopters of note included the Fine Arts Museums of San Francisco and its "Thinker" image base online in 1997 with over 60,000 images; the Exploratorium and its early interactive education "snacks" features; the United States Holocaust Memorial Museum and its extensive library, archive, and photographic databases; and the Smithsonian National Museum of American Art HELIOS American photography site, where visitors could email comments about photographs that they viewed. This list is by no means inclusive of all the early adopters and their pioneering work.

Museums were transformed by these technological advances that enabled the use of digital information in education and exhibition within the museum and online. Hilde Hein (2000) suggests that objects are "placeholders for stories" (p.12); by creating digital placeholders for objects, museums were able to continue to tell the stories that had been available only in the physical space of the museum or in published catalogues. With the advent of these technologies museums could tell the story to larger and different audiences and in new venues.

New media had not just become ubiquitous in the museum community, but in the art collected and preserved by museums. In 1998, the Walker Art Center and Franklin Furnace were among the earliest museums to curate art online and deliver digital art. In 2001, the first conference was held on digital art and preservation at the Guggenheim: "Preserving the Immaterial: A Conference on Variable Media." Collaborative projects such as "Archiving the Avant-Garde" were launched.

TRANSPARENCY OF INFORMATION:
NEW MEANS OF ACCESS, NEW AUDIENCES

The early days of the Internet for museums may have seemed like the Gold Rush, but the activity is consistent with the phases of hype cycles as defined by the industry analysis firm, Gartner, who defines five phases in the lifecycle of any emerging technology: Technology Trigger, Peak of Inflated Expectations, Trough of Disillusionment, Slope of Enlightenment, and Plateau of Productivity (http://www.gartner.com).

We could use the idea of hype cycles to explain the natural evolution of technical transformations in museums. We could say, for example, that the database cycle peaked somewhere around 1987 and suggest that a marker of the plateau was the conference held by the Museum Documentation Association (MDA) sponsored by the Getty. This conference was entitled "Terminology for Museums," but it was also a watershed event for the maturation of museum collections management systems (Roberts, 1990).

The use of Internet technologies, especially the Web, reached the "Slope of Enlightenment" somewhere around 1999. We began to see online collaborations like ArtsNet Minnesota and ArtsConnectEd between the Minneapolis Institute of Art and the Walker Art Center. Museums in general continued the evolution of museum Web sites with the development of deep and rich content. Museums followed the lead of several including the National Gallery of Art and created online access to collection information including high quality images.

We are, perhaps, on the "Plateau of Productivity" with amazing new online exhibitions, educational programming, and collections databases accessible to new global audiences. Information about the museum and its collection are more accessible than any other time in the history of museums (see Figure 2.1).

What Digital Natives Expect of Museums

Are we ready for the next transformation? What seems to be on the horizon is a transformation of museums, collections, and museum information to meet the needs of the new generation of "Digital Natives." Marc Prensky (2001) suggests that this group thinks differently than the preceding generations, i.e. the group he calls "digital immigrants." His theory is that this group lives at twitch speed due to the gaming experiences of many of them

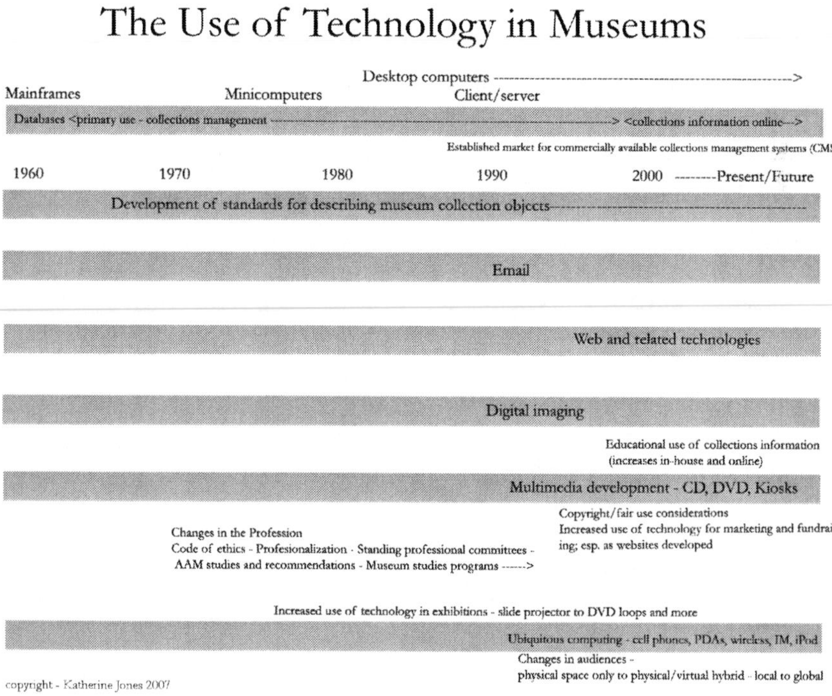

The Use of Technology in Museums

Figure 2.1 The Use of Technology in Museums (1960 to present)

during early developmental years. He suggests that there has been an actual change in brain patterns during this developmental time. How will museums respond to this new way of thinking and reacting?

Digital natives seem to be more willing to experiment online, giving up personal information on Web spaces like myspace.com, for example. Digital immigrants, on the other hand, seem to value and guard private personal information. The communications of digital natives are instantaneous, wireless, and almost constant. Their online research is more likely done using Wikipedia than in online libraries or museum Web sites. What is the next digital shift and how can museums respond? To answer these questions, museum professionals need to look at technologies that have emerged and those that are just, possibly, on the horizon.

ENDNOTES

1. From the early 1970s Chenhall was a leader in the area of standardization and museum computing. His efforts at the Arkansas Archaeological Survey were a model for the Florida Bureau of Historic Sites and Properties to begin similar electronic inventories in 1973. These inventories served as the basis for the continued development of the Florida Site File by the Florida Department of State, an effort which is ongoing.
2. See the Museum Data Bank Research Reports, Strong Museum, Rochester, NY.
3. Both projects were the result of work by numerous scholars from various subject fields. A revised edition of *Nomenclature* was edited by James Blackaby in 1990. Development of the AAT continues through the efforts of the Getty Research Institute (http://www.getty.edu/research/conducting_research/vocabularies/aat/).
4. "The study started as a student internship project to meet the requirements of the Museum Studies Certificate program at Harvard University Extension School, and continued well after graduation. It was done in the Office of Information Services and Technology of the Peabody Museum of Archaeology and Ethnology, Harvard University, and under the direction of Katherine Jones, assistant director at the museum and instructor of Information Technology for Museums and Collections in the Museum Studies Certificate program" (Loran, 1999, p. 2).
5. This program now offers a Masters of Liberal Arts in Museum Studies.
6. Unique in that one needs an understanding of technology in support of the museum's work and grounding in museum studies.
7. In larger museums, this role would have been held by individuals in the rights and reproductions office or in the photographic archives.
8. CAMIO™ managed by the Research Libraries Group (RLG) is the closest successor to AMICO. Information on CAMIO can be found at http://www.rlg.org/en/page.php?Page_ID=20638. Information on the Art Museum Image Gallery is available at http://www.hwwilson.com/databases/artmuseum.htm. Information on ARTstor is available at http://www.artstor.org/info/.
9. For example: "Gopher: a system that makes available hierarchical collections of information across the Internet. Uses a simple protocol allowing a single client to obtain information from any accessible Gopher site" (Wallace & Jones-Garmil, 1994, p. 34).

10. A revised version of this article later appeared in *Museum News*.
11. According to http://www.finalchapter.com/museum-l-faq, "Museum-L is a general purpose, cross-disciplinary electronic discussion list for museum professionals, students, and all others interested in museum related issues. All museum related topics are acceptable for posting and discussion at this time."

Section 2

Information Resources in Museums

3 Information Representation

Paul F. Marty

Florida State University

Museum professionals are responsible for large numbers of information resources, from physical museum artifacts to electronic documents about museum collections. As managing such resources in their original form can be difficult and time-consuming, museum professionals tend to rely on principles of information representation to create information surrogates or aggregates that can be manipulated more easily. Surrogates arise from the process of taking information entities and making them physically or informationally smaller (e.g., creating catalog card records), while aggregates arise from the process of creating a single resource that represents groups of information entities based on shared data (e.g., making a list of all artifacts accessioned in the same year).

Naturally, museum professionals have been representing information about their collections for thousands of years. Without information representation, the users of museum resources would have a difficult time finding, searching, sorting, or manipulating those resources, especially considering the number of information resources found in any given museum. Over the years, museum professionals have used a wide range of tools (such as ledgers, card catalogs, computer databases, and digital collections management systems) to organize and provide access to the museum's information representations.

As the amount of information found in museums continues to increase, it is important to understand how decisions made during the information representation process can affect the ability of museum professionals to meet the needs of users of museum resources. Consider the hypothetical example of an Egyptologist traveling the world looking for very particular types of black-topped pottery. While many museums will have related collections, few will have pots that match this scholar's particular conditions. In a museum without suitable information representations, meeting this scholar's needs may place unnecessary burdens on the museum's staff and artifacts. Some individual would most likely have to examine each pot in the museum's collection in order to determine whether it would be useful to this scholar. Given that few museums have access to qualified Egyptologists, it is even possible that the scholar in question would have to physically handle each artifact to make this determination.

On the other hand, if suitable information representations were available, the Egyptologist would potentially be able to review the museum's collections remotely, troubling neither the museum's staff nor its artifacts. By evaluating and analyzing these representations, the scholar could determine the appropriateness of each artifact long before arriving at the museum in person. In this way, information representation makes it easier for scholars to conduct research, especially from a distance and particularly if they need access to objects in storage; makes it easier for museum professionals to assist the users of their resources, especially given the time-consuming process of providing access to artifacts; and finally makes it easier on the artifacts themselves, minimizing the risks involved any time an object is removed from display or storage for study (Marty, 1999a).

The difficulty with information representations, however, lies in determining what qualifies as a "suitable" information representation for any given situation in any given museum (Müller, 2002). Naturally, no representation can duplicate the physical artifact in its entirety, and there will always be situations where nothing but access to the original object will serve. Nevertheless, in a growing number of situations, access to a sufficiently-detailed information representation can meet the needs of many users, including researchers, scholars, teachers, students, and the general public. Several authors have discussed the meaning of information in the museum context, identified the different users of information resources in museums, and explored ways of making information in museums more accessible (Bearman, 1988; Orna & Pettitt, 1998).

There are important differences between museum artifacts, museum collections, and other types of information resources available in the museum, such as research materials, educational materials, archives, or information about exhibits, visitation patterns, guest lectures, and other activities. There are also significant differences between information that provides basic collections data and information that places the museum's collections into specific contexts. Orna and Pettitt (1998), for instance, place museum information resources along a scale of raw, refined, and mediated information. When viewed from this perspective, different types of information resources in museums progress along a steady scale from basic facts about each artifact (what is it? where did it come from? how big is it? what is it made of? etc.) to more narrative or interpretive data about collections (why is it important? why did the museum collect it? what does it mean? etc.). Representing information resources in museums means not just accurately describing what one owns, but also supporting interpretive analyses and active scholarship over the long term.

Given the problems most museums have in simply maintaining an inventory of their collections, however, many museum professionals find themselves forced by time constraints to focus on identifying what they have, often at the expense of analyzing or interpreting the importance of their collections. Yet these interpretive data are just as valuable to researchers, scholars, and students, if not more so, and many museum professionals are building information resources that add value to collections catalogues. A museum with a sizable collection of cuneiform tablets, for example, could provide online access to thousands of database records containing basic information such as measurements, date created, and location found, but such information, while valuable, would likely be of use to only a handful of scholars. Most visitors, especially students and children, will be more interested in learning about cuneiform writing in general, with a few chosen artifacts selected as prime examples. Building such interpretive, educational information resources, however, can be far more difficult and time-consuming than simply providing access to database records.

Maintaining a balance between information representations that identify museum artifacts and those that provide interpretations of collections can be very challenging, as even the simplest forms of information representation can be problematic (Marty, Rayward, & Twidale, 2003). Consider the act of assigning an accession number or identification number to an artifact. Most museums employ some sort of standard practice for assigning accession numbers, often including such information as the year of accession, the accession lot number within that year, and the object's number in the accession lot (Buck & Gilmore, 1998). The number 1982.12.0045, for instance, might refer to the 45th object in the 12th collection of objects accessioned in the year 1982. The simplicity of such a numbering scheme often leads those not versed in museum information management to assume that if one simply added a fourth number, a code for identifying a particular museum, to the front of this number, then one could easily identify any unique object in the world.

The difficulty with this idea, of course, lies in the fact that not all museum professionals use the same methods of identifying their artifacts; there are any number of competing, yet equally valid approaches for accessioning objects into museum collections. Making matters worse, even museums that use similar numbering schemes often have subtle differences that can undermine attempts to share information across organizations. Even something as simple as dropping place-holding digits (representing the above number as 82.12.45, for instance) can have serious repercussions. When one looks at all the other ways that museums can complicate the process of numbering their collections (for instance, groups of small, similar objects might share the same accession number), even a relatively simple problem seems much more difficult.

Museums have long wrestled with the inherent difficulty of cataloguing museum artifacts, and most have had to solve these problems on their own. The simple fact that museums have unique collections of artifacts means that it is virtually impossible for museums, unlike libraries, to have their own version of the Online Computer Library Center (OCLC). When one imagines a central museum organization dutifully filling out catalog cards for each museum artifact and sending a copy of that card to every museum with that artifact, one sees the futility of the approach. The uniqueness of museum collections means that each museum has historically been responsible for documenting its own collections; even when one considers reproductions of museum artifacts, the uniqueness of each reproduction is usually such that museum professionals will want to maintain their own records on their own reproductions.

There have been many attempts to create information representation schemas that can be applied across museums for data content, structure, and values. Some cultural heritage institutions, for example, classify their collections of fabricated objects using *Nomenclature* (Blackaby & Greeno, 1988; Chenhall, 1978). The Getty Research Institute has developed a series of structured vocabularies, such as the Art and Architecture Thesaurus, specifically for the use of museum professionals at a variety of institutions (Lanzi, 1998; Peterson, 1990). The MDA offers information standards in the form of SPECTRUM, a guide to electronic collections management (Cowton, 1997). The Visual Resources Association (VRA) has established data content standards for cataloguing cultural objects (CCO), fulfilling much the same purpose as the Anglo-American Cataloguing Rules (AACR2) in library and archives communities (Visual Resources Association, 2005). Despite these efforts, however, it remains very difficult for museum professionals to agree on a given standard and even to use that standard consistently in their own institutions.

In the past few decades, many have hoped that the push to create digital information representations (including but not limited to digital images) of museum collections would help solve these problems. Digital surrogates offer faster access to information, with more access points, faster searching and sorting, and the ability to compile and print lists quickly (Abell-Seddon, 1988; Rush & Chenhall, 1979; Vance, 1988). Yet problems of information management continue to trouble museum professionals, despite efforts, some going back many years, to encourage digital data sharing across organizations (e.g., the Canadian Heritage Information Network, the Art Museum Image Consortium, the Colorado Digitization Project, and so on). Digitization is merely information representation in electronic form, and electronic information representations are no more immune to information organization problems than physical records.

In terms of creating and managing information representations, museum professionals historically have been on their own. Even when groups of museums collaborate to build a centralized repository of museum information resources, individual museums are generally responsible for representing information about their own collections. The inevitable result, therefore, has been that each museum tends to have its own approach to information representation, even when they belong to museum consortia, leading to a number of problems in terms of information management and the sharing of data across multiple institutions. Despite early and ongoing efforts to create shared databases of digital surrogates—from the centralized client–server systems at the Smithsonian in the 1960s, to the efforts of the Museum Computer Network in the 1970s, to the research of the Consortium for the Interchange of Museum Information in the 1990s, to the Open Archives Initiative today—most museums are still using their own unique systems with their own unique record structures, and even museums that use common standards and controlled vocabularies can face problems sharing their information resources. It is against this landscape of unique artifacts and common difficulties that museum professionals work to create digital surrogates of their artifacts.

Despite the challenges, there are many reasons for them to proceed, not the least being the growing number of museum visitors, donors, researchers, and other constituents who now expect museums to provide access to their collections in digital formats. As the technologies required to build a digital collection become easier to use and cheaper to acquire, more and more museums have the opportunity to embark upon digitization programs. It is important that museum professionals understand the consequences behind their digitization initiatives. The digitization process raises many difficult questions. What is a digital representation of a museum artifact? Why would someone want such a thing? Are there things one can do with a digital surrogate that one cannot do with a physical surrogate? What advantages or disadvantages do those differences bring to the museum? While it is beyond the scope of this brief essay to describe all the potential advantages and disadvantages of digital information representations, the following issues underscore some of the challenges.

The advantages of digital information representations include the ability to make a virtually infinite number of perfect copies of digital surrogates, and transmit them great distances with no loss in quality; to offer new levels of interactivity between objects and users; to take advantage of hypermedia and multimedia to remove objects from the constraints of physical space and present arrangements impossible in physical galleries; to provide remote access to information resources for visitors, scholars, researchers, and students; and to target unique information needs, by either broadcasting information resources to wide audiences or narrowcasting information resources to individual users. The disadvantages of digital information representations include concerns, worries, or fears over such issues as copyright and intellectual property; the potential lessening of the "aura" or authenticity of museum artifacts; the blurring of individual museum identities online; the risk of losing individual details and lessening a sense of reality; and the potential impact of access to digital surrogates on physical museum visitation.

When one examines these issues from the perspective of information representation, one sees that digitization in museums is just the latest in a long series of information science and technology advancements dating back thousands of years. And yet, while the underlying principles of information representation may still be the same, the challenges of digitization have brought new expectations and new opportunities. As museum researchers and professionals continue to explore new ways of representing information about museum resources, they are radically changing the way museum professionals, visitors, and all users of those resources work with museum collections. When examining these changes, it is all too easy to be captivated by their novelty and potential. It is important to remember, however, that these changes are built upon a solid historical foundation of information representation in museums.

4 Representing Museum Knowledge

David Bearman

Archives and Museum Informatics

INTRODUCTION

Collectively museums hold the universe of all objects and ideas and all their relations. They have assumed responsibility for preserving what is now thought about them and what has been thought in the past, and for representing and interpreting that for the present. Because museums are collections, they have made even nature into a cultural artifact (Buckland, 1997). Thus, representing museum knowledge is potentially a task as comprehensive as the representation of all human knowledge.

The act of collecting has privileged those attributes of the object around which the collection is constructed and deprecates others, but decisions about representation should enable the object to be re-incorporated logically into many collections and contexts, including their original context, to support the work that those in and outside museums do with museum objects.

Museum knowledge representation has acquired additional requirements as a consequence of the computerization of much museum work and museum relations with visitors. Though computers were used to inventory museum collections from the 1960s on, computer representations of museum holdings evolved in sophistication from the mid-1980s to mid-1990s, as computing systems became increasingly capable of holding extensive museum data and data models were developed to support more and more museum work processes. Since the mid-1990s, the advent of the World Wide Web and networked computing has radically transformed the task as it was previously understood, in particular by redefining its audience, and thereby forced museums to rethink the purposes and ways they represent knowledge. This chapter proposes some guidelines for the present that can be gleaned from prior museum practice and other frameworks for representations. It illustrates how radically our concepts of what it is crucial to represent have changed over the past three decades, suggesting that today's view will be found lacking soon, but nevertheless attempting to guide current knowledge representation practices.

OBJECTIVES OF MUSEUM KNOWLEDGE REPRESENTATION

While museums do strive to collect every natural and human-made thing that has ever been and to interpret everything that has ever been thought about them, museum knowledge representation supports museum missions.

Museums make and record meanings. It has been argued that information management is, therefore, the central purpose of museums (Washburn, 1984). Certainly, the knowledge model of museums is much more robust than the actual accumulation of knowledge about any given object. The model can serve to highlight lacunae in our knowledge, directing research and documentation. Through research museums seek to document the material and ideational world as they, the original discoverers, and the creators of the objects and specimens they acquire, understood them. These are, of course, diverse and potentially conflicting, perspectives.

Museums seek to convey their understandings of their collections to scholars and lay people, experts and the naïve, adults and children. As such, they strive to articulate their knowledge in many different ways, to different depths and at different levels of sophistication. The museum must hold all these representations at one time if they are to be presented to their desired audiences.

Museums seek to preserve their holdings and knowledge, not just over time, but also from one product or process to the next. The information that is recorded should, therefore, serve the purposes of each of those museum activities that need to use it, requiring that it be encoded in a way that makes it efficiently suited to the purposes for which it will be used.

These three goals—making, conveying and preserving meanings—will be explored in this essay as we delineate the domain of museum knowledge representation and suggest guidelines that might govern how we can best describe, explain and control museum objects.

THE CHALLENGE OF KNOWING

To understand what it means to represent knowledge in the museum, we need the humility to appreciate that our knowledge of the world is socially constructed. Museums strive to represent what they know, but what they know was conveyed to them by someone who first made or discovered the object, collected or analyzed it, or acquired it for the museum or managed it in the collection. When we say what something "is," it cannot be said to be "true" though it may be preferable for particular purposes or "correct" from a specific perspective. To the explorer who first collected the artifacts, the people he had encountered were Eskimo; to themselves they were "Inuit," meaning, "the people." One informant may be as certain that an object is a jaguar as another is that it is the spirit of his grandfather. The fact of the social construction of all knowledge means that in museums, all knowledge should be sourced. Yet this is one of the least observed requirements of knowledge representation systems in museums, and results in one of the most criticized aspects of museum interpretation, its adoption of an authoritative, unsourced voice (Walsh, 1997).

We must not lose sight of the fact that objects in museums have been collected from some natural or cultural context in which they originated or have been used. The museum is a storehouse of things that were consciously gathered and placed in the context of other things also gathered. We need to be aware that the representations we have were made for a purpose. Thus we might know that an object was acquired at dusk, or in the spring, or on the birthday of the collector, or in 1842, all reflecting quite different purposes and assumptions about what is significant to the object, to the act of collecting and to the different social constructs in which we make sense of such things. Any museum object has several stories to tell: the story of having been collected might be thought of as their stories as told by their original collectors, while other stories are those told by subsequent curators or researchers. Yet too frequently our abstract frameworks for representing what we know assume a singular point of view about what is worth recording and how. For example, if the day, month and year that an object was collected are the only form of "time of collection" supported, we are deprecating other perspectives that in other contexts might be more relevant.

What we know is further qualified by why we know it. We might know of an object that it was given in tribute because what we know about it was recorded by the recipient; what might have been said by the "gift giver"? In addition to lacking information from all possible sources, we are always at risk of substituting our cultural perception for that of others. We might conclude that an item was acquired by theft, for having been found in a "hoard" of objects seemingly pillaged, but it might upon further study be a kind of bank, to which voluntary deposits were made. We might "know" that a stone to which magical properties were associated by the peoples who owned it is a strong lodestone, but they did not "know" that and recording our knowledge does not alter theirs. By chemical analysis we might know that a pigment on a famous masterpiece which "hung on the mantel of a major local landowner since the 16th century" was not invented until the 18th century, yet this "fact" would not change the role this object's place in family history at all. Collections of facts, like collections of the objects to which they relate, are built up over time and have a life of their own; when we represent both a thing and the knowledge of a thing, we must be prepared for divergence between the two.

What we represent and how we encode it should be faithful to the evidence that we have. Thus if the testimony accompanying an object dates it from "the third year of the Depression," recording it as 1932 loses significant information given us by the informant and privileges the curator's interpretation of the dating of the Depression. Not recording 1932 makes it hard to correlate the object to other objects whose dates of creation are known years, so we may need to keep that representation too. In documenting the object in the museum, we are recording the history of its interactions with other objects, and with people, places, events and actions. A "hard" fact, like the date an object was first created or used, may involve documenting the testimony of the donor, the interpretation of the curator and the evidence of a scientific analytic tool, in each case recording the "date" and who claimed what, when. In the museum we need to record many forms of date as they convey different information, reflecting their different relationship to the source of the data, even though they may not refer to different years.

THE CHALLENGE OF RECORDING

Some ways of representing information will be better suited than others to particular subsequent uses. When the facts about objects are recorded, we tend to record them from the perspective of the activity in which the act of recording takes place. For museums, these activities might be collection, acquisition, conservation, exhibition, interpretation, or research for example. In each case, there is a time and place in which the documentation occurs; there is an observer who is recording, possibly with tools to assist, and with ideas involved. All these facts about the context of the activity and its bearing on the object need to become part of the museums' knowledge. Later there will be reasons, of accountability or of curiosity, to recall these facts.

Time, for instance, is an omnipresent aspect of the knowledge we have of objects, since the museum gathers the past. The periods measured in museums range from nanoseconds to astronomical units of time, and the schemes employed include all those ever used by man. For instance, if we are concerned for conservation to determine if an object in our storerooms was subjected to a specific danger, we will want to know the date and time of day it was moved there. If we are dating a geological specimen in our collection for researchers, a date expression of $3.2^{-6} \pm 5^{-3}$ will better support calculation and reflect what our carbon dating instruments actually told us. The reading from our instruments has the advantage that it will never change, but the name of the geological era, like any interpretation, could be reconsidered and the object we own could "move" to a different period. Nevertheless, if we are going to prepare a label for an exhibition, the geological period will probably be a better representation since it can be used with appropriate illustrative wall labels to locate the object in geologic times and most of the public finds mathematical expressions off-putting. The "correct" scheme with which to represent time in each *display* is the way that will be useful to the tasks at hand, but at the same time we need to *preserve* the primary form and source of data intact.

But selecting a representation scheme is not the end of the task. We must still choose how best to encode time. In our example of the time of storage, we'll no doubt want to choose "ANSI-time" or YYYYMMDD (which can be extended with HHMMSS etc.) over the prose expression June 15, 1999 because it will be easiest to manipulate in our computing system. But if we are recording the time that an artifact in our collections was created or used, we might be confronted with a more complex decision. Rather than using ANSI time to record a date range from the Gregorian calendar with which we are familiar, we might wish to represent the time of creation in terms that the creator recorded them. Prior to the mid-16th century in the West, this would have been the Julian calendar, but elsewhere it might have been based on the Zoroastrian, Jewish, Chinese, or Mayan calendars, each of which employs quite different units of time and "begins" at different dates. Or it might be a more accurate reflection of our knowledge to use culturally referential periods of time such as "the Depression" or "the Ming Dynasty" or "the time of Abraham" if we are reasoning about the time of creation and use based on other observations that we associate with these periods rather than based on recorded dates.

One more example should suffice to make the point. In our administrative recordkeeping within the museum, personal names of employees or donors for example might be represented with discrete elements for Titles, First Name, Middle Name, Last Name, but could be displayed in inverted order as "Last, First, Middle, Title" or in direct order "Title, First, Middle, Last." Each conveys the same information, but the latter is preferred for sorting lists by last name, while the former is the preferred form of address; both are enabled by representing the units discretely. But for other purposes the museum needs to know that this form of representation and encoding is very culturally specific. To avoid that, the museum could keep the same representation, but label the elements "surname" or "patronymic" and "given names" or "personal names." But if the museum needs to manage names from many cultures and periods it needs to consider whether "toponymics" might be importantly different from other "last names" and realize that names of individuals, as a category, is probably too narrow a construct to represent the makers and creators of artifacts in a universal collection. In our society corporate entities are often creators, and in other cultures and times the creation of cultural artifacts would not be attributed to individuals at all, but to villages or totemic groups or other cultural units.

The choice of how to represent information, whether it is a personal name, time or any other attribute, should not be arbitrary. There are criteria by which we can distinguish useful from less useful representations.

How we encode what we know should be useful for the purposes that we have, and as useful as possible to purposes we may have, but do not yet know. Always the information should be expressed in a way that is explicit and can be used as flexibly as possible in the tasks that we know it will be involved in, within the museum and by others. Thus, even if the museum "has always used cm height × width × depth in its measurements," a measurement that explicitly enumerates the measurement units and dimensions measured—14 cm h × 3 cm w × 2 cm d —is preferable to recording 14 × 3 × 2 as NASA recently learned at the expense of losing a multi-billion dollar planetary mission (Lloyd, 1999). In addition, the structure of recording the quantity, the unit and the dimension permits us to add our measurement of the base as 4 cm diameter and the mouth as 8 cm diameter, at some point in the future as our knowledge of the object develops.

Increasingly as we consider representation of knowledge we need to remember that we can record facts and opinions about objects not only in words or in measurements and scientific analyses, but also in images, in sound and in multimedia. When we do this today, we will encode all these diverse representations in binary form. Thus we have at least four levels of decisions that we must make about documentation—what features to document, what about those features to represent, how to express what we want to record, and how to digitally encode it. Each of these decisions will impact on what the representation that we make can be made to "do." Metadata about how these representations were made and how they are encoded is essential to their subsequent reuse.

Finally, effective encoding should aim to be of continuing value. This means we need to consider not just the purposes for which it was originally recorded, but future purposes as well. Historically museums have been intellectually profligate. They have expended vast quantities of the time of highly trained staff creating numerous representations of small parts of their collections and throwing each subsequent representation away when its work was done. Consider the exhibit, in which each of a small number of objects is re-researched, re-described and re-interpreted only to have these expensively constructed knowledge components discarded along with an equally expensive construct that holds them at the end of a brief public run. Yet to use these representations, we need to not only document who made them, why, and what purposes they served, but to do so in a manner that will be consistent with representations we have kept from other sources so that we can efficiently decide on their re-use. This places a huge burden on our choice of encodings.

A BRIEF HISTORY OF MUSEUM KNOWLEDGE REPRESENTATION

Object-Centered Data Models

The creation of museums and the keeping of records about collections of artifacts and specimens are historically intertwined. Field notebooks of naturalists, ship logs from expeditions of maritime exploration, accounts by the bookkeepers of princely hoardings, country house inventories, and registries of wills supplement museum ledgers recording acquisitions over the first few centuries of museum history. In each case, these are records of transactions, organized around the activity of recording, in chronological sequence.

Card catalogs and vertical files augmented these transaction records during the early 20th century, providing some access by object, though the principal means of accessing individual objects in museums remained the organization of their storage and the records of their acquisition. Card systems tended to reflect the idiosyncratic interests of curators and were often abandoned after their retirement, or replaced by new systems with different orientations. Projects external to museums, such as the Princeton University Index of Christian Art or the ICONCLASS classification system invented by Henri van del Waal at the University of Utrecht, evolved "encodings" for recording and searching museum documentation, but these were secondary resources. Until the arrival of computers, museum documentation was not generally organized around object-centered knowledge representation models designed to expedite retrieval and use based on properties of the thing, rather than on the history of its relation to the museum. The history of computers in museums is addressed elsewhere (Jones-Garmil, 1997) and in Chapter 2 of this volume; here we will look only at aspects crucial to understanding museum knowledge representation.

The registrars and computer scientists who formed the Museum Data Bank Committee (subsequently MCN) viewed the computer as a tool to create a catalog of the collection, just as librarians at the same time were imagining it as a means of compiling bibliographic catalogs. The systems they developed were collection inventory files processed by mainframe computers fed initially by punch card input. In the 1970s, the literature on museum computing consisted in large part of the data dictionaries of systems that were designed to serve as centralized repositories of the metadata from many museums, and very basic presentations on the nature of museum computing systems by their developers (Bergengren, 1979; Chenhall, 1975, 1978; Gautier, 1979; Porter, 1979; Roller, 1976).

Prior to the mid-1980s, data dictionaries of implemented museum systems which circulated in loose-leaf binders, and assembly language object code distributed by the agencies promoting these systems, were the primary sources of any information there was about how museum databases were or ought to be constructed. "Published" dictionaries that were influential included those of the Smithsonian Institution (SELGEM system), the Metropolitan Museum of Art and Museum Databank Committee (GRIPHOS system), and the Canadian Heritage Information Network (CHIN Data Dictionaries, Sciences and Social Science/Humanities). The secondary works focused on the nature of computers and project management and did not treat knowledge representation (Chenhall, 1975; Orna & Pettitt, 1980, 1998; Van Someren Cok, 1981; Williams, 1987).

Although intended to be normative, these data dictionaries reflected local representation and encoding practices. All represented museum objects as records of very limited length in flat files created for mainframe computers. None of them reflected explicitly on alternative choices in knowledge representation or discussed the limitations of the approaches they had taken. The data values were all presumed to be objective. The data were unattributed, and discussions of it were utterly non-reflective about their methods of recording. Limited lengths of fields encouraged abbreviations and look-up tables, and made prose impossible. Pre-relational systems restricted description of other entities to whatever characteristics could be attributed to the collection item.

In the mid-1980s, the author introduced formal data modeling and relational databases to the Smithsonian Institution, and elsewhere in the world these new computing methods, the spread of micro-computers and cathode ray displays, and the falling prices of storage, had a dramatic impact enabling choices in representation of museum data and its active use in support of museum missions. From the mid-1980s on, the Museum Documentation Association in the United Kingdom and Archives & Museum Informatics in the United States moved beyond the dissemination of data models to the promotion of museum standards for recording and handling data content and a focus on the functionality that should be associated with different types of museum information systems (Bearman, 1987, 1990a, 1990b; Light, Roberts, & Stewart, 1986; Roberts, 1985, 1988, 1993).

The rise of commercial software which spread with the mini-computer (and later the micro-computer) was an alternative to custom developed mainframe applications. But to exploit this opportunity, museum professionals needed to be able to compare different systems, and assess them based on criteria including their interoperability with other systems that were imagined (in imitation of the emerging "integrated library systems"), as functional modules of a to-be-realized integrated museum system. These comparisons were published in bi-annual volumes of the Directory of Software for Archives and Museums (Bearman & Cox, 1990; Bearman & Wright, 1992, 1994), accompanied by essays by the editor highlighting important knowledge representation and functionality developments.

A new generation of more critical analyses arrived in the mid- to late-1980s (Abell-Seddon, 1988, 1989; Bearman, 1987; Chenhall & Vance, 1988; Roberts, 1985). All these addressed knowledge representation as an issue in its own right, rather than as simply a question of encoding, storage efficiency and data preparation. All took the then relatively new perspective of the relational database to explore relations between entities beyond the collection object in itself and suggested departures from flat files. Each distinguished between issues relating to syntax of representations, encoding rules and the semantics of representation.

By the 1990s it was possible to refer to a number of guidelines for integrated museum systems functionality, and a sophisticated, community developed, relational data model from CIDOC, the International Council on Museums Committee on Documentation (International Council of Museums). The CIDOC relational model, the culmination of years of consensus building, remains the best single statement of the relationship between structured elements of information about an object in the context of museum practice.

Efforts by humanists and scientists over many years have yielded insight into the many complex relations of scientific specimens (Allkin, White, & Winfield, 1992; Chavan & Krishnan, 2003; Graham, Ferrier, Huettman, Moritz, & Peterson, 2004). A multi-year funded effort to map the knowledge structure of art and architecture artifacts yielded the Categories for Description of Works of Art (Baca & Harpring, 1996; Trant, 1993), an important conceptual mapping that some have unfortunately tried to implement as a data structure rather using the reasoning that went into it to better understand the complexity of concepts in the field.

Process-Centered Data Models

In the early 1990s, the Museum Documentation Association under Andrew Roberts decided that the shift to object-centered recording in museums was creating a new documentation requirement, but not taking advantage of ongoing documentation activity which took place throughout the museum in all of its processes. They began to map the relationship between data about objects and the events within the museum that gave rise to that data. In 1994 the MDA issued SPECTRUM, a data standard organized around common procedures within museums and designed to re-integrate object documentation and museum workflow. The standard has undergone revision since (Museum Documentation Association, 2005), but remains the best single source of information on the way in which museum data is employed in museum processes, and indeed, of how museums actually work.

By the mid-1990s, pressure was building from other quarters in technology that would make the comprehensive statements of purely object-centered data relationships obsolete. First, computers were increasingly being used to store unstructured text, still images, and, ultimately, multimedia, not all of which represented the museum collection item, as reflected in the Proceedings of the ICHIM conference from 1991 to the present (Bearman, 1991, 1995c; Bearman & Garzotto, 2001; Bearman & Trant, 1997, 1999, 2003, 2004, 2005; Lees, 1993). Secondly, computers were increasingly used to communicate over networks, ultimately the Internet, which led to a growing interest in standards for data interchange and interoperability (Bearman, 1992b, 1995c; Bearman & Perkins, 1993), and also to a growing audience of non-specialist users, culminating in the general public. Thirdly, alternatives to the relational data model, driven in part by the need to exploit both these developments, and in part by a vision of object life-history that did not privilege the museum context over other periods in the objects' life, led to a proliferation of interest in object-oriented models and methods (Bearman, 1992a; Bearman & Vulpe, 1985; Research Libraries Group, 1994; Vulpe,

1986) and in semantic linking models (Beynondavies, Tudhope, Taylor, & Jones, 1994).

In 1994, these trends came together in a profoundly new implementation—the World Wide Web—which within a couple of years fundamentally transformed the methods, the audiences, and finally many of the purposes of museum knowledge representation. To an extent that has still not been fully understood, it brought museum representations into the same arena as those of other cultural institutions, where the exhibition and interpretation traditions of museums were highly successful paradigms. And with its popularity, and the spread of inexpensive networked computers, technology assisted workflow related computing requirements began to influence knowledge structuring practices in museums (Carliner, 2003; Marty, 1999).

In fact, over the past decade, while there has been much attention devoted to how museums can use the Web for a wide range of outreach purposes, as documented in the annual proceedings of the Museums and the Web conferences (see http://www.archimuse.com/conferences/mw.html/), there has been little explicit attention there or elsewhere to the implications of these developments for museum knowledge representation. In the remainder of this chapter, we will look at what was learned about museum knowledge representation from systems prior to the mid-1990s, examine the impact of the Web and the issues it presents for museum knowledge representation, and then hypothesize about requirements for future systems development and implementation.

LESSONS FROM PRE-1995 MUSEUM SYSTEMS

Formally Declared Data Models with Maximally Disaggregated Data

The relational model was built on a formal method of data normalization. Fully normalized data (fifth normal form) is maximally disaggregated, and while inefficient for any specific purpose, less normalized representations can be derived from it. For the first time, the choices of knowledge representations made in any given implementation could be explained with concrete reference to a logically derivable form of the data.

By naming data elements consistently, using entity-process-property-encoding conventions, it became possible to begin to map data within, and ultimately across, systems. Maker-birth-date and museum-acquisition-date or object-collection-place-geopolitical name and maker-death-place-geopolitical name, are two pairs of data values that will be expressed the same ways. This meant that we could define common routines to manipulate them and common indexes to search them.

By disaggregating, and by using formal decomposition methods and naming conventions, we were able to discover and then exploit common usages and usages that while different could be formally translated to be the same. Thus we were able to relate object-creation-geopolitical era to object-creation-date and object-creation-place and recognize that Ottoman Empire was both a time and a place, or rather is a place which has different boundaries at different times.

Experience in building museum systems had begun to convince some of its practitioners of the potential value of standards, in particular knowledge representation standards, which would govern how particular pieces of information, if present, would be expressed. The goal (though the term was not yet in use) was interoperability and exchangeability of data.

Multiple, Independent Authorities

Museum practitioners looked to libraries, which had been down the path quite successfully by the mid-1980s, for guidance in how best to represent knowledge for computer applications. In the context of 1980s information retrieval systems, which depended on pre-coordinated indices, libraries had demonstrated the benefits of "authority control" in searching for "known items."

Since then, museums have been on a quest for authority control, seeking the same benefits while neglecting the overwhelming differences in their holdings and of searching in the museum context. The results have been disappointing because of the significant difference between the presentist and retrieval orientation of libraries and the historical and contextualizing orientation of museums.

The most important difference between libraries and museums was so obvious that it was typically overlooked. Museum artifacts and specimens lacked "title pages" from which descriptive catalogers could "transcribe" the computer record. This difference in practice between transcription and attribution remains poorly understood in most comparisons of library, archives and museum documentation practices. Because publications almost always have known authors, titles and dates, searching in library catalogs is designed to retrieve known items. Artifacts and specimens almost always lack all these recorded metadata, so "known item" searching is quite atypical in the museum context.

Library catalogs, therefore, maximize the effectiveness of searching by collocating all items associated with a particular person, organization, or subject, by substituting "authorized" or "preferred" terms for data values that might otherwise be in the descriptive records. Hence, when persons were associated with artifacts or specimens in museum records, as their makers, designers, discoverers, owners, etc., they were frequently not in the Library of Congress Name Authority files. Often, of course, these people had not authored books, or they had under an assumed, literary name, which the Library of Congress "preferred," but even when the person was known to exist in a library authority, variations on names were considered "not preferred" and would never be used in the library setting. In contrast, museum practice would have dictated that all names by which someone had ever been known ought to be retained as each would be found in some documentation and reflected the point of view of the person using that form of the name. In their documentation, museums would normally use the name by which the person was known at the time and in the place that the attributed relation occurred. Similarly museums found that corporate creators of material culture tended not to be in the LC Corporate Names list because the names used by librarians reflected changes of ownership and not the name at the time of creation. Finally, name authorities were of little help in attributing works in the museum that had been made by anonymous individuals reflecting the methods and techniques known to their social groups. This applied to a large percentage of material objects, because attribution to individuals, rather than to groups, is a relatively modern and almost exclusively Western phenomenon.

An effort to develop a museum specific name authority for the arts, with more open ideas about the multiplicity of names an individual might have used during their lifetime or been referred to after their death, was undertaken by the Getty Trust (Bower, 1993; Siegfried & Bernstein, 1991), but it did not fully escape the limitations of prior practice. The practice of employing authorized or preferred names to reduce the particularism of the references to individuals within museum records erased the historical context of the associations between people and artifacts, and violated much museum practice and the architectures of searching through authorities to the object record that would have enabled the benefits of collocation without supplanting context, were insufficiently developed.

Similar difficulties bedeviled the use of other library authorities. Non-literary cultural works tend not to have "titles," and when they do these are likely to be multiple and not well known. Artifacts do have common names, and the names of everyday things were compiled into a valuable "nomenclature" by Robert Chenhall in 1978. This nomenclature had numerous properties of a thesaurus and could be used in part to locate items of similar types. A more rigorous thesaurus, with a more focused topic, was developed by the Art and Architecture Thesaurus projects from 1979 to 1990, but it limited its terminology to that which had warrant in the published literature rather than drawing more widely from the realm of museum curatorial and scholarly terminology within the realm of practice (Petersen, 1990).

Subject terminology from the Library of Congress subject headings proved to be difficult to apply meaningfully. The Library of Congress Prints and Photographs Division Thesaurus for Graphics Materials incorporated "topical" subject terms that were more appropriate, but could not overcome the fact that museum specimens and artifacts do not "have" subjects (Parker, 1987).

Thus, before the advent of the Web, it had become evident that the library community's approach to vocabulary control was inappropriate for museums. Whereas libraries decided that certain terms would best represent specific ideas to meet the information retrieval needs of their users today, the museum had three additional problems to contend with. First, the historical terms actually given to an object or concept over time were important properties of that object from the perspective of scholarly interpretation; more so than the information retrieval requirements of using today's language (Bearman, 1995a). Secondly, scholars almost by definition, do not agree on what constitutes reality; indeed one of the most important things to preserve in museum documentation was the on-going debate between scholars, and therefore multiple independent truths (Bearman, 1988; Bearman & Szary, 1986; Doerr, 1997). Finally, the perspectives of various users were all equally legitimate and quite different; the terminology that quite accurately describes a thing for one kind of user is not appropriate for another (Sledge, 1995).

Museums found themselves redesigning the purposes of authority files. Exemplifying a paradox of humanities computing, that it valued nuanced prose and a distinctive voice with which to address each clientele, museums resisted using authority control to engineer commonality. Instead, museums recorded situationally correct terminology and located it where and when it had been assigned, as well as by whom it was used and how it evolved over time. The knowledge representation needed to support this did not use controlled vocabularies to limit the terminology assigned to an object in the collection, but rather to expand it. By representing the range of what was known about people, places, actions and ideas without assuming the privileged position of the museum object, museums could architect search systems that positioned thesauri between users and the database so as to expand their search language (Bearman & Trant, 1998; Sledge, 1995). This meant that museums needed to learn the limits of term expansion both up and down thesaural hierarchies so that if a user searched for wood occasional tables they would find birch end tables and maple coffee tables (Bearman & Peterson, 1991).

Dozens of specialized vocabularies were developed from particular realms of curatorial practice at this time, but they in turn lacked both "literary warrant" and use outside their domain. Some were complex classification systems that could be applied by a trained observer (such as ICONCLASS) while others were ontologies specific to a narrow domain (such as the Railway Thesaurus). Discussions of these in the 1980s led nowhere, as there was no mechanical way to integrate them into a universal ontology nor any politically or intellectually acceptable way to give one precedence over another (Roberts, 1990). What the museum community discovered, long prior to the advent of RDF and the Semantic Web, was that it was necessary to employ multiple domain vocabularies since they served their communities uniquely, and that these could not be integrated in some universal ontology (Bearman, 1994). At the same time, a degree of integration between vocabularies could be supported by explicit sourcing of values.

Self-Consciously Universalistic Data Typing

By the mid-1990s, some museum knowledge representation standardization was gaining ground. Although commercial software developers had resisted interchange standards (Bearman & Perkins, 1993), fearing loss of market if museums could easily move data to other systems, they were adopting universalistic data typing as a common approach to defining complex data types.

Commercial applications developers had numerous clients for their software. Since they wanted both to make as few changes to their code as possible and yet name data in the ways their clients viewed it, they needed to be able to define their data "under the hood" in ways that could allow different museums, and indeed different curators within the same museum, to label and record it differently while maintaining a common knowledge architecture. They found that by supporting complex data types that were explicit about the intellectual system being represented, the tools and techniques of measurements, and the degrees of certainty of the recorded facts, and disaggregated the components of the attributes as fully as possible, they could support multiple world views.

Application systems in general use would support metric or Imperial measurements using two different hard coded routines, but museum software designers learned that museum application would want to record linear distance measurements using systems from other cultures and eras—such as the Greek (stadions and plethora) or Chinese (chi and li). In addition, some museums will want to record vast distances in astronomical units and cellular and atomic distances in angstroms. The range of possible measurement units required the disaggregation of a measured quantity, the unit, the system of measure, a dimension, degrees of certainty, and measurement methods (tools and techniques); in other words, the museum software developers learned to adopt self-consciously universalistic data typing practices.

Earlier we discussed expressions of time and periods of time. Self-consciously universalistic data typing required explicit recording of the dating system, period/era common name, early date year, early date month, early date day, early date expression of time, late date year, late date month, late date day, late day time, degree of certainty, instrument, method, when time was expressed numerically. It also required that other concepts of periods of time—such as political time-periods (World War II, the Ming Dynasty), cultural time-periods (the Victorian era, the Reformation), geological time-periods (the Pleistocene), personal time-periods (adulthood, pre-pubescence), each requiring different systems of measurement and different data structures for expressing values, be accommodated in the knowledge model.

Whereas place, or location of origin, in library descriptive cataloging means place of publication, and takes the values of geopolitical place names for a city and country, the location of origin for objects in museums can be geo-morphological, geo-cultural or geo-linguistic, and geo-religious locations as well as geo-biological/botanical regions or even extra-terrestrial (oceanic and outer-space) locations.

In sum, in the museum, the "properties" of things are consequences of acts of knowledge declaration, and the tactic for making such declarations work over time is to adopt a self-consciously universalistic approach to data typing. In formal structures to represent time, space, events, and physical descriptions systems of representation are always explicitly declared. Individual data values—"x"—are replaced by tables that can qualify the data value by answering: "x" by what calendar? By what projection scheme? By what measuring instrument? And, by whom, where, when?

CONTEMPORARY MUSEUM KNOWLEDGE REPRESENTATION ISSUES

From 1985 to 1994, we saw the rise and spread of the personal computer which changed the character of office work worldwide and penetrated the household to some extent. But its effect on everyday life was trivial as compared to the influence of the World Wide Web and telecommunications-based computing since 1995. Indeed society as a whole, worldwide, has been significantly shaped in the decade since 1995 by the evolving nature of computing. The reasons are probably very simple—computing was awaiting the full integration of multimedia before it could be a populist medium.

Since the invention of the World Wide Web and http, we've seen a huge number of specific innovations in computing that have shown potential for museum applications. But the fact that vastly growing numbers of people are connected to the Internet every year and that they are spending vastly more time searching for things of interest to them (Fallows, 2005), explains why museums are trying to make themselves known to this audience and to compete for its attention rather than any particular technical synergy.

The success that museums have registered in increasing the size and variety of audiences visiting them on the Web has in turn promoted interests within the museum in further extending access. Most museums have taken accessibility of their Web sites quite seriously and are implementing W3C standards to further extend audiences. Active broad- and narrowcasting using RSS newsfeeds, blogs, and Webcasts are becoming more common. A few museums have begun to invest in embedded computing and smart buildings and are using these to individuate the information provided to visitors and to support a range of visitor information gathering studies. Each of these has further implications for knowledge representation, as do the creation of collaboration environments, intranets, and multi-platform, handheld consumer devices.

Re-Usable Metadata

The availability of a wide range of communications outlets (content platforms) all fed from digital data representations has increased the possible payback from making re-usable data. At the most basic level, museums are finding reasons to adopt XML as their standard for representation of text. As a practical matter, this means that textual components have explicit data types (albeit labels and behaviors are still based on local schemas, but at least they are identified) and that metadata can be directly displayed on the Web.

Arguments to adopt data standards for more complex objects are being considered by museum management. Object-oriented multimedia and calculated visualization methods including 3-D virtual reality are additional trends that require new knowledge representation. Location-aware data, whether to take advantage of mobile customers or GIS displays, are particularly demanding in this respect. As the audiences we reach grow and change, customer relationship management becomes important, which requires new metadata about new entities.

In sum, though neither multimedia nor telecommunications were new to museum computing, the advent of the World Wide Web, which marked their unification and extreme popularization, effectively transformed museum computing and knowledge representation.

Metadata for Management

The Museum and Its Programs

The museum Web site is a source of information about the museum and everything that it does. As a consequence, the data published to the Web from museum computing systems now has to include data about the museum itself, not just its collections, but its hours, its staff, and its services. This means not only a massive expansion of the quantity of data, but the involvement of virtually every department in publishing that data and keeping it up to date (Booth, 1998). Ultimately this means the scope of museum metadata expands to include process and responsibility, as well as audience appropriateness, language, and accessibility metadata. And it implies a need to keep track of versions of documents updated to the Web site at different times and by different departments and individuals.

The Museum and Its Societal Obligations

The museum is a legal institution and occupies a respected place in society which subjects it to various legal and ethical obligations with implications for knowledge representation and metadata management. In addition to standard recordkeeping requirements imposed on all corporations, as a publisher the museum may have legal obligations to make its online presence accessible to the handicapped (in the United States under the Americans with Disabilities Act), or to protect minors from material that is considered inappropriate to them (in the United States under the Children's Online Privacy Protection Act). It is likely to have further obligations that are specific to its role as a collecting organization such as requirements to document the origins of human remains and grave findings (as in the United States under NAGPRA) (Grose, 1996), and will be required to document provenance of acquisitions of art that might have been looted or sold in contravention of laws protecting movable cultural properties (UNESCO convention and acts requiring return of Nazi- and Soviet-looted art). If it holds natural specimens, it will need to consider laws relating to endangered species and the documentation needs of the biodiversity community (see the Consortium for the Barcode of Life, http://barcoding.si.edu/index_detail.htm/). All these legal obligations imply requirements to represent knowledge about object provenance and interpretation in ways that support compliance with legal and reporting requirements.

Collections

In the 1980s, archives had substantial success in placing their metadata into library systems by describing their holdings "at the collection level" using MARC. Since then, EAD has been nominally accepted as a way of representing archival descriptions, which had a quasi-standard life as a prose genre, in XML. Museums also consist largely of collections, and some museums have been influenced therefore to represent their holdings in EAD. However, unlike archives, museum collections are largely artificial constructs (though some will have had an existence as collections prior to coming into the museum), and they do not have a pre-existing genre of prose descriptions to convert to XML. Representation of museum holdings at the collection-level is, therefore, not likely to be of any great benefit.

Public Interactions

Data placed on the Web will be interacted with, and can be made interactive or even open to community editing. Evaluating the interactions of the public with the data, and managing any annotations and uses made by the public, involves another layer of metadata. When the general public can search museum databases, museums discover the limitations of the data their collections systems hold, how inconsistently each facet of possible description is actually construed, and how much technical language is used in the data values. They often find that the public needs facets of description that are not usually employed by their curators. Some knowledge representations may be specifically oriented to the public, or even special age or interest groups within the public, while the same knowledge might be represented in a different way for internal use. For example, a botanical collection that curators use by searching scientific names of plants might need to be made accessible by common names, and an art museum collection which curators search by artist, title, date and genre might need to be made accessible to users who search for what they remember as the subject content of the images, something previously not catalogued at all (Bearman et al., 2005).

Because museum data on the Internet can enable two-way communications, museums find that they use it to build communities, which implies collecting metadata about possible clients. For example, the museum might create a Web form to gather memberships or reservations for a lecture series. Once the museum has experience with Web forms, they could inaugurate online forums, build collective documents, participate in interactive lecture dialogues or engage online chats. However, if museums seek to attract different audiences with their own specific requirements for museum content, and/or create mechanisms that permit people other than museum staff to add data to museum knowledge-bases, they will need to adopt sourcing for all their data. In other words, every piece of information will need to have metadata associated with it to say by whom it was created, when and under what authority (if any) and who owns it, and who can change it. No longer will it be acceptable that the contents of the museum databases speak "for the museum" and with that anonymous authority. Now it will be necessary for individuals to sign contributions to the database and speak with their own authority. By definition this reduces the abstract authority of the museum and brings it closer to the level of other institutions which can then articulate their views more equally.

Publication and Its Representation

As soon as the museum becomes actively involved in recruiting public attention, its publicity machinery must ensure that the public knows enough to participate in its programs. Information about where publicity has been released and what kinds of data have gone to which newsfeeds, must be kept along with the various releases. Relevant information, at a considerable level of granularity, must be keyed to where it has been made publicly available and to whom it is targeted. One potential strength of online information provision and online exhibitions is that the statistical preferences of prior visitors can be useful guides for subsequent clients. In order to construct systems that reflect what others have selected, whether using crude counts or sophisticated profile-based preference weighting, each "page" of display information, even those generated on the fly in response to queries and user-profiles, must contain its own history of use metadata. Recommender systems are constructed on such retained links and nodes. They can both support museum evaluation needs and assist users to follow paths that others have found useful.

Location-Based Knowledge

Once they are liberated from wires, users could be inside or outside the museum and their location, their paths, and even the direction of their gaze, become meaningful criteria in judging what information would prove relevant to them. Museums will need to make interpretive information sensitive to the location of users in order to meet wireless needs. Once knowledge representation begins to consider the location of the user, it adds value to reference the geo-location of a variety of facts in the life of the objects themselves—their creation, acquisition, and exhibition. This in turn makes it possible to deliver to users in those locations information about the events in the life of objects in a potentially remote museum. Knowledge representation that takes advantage of the spatial location of the user and the object throughout its life cycle will make museum information more relevant.

Experiences

Museums are dependent on making museum visiting a life-long habit, reinforced by good experiences. Museums are therefore at least as interested in building loyalty as any brand would be, and are beginning to realize that brand loyalty is reinforced by their using knowledge of their customers to satisfy known needs. Online visiting is no different, and in fact there are many advantages in the online environment since it is relatively easy to keep data about what people do and tell us, to recognize the customer instantly on arrival and to provide feedback about matters of interest that might have unfolded since their last visit or advise them about what other visitors with similar interests have been doing. Of course any such program requires that the museum create and maintain knowledge about its customers. This knowledge can't remain within the museum shop or files of the development office either; it must be fed into the delivery of online and onsite interactive and interpretive experiences. The knowledge representation requirements about customers can be considerable—involving tracking user behavior through exhibits and interaction events and building profiles that can individuate future experiences.

Multimedia Assets

As discussed elsewhere in this volume, metadata regarding surrogates has a growing place in museums. Data on the Web is multimedia, and the multimedia content is not just still images and graphics linked to text, but increasingly includes multi-modal, time-based data, such as animations, sound, video, and complex games, that have an elapsed performance time. The absence of universally accepted standards for encoding of multimedia has led museums to keep the same data as multiple MIME types. This, and the growth in absolute numbers of files, has resulted in implementation of media or asset management systems with huge overheads of metadata about media objects added to the knowledge representations. In the evolving jargon of museum systems, this is reflected in the use of the terms "content management systems" for the Web presence and "digital asset management system" for the in-house data objects in addition to "collections management systems" which keep track of the "real thing" as well as its surrogates and occurrences of its representations.

As the museum's investment in creating rich data grows, its need to keep that information available and accessible increases over time. The museum must consider life cycle information asset management as a priority, and this in turn means further investments in knowledge representation. If the museum obtains an image of a work in the collection by photographing it in response to a request, it needs only concern itself with keeping the request it received. But if the photograph is to be kept and digitized for future use, it will need to keep information about the digital asset (how it was made, how it is stored, etc.) in order to address version control and format obsolescence, along with the digital photograph itself. Archival issues, such as how long records are to be retained and under what authority they might be disposed, as well as metadata about format dependence needed to prevent obsolescence of formats while their content is still needed, must be addressed, and these in turn call for more knowledge representation about the authorities responsible for the content.

Sharing Knowledge

Museums have historically not been interested in the computer interchange of knowledge with other museums, though scholars and other cultural institutions see museum metadata as an attractive resource. Although many efforts have been made for twenty years, museums have not succeeded in finding a compelling reason to communicate with each other. Surprisingly, even the most obvious and best funded cases for benefits in such data interchange—the requirements of biodiversity researchers—has not generated implementations of museum to museum (M:M) data sharing. Specific projects to aggregate museum metadata have been forced to develop their own approaches (Bearman & Trant, 1998a).

While the objects museums hold are unique, and much of the knowledge that museums have about their collections and activities is likewise specific, an important component of museum knowledge is logically shared by other institutions. Indeed, the aspects of museum information that are shared by others are the most important properties from the perspective of anyone wishing to understand the world beyond the boundaries of the museum itself. It is through common data about people, places and things, methods and events that links between the particular and the more universal are made. There are times when it is preferable to link to, or incorporate, knowledge that is stored elsewhere rather than to replicate it redundantly within museum databases. In these instances, structural mechanisms replace data content to bring in related facts. For example, if the museum acquired an object at an auction on July 17, 1953, the auction catalog might be incorporated by reference. If it was, the description of the object as it was known then, and as it was represented for sale, would add depth to the fact of the acquisition on that date. Rather than recording in museum databases popular explanations of the differences between lithography and other printing techniques, we could create a link to the Museum of Modern Art exhibition which is devoted to this topic (http://www. moma.org/exhibitions/2001/whatisaprint/flash.html). Of course, before the museum uses links for any information that it deems essential to its operations, it will need to ensure that the links remain active. Pointing, linking and incorporating by reference data that resides elsewhere is a method of knowledge representation that builds collective knowledge and links the particular to the more universal. The "info URI" scheme (http://info-uri. info/) developed to reference analogue object identifiers—such as ISBNs or LC call numbers—holds much promise for the many obsolete identifiers found in museum data repositories (National Information Standards Organization, n.d.).

Changing assumptions about the architectures that will support future information use have influenced the museum community, like others, in its choice of metadata packaging strategies. In 1993, the CIMI Standards framework proposed using a MARC-like interchange standard and querying remote museum databases using Z39.50 (Bearman & Perkins, 1993). This assumed that distributed resources would be brought together in central databases or that they would be queried in a targeted search. An alternative approach available at the time would have been to specify a standard structure for museum databases, which was rejected as impractical. Since 1995, arguments have been advanced for "industry-standard" cross database solutions like SQL and ODBC (Open DataBase Connectivity) as mechanisms for data interchange, though they don't address the need to agree to common schemas. CIMI (in collaboration with the Dublin Core Initiative) argued for extending the Dublin Core as a model both for content interchange and query, or "discovery," but they could not agree on which extensions would be required. These approaches should no longer be considered.

The preferred approach at present is federated architectures based on metadata harvesting using the OAI-PMH protocol (http://www.open archives.org/OAI/openarchivesprotocol.html/), which the digital library community adopted in 2001. If the museum community was willing to adopt a basis of unqualified Dublin Core description, it could create implementation guidelines specific to its needs that would support federated harvesting using OAI-PMH and participation in harvesting efforts of other communities such as the Open Language Archives, or eventually those using LOM and METS standards. Tim Cole of the University of Illinois has been active in NSF and IMLS projects that are promoting this approach, but unfortunately little beyond conference presentations has been written that makes the rationale for this advocacy clear or demonstrates its utility.

The experience of the Web suggests that looser agreements between parties may still support a degree of integration that didn't seem practical without adherence to common standards in the past. But loose linking falls far short of interoperability. Metadata registries seem to be preferred by European initiatives, though the precise mechanisms by which they are supposed to work are still unclear (http://www.ukoln.ac.uk/metadata/). The W3C is promoting the virtues of RDF and the potential of the Semantic Web, which when deployed using a name space based schema declaration model does not require adherence to any given reference model. The Semantic Web model is appealing to some museums, though its full promise is still quite distant (Hyvönen et al., 2004).

Action-Centered Data Models

It seems likely that the Web will erode the splendid isolation of museums; if not because the museums find that they want to be part of a larger universe of information, then because the players in that larger universe increasingly appropriate museum knowledge and find ways to integrate it with their systems. Action-centered data models privilege a view that integrates resources of many cultural institutions (Bearman, Miller, Rust, Trant, & Weibel, 1999). By reinterpreting all facts as statements made about events, they unify time, place and ideas, the three remaining facets of Ranganathan's five elements other than objects and energy, and thereby link things with processes (Ranganathan, 1933). Together with the CIDOC Conceptual Reference Model (http://cidoc.ics.forth.gr/), this way of looking at relations as action linking entities emphasizes the unity in diverse objects that have been the subjects of actions such as discovery, invention, creation, publication, interpretation, analysis and presentation.

Although the development of an object-oriented version of the CIDOC relational data model was a logical next step for the ICOM Committee on Documentation to take, museum practitioners within that community did not take the lead in its development. Instead, academics took a lead, and ultimately brought the model to the International Standards Organization without much museum community input. Object-oriented models have had little influence on museum knowledge representation practices, but the CRM, and especially its support for event-centered views, has had an impact on knowledge models in the broader information community (Doerr, Hunter, & Lagoze, 2003; Hunter, 2002; Lagoze & Hunter, 2001). It may be ultimately that the model helps integrate views of heritage by serving as common frame of reference for understanding specific schemas (Lee, 2004).

CONCLUSIONS

Describing everything for every purpose and managing the data intelligently over time is not easy. It requires a great deal of self-consciousness about purposes. Explicit rationales for knowledge representation will be required to ensure others preserve the data properly. The ultimate payback will come when museum knowledge can be readily integrated with knowledge from other sources. This larger goal will require not just good knowledge representation practices, but the political and economic commitment of museums to cooperate.

Ultimately the environment itself will need to be built to support intelligent processes on the objects represented in it. We require a cyberinfrastructure of knowledge, with the toolsets and underlying methods to support connections between ideas and the objects that embody them. This vision is what drives the Semantic Web efforts, though I suspect that it may require less of a breakthrough in the means of representation and more investment in knowledge engineering.

5 The Information Revolution in Museums

Darren Peacock

University of South Australia

INTRODUCTION

Digital information has become ubiquitous in our lives. In the post-industrial information age, we produce more information than anything else. More than half of the population is employed in information related activities. More than ninety percent of the information that we create is in digital form. The digital deluge is seeping or flooding into every corner of our lives.

For individuals and organizations the "information revolution" seems to mean an inexorable increase in the volume and complexity of the information we create, consume and manage. The formats, devices and contexts in which we use digital information are also rapidly changing and proliferating. Our ability to author, access and exchange information grows apace. Every day, with our keyboards, voices and pointing devices we are tapping, speaking and clicking more digital information into being.

The ways we talk about information have also changed. Daily we are reminded of the emerging "knowledge economy" and the "information society." Information and associated concepts such as "knowledge" are much discussed, albeit from often divergent understandings. The value of these abstract, intangible entities for individuals, organizations and economies is increasingly recognized and frequently reiterated.

In many organizations, information is now recognized as an asset of value equal to or greater than their more tangible assets. Some argue that information is the key asset for any organization (Drucker, 1994); others say it is the only asset that matters. As an asset of recognized and growing value within organizations, it has become something which must be actively managed.

The recent discovery of the value of information does not surprise many museum professionals. We are, after all, used to seeing ourselves as information providers and knowledge specialists. Even before the explosion of digital technology and information in the 1990s, museum leaders have clearly understood that museums are in the business of information. Lytle of the Smithsonian is quoted as declaring in 1981 that "The Smithsonian is Information" (as cited in Orna & Pettitt, 1998, p. 29). In 1988, anticipating the electronic "global village" created by the World Wide Web, George Mac-Donald described the advent of the "information institution."

Information management is familiar and comfortable terrain for museums. Museums are typically early adopters of new information and communications technologies. This is evidenced by the uptake of database systems for collections management, the innovative use of multimedia and interactive digital technologies and the rapid proliferation of museum sites on the World Wide Web in the 1990s. More than most organizations, museums are likely to have considered how information is created and used, to value it, and to have established information policies and procedures to manage it.

Yet information management practices are not solely driven by changes in technology. They are social practices as much as technical ones. They occur in an organizational context of people, relationships and ideas. To understand their evolution we need to understand the evolution of organizations—their goals and behaviors—as much as we need to understand new technologies. As John Seely Brown and Paul Duguid (2000) remind us,

> The ends of information, after all, are human ends. The logic of information must ultimately be the logic of humanity. For all information's independence and extent, it is people, in their communities, organizations and institutions, who ultimately decide what it all means and why it matters (p. 18).

And that's where the real complexity begins.

The information revolution is not just about devices, data and connectivity, but about how we think of information. Museum information management is being transformed not just by technology, but by a paradigm shift in our understanding of the value and ends of information. Our earlier preoccupations with information-based technologies as tools for automation are giving way to the idea of information-based technologies as knowledge enablers.

What is emerging is a user-centered view of information as a service, rather than an end in itself or as an aid to manual processes. Content management is a new paradigm for information management that embodies this shift. It recognizes that the value of information lies in creating useful meanings for people. A content management approach provides a strategic, policy and technology framework for achieving this new service-oriented approach to information.

This is the story of one museum's journey towards that new paradigm. It represents an attempt to come to terms with the new challenges and opportunities of managing information "after the revolution." As this particular revolution is far from over, what is presented here is a moment in time, showing a new alignment of priorities and practices rather than a fully resolved conclusion. Adaptation is an ongoing process, but it is often useful to take account of the distance traveled, not least to light the way ahead.

New Museum, Old Problems

The National Museum of Australia (NMA) is a history museum established by the national government for the purpose of collecting, researching, exhibiting and disseminating information in relation to Australia's social and natural history. Although the idea of establishing a national museum in Australia is a century old, the NMA was not formally established until 1980. Not until 2001, as part of celebrations to commemorate the centenary of Australia's federation, did the National Museum of Australia open its doors to the public.

Despite its protracted gestation, the NMA shares a common information management history with similar institutions. In the 1980s it established computerized records of its collections using a series of homegrown and proprietary database systems. In the 1990s it eagerly embraced the Internet as one of the first Australian cultural institutions to establish a presence on the World Wide Web and as the original host of Australian Museums on Line (AMOL), a Web portal to Australian museum information. When the NMA opened in 2001 in a spectacular new building in Canberra, the national capital, it boasted a range of innovative interactive multimedia installations and a highly advanced technological infrastructure, including a broadcast studio for producing television, radio and Web content.

Like many institutions, the National Museum of Australia had progressed strongly with a range of information-based technologies, but the information and systems it used were poorly integrated. Under the pressure of constructing an entirely new museum in less than four years, the organization had achieved a remarkable amount, but had not yet established the foundations for effectively managing and integrating its information resources. After the celebrations to mark the opening of the new building, some more hard work began to build an information architecture to support the museum's future information needs.

Information management practices tell us a lot about the history and culture of an organization. Change proceeds at a variable pace, sometimes rapidly, sometimes not at all. In a typical pattern for museums, new information-based initiatives at the NMA were often driven by a particular project requirement or an exciting new technology. Although the museum had completed some successful projects, it lacked a strategy, policies and procedures for directing its information management practices and for managing its information assets.

In any organization where information practices are not coordinated and are left to evolve in divergent ways, using different, sometimes incompatible technologies, fragmentation of information content and management will almost certainly occur. It then becomes hard to gain a whole view of the organization, its information assets or activities. Communication across the organization becomes strained or breaks down and productivity and creative exchange are impeded. Ultimately, this leads to the "silo effect" where information is held in separate locations and access is diminished. At various stages in the life of any organization, the silo syndrome may emerge. In a period of rapid change, such as that experienced at the NMA, holistic, long-term thinking about information is often an early casualty. Getting back onto a convergent path requires vision, consensus and commitment, not just technology.

Integrating information management practices within museums is as much a cultural process as a technological one. Debate about information management issues and priorities often exposes fundamentally different perspectives within organizations. Established practices reflect and reinforce entrenched beliefs, customs and power relations. True to the professional type of organizational culture identified by Mintzberg (1989), museums value consensus decision making about strategic issues. Building a genuine consensus on the future requires a reconciliation of divergent perspectives that are often strongly held.

Experts in the field of information management such as Elizabeth Orna and Charles Pettitt (1998) suggest that a common barrier to the integration of information in museums is a perceived conflict between "outward-oriented activity designed to attract the public and provide exciting forms of access, and the work which is carried out entirely within the walls of the museum to record essential information about the collections and make it accessible" (p. 71). Yet effective information management needs to do both, often with the same information. According to Orna and Pettitt (1998),

> This should be one of the main issues addressed by any strategy for using information. It is not an easy task, and carrying it through is harder still, but there are enough hopeful developments to show that the aims of managing collections information and providing access are not mutually antithetical, that you can't have one without the other, and that they can actually support each other to the benefit of museums and their users (p. 73).

This is echoed in the observations offered by Ingrid Mason (2002) about competition for internal attention and resources between the very public presentations of the World Wide Web and the back end information systems supporting everyday business.

> The slick formats and whirligigs seen on museum Web sites and in exhibitions are not equaled in their sophistication by the information practices behind the scenes driving museum business (p. 16).

There is often a gap between outward appearance and internal reality.

Howard Besser (1997a) offered another explanation of diverging approaches to creating and managing museum information. He observed how technological constraints shaped the development of two distinct and separate "camps" of museum information practice—represented on the one hand by the collection database "driven by the need for record-keeping and inventory control" and on the other the multimedia exhibit "designed for explanation and access" (Besser, 1997a, pp. 160–161). At that time, Besser envisaged a growing convergence of these two traditions through the medium of the Web. Nearly a decade on, there has been some progress, but the cleft is still discernible in the practices and products of museum information management. Collections data is typically managed separately from Web content. Few of the rich data resources online are presented with the interactive flair of the multimedia tradition.

Certainly, technology itself has played a role in shaping and constraining the directions and focus of information management practice. Software incompatibilities have limited the possibility of data exchanges; processing, storage, bandwidth and other hardware constraints have all narrowed our horizons and cribbed our views.

Ultimately, a mix of socio-cultural and technological factors may account for this common lack of integration of information in museums. Whether silos emerge from organizational structures, competing priorities, or technical constraints, the effects are the same—poor access to information and barriers to integration. In the information age, the silo approach is unsustainable for any organization that wishes to participate effectively in the digital knowledge economy. Integration of information management objectives and practices is the only sound foundation for any organization. Those foundations must be built on shared assumptions, consistent terminology and common goals.

Of course the nature of an organization's information needs will be shaped by its unique origins, information assets and understanding of its role. A natural history museum will have more scientific data, an art museum more images and a history museum more text-based material. Most museums will have a mix of these and other types of information. Some museums conduct active external research programs generating further data through fieldwork; others will give priority to publication or to educational activities. Each difference in emphasis changes the ways in which an institution creates, values, manages and utilizes its information resources. As Orna (1999) observes, "Information means something special and different for each organization, so each needs to formulate its own definition of information in the light of what it wants to achieve" (p. 38). The information history and current situation of any one museum is unlikely to be the same as any other.

At the National Museum of Australia, in developing our information strategy we wanted to ensure that we took a view that was holistic, integrated, cognizant of the past and directed to the future. To develop the strategy, we accessed external expertise from the academic and government sectors, the multimedia industry, as well as peer institutions. Most importantly of all, we consulted broadly across our own organization.

In a museum, perhaps more than other organizations, information is everybody's business. Any attempt to improve our information management and integration across the organization's internal boundaries, competing priorities, and varied hardware and software platforms would require widespread involvement and support.

> If a museum is to make productive and profitable use of information, it needs not only to define what information means for it, but also to understand itself as a community of users of information, to recognize the 'stakeholders' in information, and to provide them with the means of negotiating over information (Orna, 1999, p. 19).

We ran internal workshops within and between different business units to listen to what staff needed to do their own jobs and what information they needed to share with others—both internally and externally.

This process recognized that museum staff were both producers and consumers of information. Often they were brokers too—mediating, filtering or packaging information from within the organization for external users. In fact, one of the key findings of this process was the interdependence of the information flows within the organization. Most key business processes depended on the flow of information. While this might seem obvious, the depth and extent of that interdependency was surprising. From the exhibition gallery to the collection store, crisscrossing the administration from finance, to marketing and development, supporting education, publishing and public programs, information made the organization tick and hum. Poor information flow is likely to stop or at least stall the business of the organization. Yet vital as it was, the ways in which information was made and exchanged were poorly understood and not effectively managed.

This discovery—that information is what makes things happen in organizations—has animated much of the recent literature in knowledge management. An axiomatic truth of knowledge management is the need to identify and strengthen those flows of information and knowledge within organizations. Before describing the approach taken at the NMA, it is probably useful to venture into the vexed area of the differences between data, information and knowledge, a dilemma that bedevils much of the recent literature about information and knowledge management. "Knowledge management" may just be one of those terms which will eventually be consigned to the trash bin of management faddism. Nonetheless, the debate that the literature has engendered does drive us to think more precisely about the value of information and how to use it.

Information or Knowledge?

Certainly, one of the themes to emerge from the workshops was the many different ideas people have about what constitutes information. The term information is used to describe a wide range of things that might also be described as facts, data, knowledge or understanding. Terms such as corporate memory, wisdom, expertise and practice all allude to a wealth of ways for describing the knowledge that exists within an organization. The various forms of "information" within an organization—written, verbal and electronic—add to the complexity and confusion. Similarly, the medium of communication (e.g., email) may be what the IT department is concerned with managing, while the content of the message is the information that concerns others. "Information" may be used to describe almost any form of record or communication. "Knowledge" and its synonyms seem to encompass something else.

Generally, most writing on the subject of information suggests a three-part continuum ranging from data, through information to knowledge. Checkland and Howell (1998) provide a good overview of some of the common definitions. Data is typically described as "raw" factual material, derived from observation or measurement. Information, strictly defined, is data that has been processed in some way to make it meaningful. Data is "transformed into information when meaning is attributed to it" (Checkland & Howell, 1998, p. 95), that is, through a human act of meaning-making. These acts of meaning-making include interpretation, reading, writing and speaking. Data exists regardless of meaning.

Knowledge, which is often conflated with information, is distinctly and crucially different from it. While information can be created, exchanged and circulated in a variety of forms such as documents, conversations, pictures and graphs, knowledge exists only within the minds of individuals, in the mind of a human "knower." There may be shared or common knowledge, but the organization does not "know" other than through its people.

Knowledge may be given expression as information that can be used for meaning-making by another person to develop his or her own knowledge, but the original knowledge remains with the original owner. I may know or have my own understanding of the same piece of information as you, but my knowledge remains limited to me, as yours is to you.

Orna and Pettitt (1998) explain this as an ongoing cycle of transformation as people communicate information—"knowledge made visible or audible" (p. 20)—back and forth by means of "information products." This process is essential for the ongoing development of knowledge. Information, they suggest, is "the food of knowledge . . . Just as we have to transform food into energy before we can derive benefit from it, so we have to transform information into knowledge before we can put it to productive use" (Orna & Pettitt, 1998, p. 20).

Knowledge is not a thing, so it cannot be "managed" in any regular sense of the word. Knowledge resides with individuals. What organizations can do is facilitate its development and exchange between people. Organizations can utilize and benefit from the "information products" which they engender, to use Orna and Pettitt's term. Knowledge making and sharing is a social process; technology plays at most a facilitating role.

At the NMA, having already experienced one failed technology-driven knowledge management project, we were skeptical about magic bullet IT "solutions" to our information needs. Maintaining a clear idea about the complex, human nature of knowledge and how it is created and exchanged is essential in avoiding the siren simplifications of some management theorists and software vendors.

Throughout the organizational knowledge debate that raged in the 1990s, people such as Brown and Duguid (1991, 2000) recognized that enabling and nurturing knowledge within the organizational context was not simply a technology problem. In their *Social Life of Information* (2000), they emphasized the importance of human learning processes and shared practice as generators of knowledge in organizations, including the concept of "'communities of practice': tight-knit groups of people who know each other and work together directly" (p. 143).

The notion of knowledge as a system of relationships and reciprocity rather than a thing to extract suggested to Brown and Duguid the metaphor of an "ecology" to explain the creation and circulation of knowledge within organizations. Thomas Davenport and Laurence Prusak (1997) had introduced the concept of information ecology to describe how organizations manage knowledge. Knowledge flows have also been likened to energy—intangible and pervasive.

For us, what emerged in our workshop discussions was the outline of our museum's information ecology, a patchwork of relations and exchanges that drew on and developed the knowledge of its participants. True to its human origins, the ecology was complex, fluid and highly variable. Most of the activity and few of its outputs appeared on the organization chart. The tangible manifestations, or "information products," were well in evidence—exhibitions, Web pages, publications, object inventories and management reports—but the processes of knowledge making were harder to discern. The information ecology that produced and circulated knowledge was also surprisingly indeterminate. Many of the connections relied on practices and processes that were undocumented either as human procedures or automated operations.

Information got managed and knowledge was exchanged, but by a largely intangible system with only rudimentary use of IT. Although there were many wells of information, there was no discernible water supply to connect them other than the human relations and communities of practice formed around particular activities such as collection management, exhibition development, school programs and public events.

There was a common complaint that the corporate information stored in databases, on the intranet and in procedures manuals was inaccurate, out of date or incomplete. The effect, but perhaps also the cause of this was a strong preference for and reliance upon the tacit knowledge of colleagues. Our IT systems were less important to managing information than our human information network.

The ecological metaphor is useful for thinking about how information and knowledge are made and circulated in organizations. However, attractive as the metaphor is, it may lead us in the wrong direction both logically and ethically, used as we are to engaging with "natural" environments in terms of hunting, mining and harvesting. Rather than viewing the information and knowledge "environment" as ripe for exploitation—as early knowledge management theorists might have—it is important to emphasize the social values of participation, reciprocity and collectivism.

Similarly, the frequently used alternative metaphor of "information architecture" often favored by technologists distorts our thinking in another direction, focusing on structure and engineering at the expense of the organic and free-flowing qualities of knowledge and information implicit in the ecological metaphor.

If neither ecology nor architecture best describes the dynamics of information management practice, another model of this system is required. To capture the complexity of the concepts involved I prefer to think of the goal of a "commonwealth of information," one that is shared, but which brings with it both rights and responsibilities. Active citizenship is encouraged and the good of the whole is paramount.

To shape our commonwealth of information we articulated six major goals:

1. Information integration
2. Collection development and management
3. Information access
4. Research
5. Commercial opportunities
6. External reporting

These goals represent a balance between outward-looking opportunities and responsibilities and the requirements of internal business processes, that is, making information easier to find, use and share.

The idea of a commonwealth implies shared ownership and responsibility for creation, management and use of information. It recognizes that knowledge in organizations is created by the free flow of information and the interaction of its members. Information is a common good for conducting the business of the organization; sharing is emphasized over ownership.

The Heart of the Problem

Whatever metaphor of organization—ecological, architectural, social—a museum uses to describe its system of information and knowledge management, the heart of that system is information about its collections. As Orna and Pettitt (1998) describe it, "At the centre of the museum's [information] requirements are the collections; all the other kinds of information which any museum requires depend on them" (p. 25). Few would dispute this assertion. What has changed however, in the wake of the digital revolution, is the end of that information, the why and the how of information management.

The desire to record, manage and optimize collections information has been the driving force behind the use of information technology in museums. The development and implementation of computerized collection information management systems has, at least until recently, been the major focus of information management practice in museums.

Katherine Jones-Garmil (1997) presented a comprehensive overview of the first thirty years of museum computing technology. Marty, Rayward, & Twidale (2003) updated and extended this chronology in a recent survey of the museum informatics field. What they and others such as Besser (1997b) have highlighted is the slow and often arduous evolution of IT systems and museum work practices to enable effective use of database technologies for creating and storing collections-based information.

Progress in implementing data management systems has persistently been hindered by a lack of agreed information standards within and between museums as well as access to hardware and the availability of software suited to the task (Marty et al., 2003). As Besser (1997b) points out, the automation of library information progressed in a more fluid way from title lists, to abstracts, to extracts and then to full text online versions. All too often museums became mired at the very first step: naming and describing an object. Lack of shared standards across museums limited the potential for data exchange, which libraries had mastered decades ago.

Describing museum objects is inherently complex because of the diverse nature and origin of collections, the overlapping disciplinary structures of knowledge in which they are described and the very fact of their uniqueness. As Marty et al. (2003) observe,

> No two museums can possess exactly the same historical object or work of art; even reproductions vary greatly in such crucial identifying features as size, material composition, and provenance. These factors not only make describing each object a time-consuming and individual task, they also make it very difficult to share this task among institutions (pp. 265–266).

Despite these obstacles, the creation of collections databases did proceed apace. By the end of the 1980s there was an established marketplace for proprietary database software designed for museum use. A decade later, a survey by the Canadian Heritage Information Network (CHIN) (1999) of IT use in more than 700 Canadian museums and heritage organizations of all sizes found that collection management was the most frequently cited reason for using information technology.

Most museums have now ridden that first wave of automation. The core of the collection information they have is stored and maintained within an electronic database, usually one of the dozens of proprietary software systems that have emerged in the past twenty years. However, the depth and extent of that core of information remains problematic. The frequently lamented lack of standardization in data sets, vocabularies and metadata usage has brought forth a profusion of localized approaches.

How this history is manifested in any particular institution may be difficult to predict. Given that most museum collections are as unique as some of the objects they include, the core of electronic information they have is likely to reflect local history and circumstance. Notwithstanding the significant achievements in standards development represented by SPECTRUM, CIDOC and CIMI, collection documentation remains balanced between art and science, highly vulnerable to the idiosyncrasies of collections, institutions and individuals.

Notwithstanding these problems of collection description and data exchange, museum automation has been able to achieve significant benefits in routine data processing tasks. After all, counting, listing and simple retrieval tasks are what such systems are ideally designed to do. As discussed earlier, data is the easiest of all information to manage. Typically, collection information systems managed a set of descriptive object information as fielded data organized into a relational database structure. Different systems varied only in terms of how and what data might be entered and the ease of searching or recombining different data elements for listing or analytical purposes.

Principally, these systems served the key functions of inventory and movement control. They were designed to improve the accountability and operational efficiency of processes for physically managing collections. Until recently, the design and the patterns of their use did not generally support the addition of extended information, although images were increasingly included as file size constraints diminished. Nevertheless, images usually served the purpose of identification rather than interpretation.

The content of these systems was tightly defined to "factual" information, (i.e., data) and squarely centered on the object. The interpretive activities of curators, educators and other researchers were recorded elsewhere, seldom in electronic form. Jim Blackaby (1997) neatly described this disconnect between the "base facts" of collection system data and the rich sources of information stored elsewhere.

Though a museum might dutifully measure and record the location of each of its blacksmithing tools, for example, actual information about blacksmithing is likely to be in a file folder in the educator's office or in the curator's reference files. An artist's name might be appropriately recorded for each of his or her works, but other information about that artist might be available in an article stored in a file folder, distributed among notes made by a researcher, typed on a wall label now removed to a storage area for old exhibit furniture, or be only directly available by reading a selection of letters stored in the archives (p. 205).

The various pieces of information that might form the building blocks of knowledge are typically scattered throughout the organization. Occasionally, perhaps for the purpose of an exhibition, the various pieces might be assembled into a greater whole and even possibly preserved together in the form of a printed catalogue. Yet little attention was given to sustaining or extending these aggregations.

Collection information systems didn't solve a museum's deeper information management problems. They certainly don't address the fragmentation of knowledge suggested in Blackaby's example. Instead, these systems achieved operational efficiencies by producing useful information for reporting and managing tasks and for answering simple queries on basic data. This was the major advance in museum information management practice enabled by IT until the early 1990s. Then, just as this first wave of IT-based automation from the 1980s had been assimilated, the next wave of innovation hit, driven by the Internet tsunami. The Internet not only changed the technology of information management, it redefined the problem.

Turning the Inside Outwards

The wave of innovation wrought in museums by digital technology in the 1990s can be seen as two waves that merged into one. The first wave—the proliferation of low-cost tools for multimedia production and playback— was subsumed and carried forward by the second wave arising from the exponential spread of networking technologies such as the Internet that enabled electronic communication and remote access to digital information on a scale barely imagined before.

The impacts of new digital technologies on museums have been extensively discussed elsewhere, including in this volume. After a decade, the impact of the Internet and in particular the World Wide Web has been strongly felt if not yet fully understood across all museum endeavors. Our ideas and practice in education, marketing, exhibitions and collection management have all been changed. Our ideas about the implications of the "digital revolution" for our information management practices have been less discussed. Yet these changes go to the heart of what the information revolution means for museums. Imperatives arising from the Internet create a new center of focus for museum information management practice. It is a profound shift not just in practice, but in the purpose of museum information management.

Stephen E. Weil (2002), elaborating on a much-quoted reflection on the changing role of the American museum, discusses the transition of modern museums from "being *about* something to being *for* somebody."[1] Weil describes this shift, "from the care and study of collections to the delivery of a public service" (p. 41), as an evolutionary transformation demanding new skills and attitudes within museums. The ripple of its effects extends far and deep. As Weil observes, "Traditional wisdom holds that an organization can never change just one thing. So finely balanced are most organizations that change to any one element will ultimately require compensating and sometimes wholly unanticipated changes to many others" (p. 40).

One of the most important manifestations of this shift from things to people, it seems to me, is the way it changes how we think about and use information. The effects of the "digital revolution" on museum information management practices seem to lend further fuel and force to the broader shifts described by Weil.

Information, like our institutions themselves, must exist to serve, not just to be. The paradigmatic shift embodied by the Internet and World Wide Web destabilizes the museum's traditional information management practices by positing the user rather than the object as the focus of our practice. Today, our information policies, procedures and systems need to be designed and geared to meet the needs of somebody or some bodies beyond the museum, not simply the administration of internal processes.

Just what those new needs are and how best to define and serve them is only slowly being discovered. The first major challenge is turning our focus and our information outwards. For museums, it seems, this is a more difficult shift than for our library and archive peers.

According to Ingrid Mason (2002), in contrast to libraries and archives, museum information management practice has had to be remodeled to meet the demands of Web technology rather than modeling it to its own design as libraries and archives have been able to do.

> Web technology has allowed libraries and archives to develop from the inside out. It has meant a simple shift to new technology—a push out onto the Web. Museums on the other hand seem to be developing from the outside in (p. 16).

Libraries led on the Web with a ready-made product—a catalogue designed for public use that could readily be transformed into an online service. Museums, in contrast, had collections data designed primarily for internal use. As a result, their Web offerings developed more strongly in other areas such as exhibitions, education and interactives. Museum collections information was generally not a ready fit for the World Wide Web.

Turning museum collections information outward to a non-expert public has been a much more difficult assignment.

After establishing a rich online presence, many museums are only now facing the challenges of systematically aligning information management practices to support sustainable online information delivery. As Ken Hamma (2004a) has recently observed, "the number of museums with Web sites is large, but the number of museums that have integrated digital knowledge management functions into their organizations is still relatively small" (p. 11). According to Hamma, there is a need for better alignment between the creation and maintenance of digital assets and the strategic goals of institutions. Managing the changes needed for integrated digital information management will be "a persistent, ongoing task" (p. 11).

At the National Museum of Australia, our experience of the Web had been like many other museums. An initial period of enthusiastic experimentation was followed by the sobering realization about the sustainability of HTML-based Web sites and the management of workflow processes to support them. The challenge for us was to move content production for the Web from the experimental periphery of the organization to the mainstream of our information activities. In adopting a content management approach, we built a sustainable infrastructure to support internal operations and external publishing from the same common base of information.

BRINGING IT ALL TOGETHER

Content management is a new approach to digital information management that holds great promise for museums. This model of digital information has arisen in response to the challenges and opportunities of the Internet age. Content management enables organizations of all types to create, manage and deliver digital information efficiently to a wide range of audiences in a variety of formats. It operates on the principles of "granular" information and flexible reuse. It is particularly effective in organizing rich and complex information into customized offerings for diverse users. According to Bob Boiko (2002), author of the *Content Management Bible*, "Content management is about gaining control over the creation and distribution of information and functionality. It's about knowing what value you have to offer, who wants what parts of that value and how they want you to deliver it" (p. 65).

This is particularly relevant to the needs of museums seeking to address the paradoxical problem/opportunity of the Internet. More than most organizations, museums are in the challenging situation of having large amounts of rich digital content and broad, diverse audiences to serve. Matching the two efficiently in the era of "mass customization" is no easy task.

Content management enables flexible organization and management of digital information so that it can be readily configured in different ways for different purposes and audiences. The same information that appears in a retail catalog could be used in an artist's profile or an education resource. The purpose of content management is to deliver rich digital content across diverse platforms—online, onsite, wirelessly, in print—to where audiences can use it. Information available on an exhibition kiosk can be published simultaneously to the World Wide Web or laid out for print publication.

Not only does this save time and effort, expediting workflows and avoiding "Webmaster bottlenecks," it dramatically multiplies the amount of content that can be produced by recombining granular pieces of information into different presentations in different formats and locations. For museums, content management offers an efficient, sustainable solution to our key problem in the digital era: how to make the rich and rapidly accumulating digital information assets we make and keep accessible to broad and disparate audiences in ways which are useful, i.e., customized to their interests, abilities and preferences.

According to Boiko (2002), "Content is information put to use" (p. 6). That use is the satisfaction of the information needs of audiences by providing what they need, where they need it, in the most useful formats. Our information is valuable only when it is used. To maximize reuse we must forget artificial distinctions between internal and external, collection and Web-based applications of information. Managing content for the Internet is not another problem separate from the problem of managing collections information. Managed in a granular way, with regard to all potential uses and users, museum information can become more useful to everyone.

At the NMA we understood that our information technology choices needed to support both internal information sharing—"the commonwealth of information"—and efficient publishing to the World Wide Web. Knowledge exchange within the museum would happen more readily if we enabled staff to create and share data and information easily and efficiently in integrated systems. Likewise, generating knowledge beyond the museum, in the minds of visitors or online users, depended on having a range of accessible information products customized to their preferences, needs and abilities.

Not surprisingly, no single IT system exists to do both tasks. Tools for Web publishing developed in the multimedia side and collection information systems evolved from the data processing tradition. Most content management software is designed first and foremost to expedite and streamline the publishing of Web sites. As a product that is still in its early stages of development, content management systems are highly variable in their capabilities and flexibility.

Any complex IT project is difficult enough, but we recognized that these were not separate problems and the synergy between the two objectives offered a substantial upside if we could achieve the alignment. From a technology point of view, the challenge was to acquire or develop software to meet the functional requirements of collection management operations and Web publishing. From an organizational point of view, the challenge was to ensure that work practices enabled the free and efficient flow of information within the organization and beyond with the minimum of bottlenecks and duplicated effort.

In embarking on our vision of integrated information management, we ended up implementing two "best of breed" systems—a traditional collections information management system and, in tandem with it, a content management system to handle publishing workflow and outputs as well as serving as the repository for non-collections information we wanted to publish online, to kiosks within the museum or to mobile devices. That information includes all the other essential information that a museum creates and disseminates: visiting information, exhibitions, an events calendar, education resources, retail and publication offerings, as well as the functionality that is increasingly standard on most Web sites: secure e-commerce, booking, subscription and other forms, discussion forums, personalization settings and personal workspaces.

Our choice of content management software was based on the need for simple authoring tools, multiplatform delivery, rich media handling and integration with the collections information system. We chose a customized version of an open source system developed by Massive Interactive, a locally-based Web and multimedia design studio. The system architecture we established dovetailed the collections system into the content management system, so that designated data fields are accessible via XML queries for presentation in various delivery formats online and onsite. The system architecture deployed at the NMA is shown at Figure 5.1.

For our collection information management system we chose KE's EMu, for its object-oriented database structure and its ability to manage contextual material, among other reasons. Descriptive object data entered into the collections information management system can be designated as "Web ready" for use by the content management system, obviating the need for rekeying and rehandling. Supplementary contextual material in the form of text, images, audio, video or multimedia files can be stored in either repository. We extended the scope of the data in the collection system to enable us to capture richer and contextualizing sources of information including images, multimedia, extended textual materials that may be only indirectly object-related.

Information that is created in Web ready fields is replicated automatically into the content management system using XML. In this way the content management system can dynamically create many different template-based presentations of the same data based on search queries, personalization tools and user interactions. The content resources of the museum are therefore effectively repurposed and presented both as users request them and in the various configurations the museum designs.

Increasingly, more people access museum information online than in person. Museum information management practices and systems have to be reorganized around this new fact. At the National Museum of Australia we

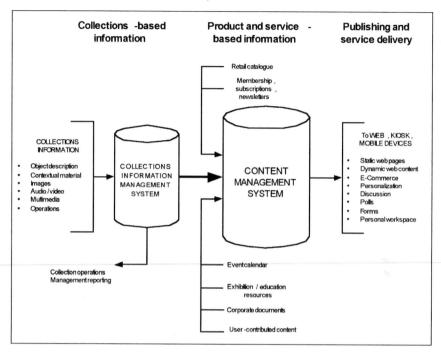

Figure 5.1 Content Management Architecture at the National Museum of Australia

have redesigned our information systems and practices in response to the needs and demands arising from the networked information society.

Firstly, we have consolidated our principal information asset—the tangible information about our collection—into a single system. This system is not only an information repository; it manages the workflow processes for all of our collection management operations. It is a living system accessible by everyone.

Secondly, we have extended the scope of the collections information system to accommodate a diverse range of contextual materials, not limited to simple object description. Thirdly, we have ensured that the data structures within our collection information system are aligned to those within the content management system so that collection information (and relevant contextual material) may be readily repurposed as online content without double handling.

Finally, we have streamlined the Web publishing process by enabling distributed authoring across the museum—from retail and events information to exhibitions and scholarly research. While the content management system provides the publishing engine, information in the collections management system will continue to provide the richest source of material. As technologies evolve the architecture will no doubt become less complex, but the new principles of information management we have established—audience focus, granular content and flexible reuse—will certainly endure.

CONCLUSION

In the Internet era, the challenges of museum information management have moved from process management to the realm of strategic management. Our information policies, practices and systems need to ensure effective utilization of our key asset—digital information—in all its forms. The access imperative arising from the digital revolution means that museum information management practices and systems have to be reoriented outwards to meet the needs of external audiences, not just internal process activities.

Established paradigms of information management and their associated technologies may blind us to the changing functions of information. The collection-centered model of information that was the foundation of museum information practice is no longer adequate in the networked digital age. We need to move from the inward-looking data processing model of the past to a focus on the delivery of digital content services.

Content management technologies based on XML and template-based Web authoring tools enable flexible and efficient configuration of information for multiple purposes and diverse users. The content management model adopted at the National Museum of Australia integrates the key information assets of the museum within a publishing system designed to provide a range of content experiences and services to users.

After the revolution, the value of our information management practices lies in their ability to deliver useful content in ever more diverse formats to growing, but ever more specialized user populations. The content management approach provides a model that is efficient and sustainable into the digital future.

ENDNOTES

1. He traces this idea through Joanne Cleaver to Michael Spock at the Boston Children's Museum.

Section 3

Information Management in Museums

6 Information Organization and Access

Paul F. Marty

Florida State University

The ability to create digital representations of museum information resources has transformed the way users of these resources work with museum collections, inside and outside of the museum. The widespread availability of digital objects (as well as digital collections of digital objects) has resulted in unprecedented levels of access, where resources once inaccessible to the general public are increasingly available for all the museum's users over the Internet. Accustomed to preserving information for use within their institutions, museum professionals are now spending more of their time disseminating this information outside of the museum. This shift in focus has led to fundamental changes in how museums are perceived and what role they should play in the information society, changes that have affected not only museum visitors but also museum professionals (Cameron, 2003; Coburn & Baca, 2004; Hamma, 2004a; Knell, 2003; Rayward, 1998).

Consider the example of collections management systems. While museums have for centuries employed paper-based information systems for managing their collections data, the recent availability of stand-alone and networked computer systems has encouraged even the smallest and poorest museums to purchase computerized collections management systems. Even museum professionals who do not wish or cannot afford to purchase "off-the-shelf" systems (which can range from a few hundred to tens of thousands of dollars) can develop "home-grown" collections management systems fairly easily using spreadsheet applications like Microsoft Excel or database applications like Microsoft Access or Filemaker Pro.

Digital collections management systems offer many advantages for museum professionals, including the ability to access, search, sort, and modify individual records instantly. The ease with which one can transfer database records over a computer network with no loss of data quality enables information from one record to be used by multiple employees, working with different applications, often simultaneously. As computerized database systems add more features and become more complex, museum professionals are able to manipulate records in ways never before possible (Coburn & Baca, 2004). A growing number of collections management systems offer features such as the ability to access controlled vocabularies or authority lists in real time, and the ability to track changes to records as different users access the system. Designers and developers are constantly experimenting with new features, such as the ability to conduct advanced queries where systems return records based on patterns found in the relationships between records (Dworman, Kimbrough, & Patch, 2000).

What is perhaps most remarkable about these systems, however, is how quickly museum employees can come to depend entirely on them. Individuals who had never before used a computer on the job can become so dependent on having access to computers that a power outage or system failure can make it impossible to perform work done by hand only a few short years before. Moreover, museum employees who are regular users of the museum's collections management system can become so dependent on the system's particular features that the mere thought of changing to another system (with subtly different features) can be devastating. The introduction and use of a digital collections management system in a museum has implications that go far beyond the technological; they can lead to a wide variety of sociotechnical problems for the museum's staff. It is important to consider the sociotechnical difficulties that can arise from the integration of information systems into museum work (Blackaby, 1997; Marty, 2000).

For instance, a new collections management system will usually mean that someone, often the museum's registrar, will need to supervise (or worse, actually perform) the transfer of data from old to new systems. The process of "crosswalking" these data, specifying how and in what format information from one system will move to another system, can be nightmare for all involved, whether one is going from a paper-based to a digital system, or from an older digital system to a newer one. Today, many developers of collections management systems can either handle these crosswalks internally or offer consultants capable of helping museum staff members get through the process; nevertheless, there is still the issue of training museum staff members on using the new system, and integrating the new system into the museum's current collaborative practices.

Most collections management systems do much more than track information about individual museum artifacts; they serve as electronic guides for all the museum's working practices: packing, shipping, conservation, registration, etc. The technological structure that governs how a collections management system stores information about museum functions—the screens one has to page through and the forms one has to complete when preparing an artifact for shipment, for instance—can have a profound impact on the social structure of the museum, and changing a collections management system can frequently mean changing the way museum staff members work. While some sophisticated information systems are capable of adapting to differences in workflow from museum to museum, the installation of a new system will usually require some degree of flexibility on the part of the museum's employees.

Sociotechnical changes can frequently occur in unpredictable ways. For instance, museum professionals moving from a home-grown collections management system (developed in something like Filemaker Pro) to an expensive, off-the-shelf system designed and installed by a museum consulting company can face many more problems than the technical difficulties of data transfer and conversion. Museum professionals implementing such changes often find that the difficulties of training museum staff members on the new system and helping employees adapt to new work practices pose more of a challenge than any technical problem. Abandoning a home-grown system can also have significant cultural or personal ramifications for the museum staff members responsible for developing that system, and other museum staff members may feel resentful or annoyed about having to change systems and standard operating procedures.

Access to digital collections management systems results in similar sociotechnical changes in the way museum visitors—inside or outside of the museum—make use of the museum's collections. As museums explore new methods of making collections information available electronically, museum visitors are frequently able to manipulate or interact with museum collections in ways never before possible. A growing number of museum visitors now expect to be able to access detailed records about museum objects on demand, whether or not those objects are on display. As the ability to do this becomes common, museum visitors continue to raise their expectations, desiring the ability to do things with digital surrogates impossible with physical objects, and to manipulate the museum's information resources in ways previously unavailable.

Recently some very interesting applications have been developed for museum visitors that focus on new methods of manipulating collections of digital objects. An excellent example is provided by Douma and Henchman (2000), who describe an online exhibit—http://webexhibits.org/feast— which allows users to examine in detail Bellini's *Feast of the Gods* and manipulate images of the painting through interactive technologies. Originally designed for art history students, this exhibit illustrates how much museum visitors are able to learn when given the ability to manipulate digital objects. Using a variety of interactive tools, users can "strip away" layers of this painting, examining earlier versions through infrared or x-ray lenses, and determining the exact material composition of any pigment used on the painting. This online exhibit, along with similar ones such as Manet's *Toreador*—http://www.nga.gov/collection/toreador.shtm—developed by the National Gallery of Art, demonstrates how offering individual users the ability to examine objects in ways previously unavailable (or available only to in-house experts) can make a museum's collection "come alive" and help museum visitors feel like active participants in the process of discovery.

Other innovative examples of users accessing digital collections demonstrate what happens when individual users are given the opportunity to explore relationships between objects, potentially discovering previously unknown connections. An excellent example of this phenomenon is discussed by Gillard (2002), who describes *HistoryWired*—http://history wired.si.edu/—developed by the National Museum of American History at the Smithsonian Institution. In this prototype exhibit, users are given the ability to manipulate a collection of 450 artifacts, looking for connections between groups of objects along temporal, cultural, and thematic lines. As users choose menu items and move sliders, groups of artifacts with common characteristics are selected and deselected, making visible such groupings as all daily life artifacts created in the 19th century, made from glass, and used for food and drink. The innovative graphical user interface employed in this exhibit makes these connections much more visible than they would be in a more traditional database system. Related systems, such as the *Revealing Things*—http://www.si.edu/revealingthings/—another production from the Smithsonian, have experimented with other methods of helping users discover unusual or unexpected connections between digital objects within digital collections.

As impressive as the technical abilities of these systems may be, even more impressive is the change in mindset such experimental approaches to digital collections management engender. Digital surrogates, digital collections, and the things one can do with them are tearing down old barriers—barriers of access, of understanding, of integration, of exploration, etc.—and offering all users new ways of manipulating museum resources. As these barriers are torn down, it becomes easier for museum professionals to meet the changing needs and expectations of their visitors, especially as their visitors become increasingly information-savvy.

Access to digital collections, for example, means that lines between the traditionally separate activities of collections management and exhibit design begin to blur. Museum professionals no longer need to maintain one source of data for internal use (e.g. by museum employees) and one source of data for external use (e.g. by museum visitors)—although such an approach does require additional metadata for tracking which data are suitable for different audiences. As an increasing number of online exhibits draw data directly from internal collections management systems, these exhibits become even more timely and interactive, representing the most up-to-date knowledge of the museum's curators and other experts (Besser, 1997b).

These kinds of dynamic online exhibits are not only the best way to meet the changing expectations of museum visitors; they offer new ways to draw people into the museum and to build stronger relationships between the museum and its visitors. When the Minneapolis Institute of Arts restored one of their paintings in 1999, they made it possible for the museum's visitors to follow the restoration online—http://www.artsmia.org/restoration-online/—in real time (Sayre, 2000). This approach was so popular and the museum's visitors were so entranced by the way this Web site drew them into the process and helped them learn more about the museum's behind-the-scenes activities, that the Minneapolis Institute of Arts repeated the project (with a different painting) in 2004. Similar projects, such as the online restoration of Sue the T. Rex—http://www.fieldmuseum.org/sue/—at the Field Museum in the 1990s, testify to the power of integrating behind-the-scenes activities and online museum exhibits.

Dynamic, interactive projects such as these that tear down old barriers of information access and use have the potential to revolutionize the way museum visitors approach museum objects. These changes can be unnerving, since the act of making things more accessible also involves giving up a certain amount of control over access to information. For many years, people have worried that easy access to information about museum artifacts might diminish or somehow threaten the sacred nature of the museum object. But along with the risks come many potential advantages, including the ability to build new communities, new interactivities, and new connections among digital collections and the users of those collections.

As the old barriers go down and new capabilities emerge, the focus within the museum is shifting from providing information about objects to providing the users of these information resources with the ability to make connections between objects (cf. Adams et al., 2001). Museum professionals can encourage their visitors to draw new connections between artifacts, store those connections on the museum's computers, and share them with other museum visitors. A growing number of museums (such as the Metropolitan Museum of Art, the Fine Arts Museums of San Francisco, and the Virtual Museum of Canada) offer visitors the ability to create their own personal collections of favorite artifacts and to share those collections with others. It is only a matter of time before museums allow users to add value by contributing their own knowledge to the museum's collections, building up the museum's store of information resources by adding new connections and new interpretations in a never-ending process of building new collections across user communities (Lynch, 2002).

In the 21st century museum, digital collections come alive thanks to the ability to create new relationships between users and objects, and to discover previously unknown relationships between digital objects in a collection or in different collections. While this focus on creating relationships and building user communities may for a time take museums away from the power of the individual artifact, the result of these changes will be unprecedented levels of information access and innovative abilities to work with digital collections and museum information resources in ways never before possible. By embracing the potentials of digital collections, museum professionals find new methods for cultivating life-long constituents for their museums.

7 Information Policy in Museums

Diane M. Zorich

Information Management Consultant for Cultural
Organizations

INTRODUCTION

It is a sad fact of history that cultural heritage often, quite literally, goes up
in smoke, disappears in conflagrations of nature, is sacrificed to the whims
of dictators or despots, or is destroyed in efforts at historical revisionism
(Battles, 2003). In ancient times, we have examples such as the burning of
ancient libraries in Alexandria and the destruction of Herculaneum's Villa
dei Papiri near Mt. Vesuvius. More recently we have witnessed the 1993
bombing and plundering of the Kabul Museum, and the looting of Iraq's
National Museum in 2003.

The fate imposed upon these cultural legacies was not of their making.
Like all social institutions, they were subject to the circumstances and events
of their times. The same cannot be said of a new threat that hovers over
today's cultural repositories: the wholesale acceptance of unstable, transi-
tory, and vulnerable new media technologies to create, store, distribute, and
access information.

It is hard to dispute the revolutionary impact that these technologies
have had upon museums. The revolution began quietly, starting often at the
grass-roots level and building into a crescendo as the technologies became
indispensable for conducting day-to-day operations. With increased use and
dependence, new challenges and risks emerged. Museums, often touted as
memory institutions, now store much of their information on media that
have notoriously poor preservation capabilities. They distribute information over digital networks that are subject to illegal (or at best, unaccept-
able) copying—of images, text, even institutional identities (as evidenced
through cybersquatting). Their information systems are subject to malicious
acts such as sabotage and hacking. The underlying technology infrastructure
can fail, leaving institutions unable to access their information unless redundant systems are in place. And the technology itself changes so frequently
that institutions often lose or ignore information in older systems as they
struggle to keep up.[1] These risks, while less sensational than those caused
by social or political turmoil, are more insidious. They are assumed with
little or any forethought, and when they suddenly surface, they can become

debilitating to a museum, threatening the foundations of an institution's information infrastructure.

Despite these risks, there is no going back to pre-new media days, nor would we want to. But the risks and challenges imposed by information technologies require a systematic approach that identifies all the issues at play and provides guidelines and governance on how they will be managed. This is the purview of policy—the creation of a "set of principles, values, and intent that outline expectations and provide a basis for consistent decision-making and resource-allocation" (Zorich, 2003, p. 16). This chapter explores the information policy issues that museums face today. It begins with a review of the nature and importance of museum information, and the history and the changes this information has undergone over the last two decades. Included in this review is a discussion of how museum information flows into and out of museums, and the role of information in the overall structure of a museum. Following this discussion is a summary of the critical information policy issues facing society and their impact on museums. Finally, the importance of an institutional information policy (and the types of issues addressed by such a policy) is presented as a strategic way to address the broader issues that information now presents.

WHAT IS MUSEUM INFORMATION?

To understand the policy issues that affect museum information, it helps to understand the character and use of the information itself. What do we mean when we refer to "museum information"? How is it similar to information in other organizations? How does it differ?

Collections are undeniably a key component of museum information, and thus the central focus of discussion and study on this topic. But the emphasis on *collections* information has overshadowed information derived from other museum activities. Critical information in other operational areas, such as human resources, facilities, finances, and general day-to-day administration are equally important to an institution's existence. So too are institutional histories, public and community relationships, and other "backstories" that offer insights into a museum's past and present roles. When considering museum information, we need to broaden our perception to include these other areas.

To do so, we must change our thinking of museum information from an object-based orientation to an activity-based one. If we consider museum information the sum total of all insights derived and recorded from museum activities (rather than objects), we are likely to encompass the fullest range of information that resides in our institutions. We are also likely to find that museum information has much in common with information found in business, industry, and other types of commercial enterprises. In the administrative arena, for example, museums are probably no more distinct than other entities. The information needed to manage personnel, finances, and facilities is consistent across a wide array of organizations. This consistency is largely the reason that "off-the-shelf" software programs for spreadsheets, word processing, databases, and email have been so successful. These programs provide functionality that suits the nature and needs of this type of information across organizations.

Where museum information does differ from other organizations is in the area of collections. Museums collections consist of objects that tend to be unique (not "multiples" of the same thing, like books or manufactured products), and this uniqueness does not easily lend itself to standardization or categorization. Collections information also is mutable: nearly every category of information about an object is subject to change over time. A further complication is that collections information is subject to a wide array of interpretations. The adjective "Persian" in a description of a "Persian carpet" can refer to a culture, a time period, a style, a technique, or even a dynasty, depending upon the traditions of an institution or a particular scholarly field.[2]

The myriad activities undertaken in a museum add another level of complexity. The model of an "object lifecycle" is often used to identify the activities that take place around objects. Starting with pre-acquisition and ending with storage or (rarely) deaccession, the in-between phases of this lifecycle can include anything from acquisition, cataloguing, conservation, exhibition, publication, education, interpretation, photography, and more. Each of these activities uses and generates information, resulting in another aspect unique to collections documentation: the information generated from museum activities is cumulative. Older and even outdated information is generally retained because it adds to the history, context, and physical and intellectual knowledge about collections. For this reason, the amount of information collected on museum objects (theoretically) has no limit.

The diversity of people involved in various museum activities also contributes to the complexity of collections information. Curators, registrars, educators, conservators, researchers, editors, and other professionals add their insights at various stages in the object lifecycle. These insights reflect different experience, knowledge, needs, and points of view of each individual and their respective roles vis-à-vis the collection. A curator, for example, may describe the materials on a necklace as "beads, bells, and cord," while a conservator may record the same materials in a more elemental manner as "glass, copper, and cotton fiber." The curator's description reflects the functional aspect of the object's materials and is more useful to her purposes, while the conservator's description represents the physical aspect more suitable for the types of technical analyses that conservators undertake.

Thus the nature, use, and interpretation of collections information make it more heterogeneous and dynamic than other types of information sets. These distinct and complex characteristics impose special demands which drive many of the information management decisions in a museum.

THE VALUE OF INFORMATION

Information is valued for economic, social, and intellectual reasons. From an economic perspective, we increasingly view and treat information as a commodity that can be bought, sold, or traded. Governments now discuss their "information industries" as a percentage of their gross national product. Information stored in databases is now made available by "subscription" or license, transactional processes that require payment in return for information.

Socially, we perceive information in the context of empowerment. Those who "have" information are at an advantage (social, economic, and intellectual) over those who "have not." The whole debate over the "digital divide" centers on the social value of information and the belief that equitable access to information, and the technologies that deliver it, confer equal opportunities for individuals in society.

Intellectually, information is viewed in the context of knowledge, and it is this aspect that underlies most museum information use. It is the raw material we collect and use to promote and foster knowledge. We put it into tangible forms that can be seen, heard, collected, compiled, and codified. While it may be interpreted differently from one individual to another, it can be shared in the same form by all. Visitors read the same exhibit label on the wall; museum staff see the same catalogue record on their computers.

The Nature of Information

Information has no inherent intellectual value until we transform it, using our cognitive abilities, into some type of knowledge. Knowledge is often described as a melding of information and experience to produce understanding or meaning. It is an intangible construct that varies from person to person. A teenager and an adult may read the same exhibit label, but can take away a different understanding from it. A curator and cataloguer may read the same catalogue record, but interpret it in different ways.

Elizabeth Orna and Charles Pettitt (1998) note that information and knowledge come together during the process of communication: "Information is what human beings transform their knowledge into when they want to communicate to other people. It is knowledge made visible or audible . . . and put into external containers like books, articles, conference papers, or databases" (p. 20).

A very basic example of information and knowledge in a museum context occurs in the creation of a Web-based educational program. Content for this program may be drawn from information located in many distinct areas: catalogue cards, research notes, publications, the curator's experience with objects, the educators' understanding of audiences and themes, etc. All the relevant information is gathered, read, translated, combined, and enhanced, and a new rendering of that information is produced in the context of a chosen pedagogical theme. This new rendering is presented to the viewer, who incorporates it with their own experience to elicit a personalized understanding or knowledge.

Because knowledge is so critical to the human condition, and information is vital to generating and imparting knowledge, it should be no surprise that humans are compulsive compilers of information. Earliest evidence of this aspect of our nature is found in Paleolithic cave paintings; the most recent manifestation exists on the Internet. What has changed over time is the scale: our accumulation of information is now so great that we are grappling with how best to organize, access, and use all the information available to us.

The Status and Value of Museum Information

Information is a key museum asset, on par with collections, staff, physical facilities, and finances. Yet it rarely merits a mention when museums list their assets. There are probably several reasons why this is so. Information is an abstract concept that is less tangible than other museum assets. Unlike a building or a collection, no immediate visual representation comes to mind when you mention "information." Museum information also cannot be easily quantified, like collections or staff, and thus may be excluded from consideration because a monetary value cannot be placed on it. Perhaps the real reason for its exclusion is also the most benign: information is simply taken for granted in museums. As a by-product of activities conducted around other assets, information is so engrained in every aspect of museum operations that it is never considered a resource in its own right.

This omission is undeserved and increasingly unwise. Consider what would happen if, by some quirk of fate, a museum lost all the information it had on its holdings. In the absence of information about collections origin, use, attributes, and aspects, the museum loses its ability to understand or impart any knowledge about the collections. Even if information can be partially reconstructed through contemporary research, the collections context, history of use, and all the other dynamics that result in the richness of museum information is essentially lost forever. Such a museum would become an amnesiac institution—unable to recall, interpret, or make sense out of the cultural history it sought to preserve.

Now consider a second scenario in which a museum's collections are destroyed, but the information about them survives intact. While the loss of objects representing our cultural heritage is devastating, the ability to impart knowledge about the collections continues because the information remains. For museums, whose mission is to promote and enhance knowledge and understanding, information is of critical importance.

Thankfully, both doomsday scenarios are unlikely.[3] But they do underscore how critical information is to the very nature of an institution: it is the building block for the very knowledge that museums seek to engender and impart. A museum that ignores information as an asset risks losing some or all of its ability to accomplish its mission should that information be compromised.

TRADITION AND CHANGE IN THE USE OF MUSEUM INFORMATION

The Users of Museum Information

The increasing amounts of information, and the ubiquity of digital technologies to access it, has broadened the user base for museum information to an extent never seen before. From ancient times until the 17th century, information in libraries, "cabinets of curiosity," and private collections was restricted to those few literate members of society whose wealth or scholarly endeavors afforded them unique access. This traditional base of users remains, but is joined now by constituencies as diverse as artists and artisans, the young and old, school groups and tour groups. The use of museum information has expanded from scholarly interests and wealthy patronage to pure enjoyment and entertainment by all segments of society.

What does this expanding user base mean for the management of museum information? Museums must now collect and deliver their information in ways that match the myriad interests of these constituencies. Fifty years ago, museums would not have thought it necessary to collect information on the gender and ethnic background of artists, but today they may do so because of an increasing number of inquiries requesting "works by female artists," "all works by Latino artists in Los Angeles," or similar requests based on gender and ethnicity criteria. Similarly, museums are delivering their information in new ways to meet the requirements of diverse audiences with mechanisms such as ADA-accessible Web sites, multilingual audio programs, streaming video of museum events, and a host of other accommodations.

The expansion and diversity of users offers museums new-found opportunities, but herein lies the dilemma: with so many audiences and so many needs, museums are torn by their eagerness to explore and assist, and their limited resources for doing so. As in other areas of museum work, priorities must be assigned and decisions made for how best to serve the information needs of today's audiences. Policies are needed to guide this process.

Management of Museum Information Resources

Before information technology became a fixture in museums, information was segregated into departments: each department would create, collect, and control discrete pieces of information relevant to its specific needs. For example, acquisition, loan, and traveling exhibit information might be under the jurisdiction of the registrar's office, while the finance office (or equivalent) would handle fiscal information, a conservation department would manage treatment information, etc.

Within this scheme, information entered each department from a great variety of sources (e.g., department staff, other museum staff, scholars, peer institutions, published resources) and went out to an equally diverse group, but control and management remained at the department level. Access was available only by contacting respective departments and following their procedures. Frequently requested information (usually collections information kept in a card catalogue or accession ledgers) might be placed in a common area where it was more widely accessible.

The separate management of information by departments was partly a limitation imposed by manual recording systems and partly a result of departments wishing to control information for their specific needs. In manual information systems, the only way to ensure information access to everyone is literally to make copies of all records *or* to have an "open door" access policy. The former option is an untenable undertaking, requiring extensive human labor and space. The latter option is also difficult, since departments could not risk lost or misplaced information resulting from an "open door" policy.

The tradition of information management in pre-automation days meant information did not "flow," but rather passed through a series of guarded checkpoints. While this appears cumbersome today, consider that demands for information were much lower in the past. The absence of automation meant activities were undertaken by different means and at a different pace. (No word processors, email, financial spreadsheets, or databases meant everything had to be typed/retyped, snail mailed, recorded by hand, etc.) It was a different way of doing business that pre-empted the possibilities of undertaking many of the types of projects and programs museums presently develop.

As automated systems replaced manual ones, a new, more expanded way of doing business has developed, but the tradition of segregating information by departments remains. A 1999 study conducted by the former Consortium for the Interchange of Museum Information (CIMI) (Sanders & Perkins, 1999) confirmed that over a decade of information technology use had done little to integrate information resources within a museum. It concluded that museums continued to use information in a fragmented, segregated way.

The CIMI study suggested that lack of standards was the issue limiting integrated information management in museums. But in the period since this study was conducted, another factor has altered the information dynamics in museums and may be driving the fragmentation of information even deeper into individual departments. This factor is the explosion of new projects that are the result of the pervasive use of information technology throughout a museum. Education departments, for example, now jointly create Web-based curricula, pre- and post-visit activities, and other pedagogical presentations with teachers and schools. Registrars' offices are digitizing collections, adding visual information to what had been largely text-based information systems. Exhibition departments routinely develop Web-based components to complement a museum's onsite exhibit. These activities result in a plethora of new resources being added to the museum's information base, but because they are most often undertaken at a department level, their information, and sometimes the resources themselves, are often not available throughout the museum.

Adding to this information explosion is the museum community's increasing reliance on external information resources to accomplish their tasks. Online tools, such as the Getty vocabularies, the Visual Resources Association's *Cataloguing Cultural Objects* guide, and the Library of Congress classification systems, are routinely used to catalogue collections. Curators and designers scour auction sites like eBay® to purchase items for planned exhibits (Kraus, 2000). Museums housed in academic institutions (such as those in universities or with in-house research institutes and libraries) subscribe to electronic information resources such as online journals, reviews, and image databases.

These external resources contain information compiled by others, and their use in museums continues the community's long tradition of developing rich, in-depth collections information drawn from diverse resources. From an information perspective, however, the entry of these resources into the museum, and the frequency at which they are being incorporated, often outpaces the rate at which they can be managed.

Another factor altering the information dynamic in museums is more sociological in nature. Automation speeds up the pace of activity in all areas of our work. We now routinely generate thousands of emails during the life of a project—a level of correspondence and contact that would never have been considered in the pre-email era. We attempt projects made possible only because of digital networks (e.g., the Nazi-Era Provenance Project, ARTstor, the Collections Exchange Center, etc.). We digitize collections at an assembly line rate because it now is possible to copy everything (once) digitally and make it accessible to all over networks.

This increased pace of activities fuels increased expectations. Museums are now expected to have Web sites, offer online database access to their collections, and provide visitors with an "offsite but online" experience. Similarly, museums expect that their programming will be made available in an online environment as well as at the museum's physical facility. And museum professionals themselves expect to locate information on the Web and are disappointed (irritated?) when they cannot find it.

All these factors—the increasing pace of activities and expectations, the proliferation of internal and external information resources, and the lack of information integration—has led to an unfettered growth of information in museums. In nearly all instances, the changes have occurred too quickly for our information management policies and procedures to catch up. We now find museums with more uncoordinated information resources than ever, resulting in expensive redundancies in time and technology, as well as lost opportunities in sharing and collaboration. It is not unusual to hear of instances in a museum where several departments are independently digitizing the same materials for separate purposes (instead of coordinating the digitization process to eliminate redundant efforts), or recording collections data in their own separate databases (because they don't like and/or know how to use an enterprise-wide system.) These kinds of costly situations abound when creation and use of information resources grow without strategic planning and oversight.

INFORMATION POLICY ISSUES

The changing dynamics of museum information has spawned a new set of internal policy issues that require better integration and management of this information. However, there also are external policy concerns affecting society at large that affect museum information management. These issues, and how they play out in museums, are discussed below.

The pervasive role of information technology in all aspects of our lives has engendered the phrase "the Information Age" to describe our era. As we move from an industrial age to an information one, societal ethics and mores have shifted accordingly, and we find that areas of little concern in the past have now risen to the forefront of debate. Perhaps the clearest example of this is the area of intellectual property, which occupied a fairly low profile in the legal community and in society at large prior to the digital revolution. Today it is the centerpiece of many key debates wrought by digital information and networks: creators' rights, the downloading of music, cybersquatting, and technological encryption all involve intellectual property at their core.

So many unanticipated issues have arisen with the digital revolution that as early as the mid-1990s, even governments, with their notoriously slow bureaucracies, began to respond. National information policies emerged to promote development in the new "information sector" and minimize the negative impact of this growth on society (National Library of Australia, n.d.). To read these reports a decade later, one is struck by how enthusiastic (naïve?) they are about the potential of information technologies to transform society for a greater good. But their prescience is equally striking: many of today's primary policy issues, such as the "digital divide," intellectual property concerns, and privacy rights, were first articulated in these early reports.

What are the major information policy issues that affect us today? How do they affect the management of museum information? Some of the key issues are grouped below by administrative, legal, and ethical categories. (Readers should note that this categorization masks the complexity of how these issues actually play out in the real world, where they tend to be entangled and overlap with one another.)

The Administrative Arena

Administrative issues affect the management of information from creation through use and sustainability. There are two broad concerns in the administrative arena: the use of information and the use of information technology.

Use of Information

Access to information resources refers to the ability to identify and use information resources. In museums it covers a wide swath of circumstances. Policy questions include:

- Who should have access to particular types of information and information resources?
  - Under what circumstances?
  - Within what timeframes?
- Under what circumstances can access be denied, or revoked?
- Can a museum adequately manage the level of access it chooses to provide?

Museums have many legitimate reasons for denying access to both their collections and information resources. Tribal communities may request restrictions for cultural or spiritual reasons. Access may be denied for purposes of donor-requested anonymity, contractual obligations, or concerns about individual privacy rights. Less clear are access restrictions that favor one individual or group over another. Should a scholar be granted her request to restrict access to collections she is studying until she has published her research results? Should museums restrict access to an information resource because it may have economic potential that the museum wishes to exploit? In practice, the concept of open access is fraught with exceptions.

Management of information resources issues revolve around resources available for information collection, integrity, preservation, and use. Some key policy questions in this area are:

- Who oversees the information resources in the museum?
  - What are the responsibilities of these overseers?
- Are the information resources effectively organized?
- What plans exist for sustaining information resources over time?
  - In the short-term (e.g., archivally securing the information with backups, redundant systems, etc.)?
  - In the long-term (e.g., migration/emulation)?

- Who deploys and trains the staff managing information resources?
- What finances are available for information management, and how are they to be allocated?
- What finances are available for information resources, and how are they to be allocated?
- What is the policy for retention of information resources?

Digital Rights Management (DRM) refers to the process of administering rights and delivering content over electronic networks. (It also refers to the software systems used to implement this process). DRM is forcing museums to consider the following policy questions:

- Should museums be charging for access (via DRM systems) to what are often public domain materials? If they don't charge fees, how will they be able fund the infrastructure that allows for digitization and delivery of this content in the first place?
- How does a museum reconcile the use of DRM systems (which technologically restrict access to content except when licensed) with the legitimate need for fair uses by the public?
- How can museums use DRM systems without compromising privacy?
- Will the DRM systems a museum uses compromise long-term preservation prospects (e.g., making it impossible to gain access to a data- or image base because it is encoded in an obsolete encryption schema)?

Use of Information Technology (IT)

Technology is the engine that drives modern information management and is an important part of an institution's information resources. The underlying technological infrastructure consists of technology products and services.

Safeguarding information technology systems is critical to the dependability of IT systems. Physical facilities must be safeguarded against physical accidents (i.e., burst water pipes, overheating) as well as malicious incursions (i.e., software viruses, hackers, sabotage, fraud). Users are one of the key issues in this area: protection can only go so far without responsible user behavior such as password use, respect for privacy, responsible email and ecommerce use, etc. Key policy questions are:

- What constitutes the information technology infrastructure in an institution? Is it computer hardware, software, and networks, or does it also include telephone systems, printing and mailing services, and digital library services?
- Who manages the physical facilities where information technology is housed?

- Who has access to these facilities? How is access gained?
- What are appropriate and inappropriate uses of these facilities?

Information systems are extremely costly, so maximizing their effective use is a key concern. "Responsible use" policies are cropping up in business and education, particularly in university environments, as a way to pro-actively streamline capacity and limit the misuse of costly resources such as storage and bandwidth. Areas of particular concern are broadcast messaging, peer-to-peer software misuse, and disruptive network traffic. Key policy questions in this area are:

- What are appropriate uses of the technology? Generally, these are defined as uses directly related to one's job, but increasingly employers allow for some non-job-related uses that, in the end, help improve employee morale and productivity, such as allowing use of company email for communication with family members during the workday (a practice emanating from so-called "family-friendly" employer policies).
- What are inappropriate uses of the technology? Clearly inappropriate uses include harassment, libel, pornography, game playing, spam, and chat rooms.

The Legal Arena

Information policy issues in the legal arena focus largely on the ownership and use of museum information. Because these areas tend to be more complex than administrative issues, they are only broadly summarized below. Museums need to consider each of these areas as issues that may affect their information management and policy.

Intellectual Property Law

Intellectual property law can place severe constraints on the use of information embodied in a tangible work like a song, a work of art, a photograph, or a database. Museums increasingly use these kinds of works in their online programming, where the ease of copying and widespread distribution poses particular risks to creators' rights.

Control of intellectual property in the digital environment is one of the biggest policy issues of the day. On the conservative side are publishing, music, and other large content industries that argue for tight control; on the other end of the spectrum are "copyleft" proponents who feel all information should be freely available. Occupying a middle ground are libraries, universities, museums, and other public institutions and individuals who support the intellectual property rights of others, but insist on a strong "fair use" provision for educational purposes.

Privacy and Publicity Laws

Our society promulgates the ethic that people have a "right" to privacy, i.e., to protect themselves against unwanted attention and any harm that may come from it. Aspects of this right are codified in our laws and usually prevent basic intrusions on a person's right to solitude, secrecy, control of information about oneself, and personal autonomy. Similarly, we cannot use a person's likeness or any other identifying characteristic for commercial profit (the right of publicity).

Museums collect and manage information on many different people: donors, collectors, researchers, creators, subjects of artworks and photographs, staff members, lecturers, etc. While museums have been sensitive to privacy and publicity concerns when using information in traditional museum activities, they are less aware of these issues in an online environment. Of late, governments and states are becoming more concerned about violations in this context, particularly in the area of collecting information from online users. U.S. legislation such as the Children's Online Privacy Protection Act of 1998 (15 U.S.C. § 91) and the Can-Spam Act of 2003 (PL 108-187), and the U.K.'s Data Protection Act (1998) are only the latest examples in what is likely to be increasing efforts to establish more online privacy laws.

Privacy issues also arise in the context of monitoring employee email and desktop computer use. The Electronic Communications Privacy Act (ECPA) (18 U.S.C. § 2510) allows employers such as museums to monitor employee email if certain criteria are met, but there are ethical and morale issues to consider should a museum elect to do this.

Cultural Patrimony Laws

The increasing activism of indigenous cultures has resulted in laws that protect cultural materials and inform indigenous groups about such materials in museums and other cultural heritage repositories. Since the enactment of the Native American Graves Protection and Repatriation Act of 1990 (25 U.S.C. § 3001 et seq.), U.S. museums with Native American collections have had to alter their information resources in ways that would allow them to comply with this legislation. For some museums (those with particularly large Native American holdings), the information requirements of this legislation made it necessary to develop entirely new information management strategies within their institutions.

The First Amendment

The critical freedoms of speech, religion, assembly, press, and petition that are guaranteed in this amendment have long been promoted and protected by museums, frequently at their own peril.[4] The tension between the idea and reality of free expression often comes to light over information. A most recent example is the USA PATRIOT Act (Pub. L. No. 107-56), which grants the U.S. government broad powers to gather information from public institutions without the knowledge of the individual under investigation, jeopardizing many First Amendment guarantees and intruding into the area of privacy rights.

Contract Law

Contracts often contain details about how specific types of information are to be used. They frequently, for example, have "confidentiality clauses" that prohibit the parties from revealing certain information. Conversely, they may specify that one or both parties are obliged to broadly disseminate certain information (for example, when promoting a jointly developed product.) The increasing use of contracts in museums to cover products and services, as well as less tangible offerings like "rights" (such as intellectual property rights), makes the information issues addressed in contracts extremely important for an institution.

The Digital Millennium Copyright Act (DMCA)

The DMCA (Pub. L. No. 105-304) increases intellectual property protections in a digital environment. Those who provide services on the Internet must set up certain "safe harbor" provisions, or risk being held liable for the actions of the users of their services.

According to Michael Shapiro and Brett Miller (1999), the DMCA will "have a significant effect on the online conduct of museums—as content providers, service providers or users of materials in the networked environment" (p. 100). The information policy implications of this Act are vast: it restricts the ability of a museum, as a user, to invoke fair use in a digital environment; it puts a museum, as a service provider, in an unwanted "tough guy" position of being an enforcer in IP infringement challenges; and it requires that, as a content provider, a museum be extra vigilant in posting third-party IP on the Web.

The Ethical Arena

The role that ethical issues play in museum information is rarely articulated, although these issues come to the fore just as often, and perhaps more often, than legal ones. Ethics are values that guide behavior and beliefs. They frequently overlap with areas of law, although ethical standards tend to exceed legal ones. In the context of information management, ethical issues often arise in the following areas.

Privacy and Confidentiality

Information about people, events, and objects can intrude on the privacy of individuals or groups. In pre-digital days, paper records offered somewhat of a natural barrier to these intrusions: the practical limitations of physical access made it less likely that private information would be disclosed. This barrier no longer exists in the digital environment, and privacy incursions are more likely to result unless other mechanisms are put into place to prevent them.

Another privacy dilemma occurs when social responsibility comes head to head with a respect for privacy, such as it does in the case of "Megan's Law" (Electronic Privacy Information Center [EPIC], 2003). For museums, this dilemma plays out in their role as public trust institutions. At what point do museums, which may receive taxpayer support and funding, have a right to keep any portion of their information private?

Confidentiality is an aspect of privacy that also is affected by policy concerns, usually in the area of access. For example, the dilemma between the public's legal right of access and U.S. federal agencies need for confidentiality led to the creation (in 1966) of the Freedom of Information Act (5 U.S.C. § 552).

Access to Information

Providing access to information is an underlying tenet of public institutions and a basic right promoted in a democracy. But access is a complex issue because it frequently abuts against other rights and pragmatic concerns. For example, most people believe that certain information, such as pornography, should not be accessible to children.

It is usually the less clear-cut access issues that pose dilemmas for museums. For example, fees often create an inadvertent barrier to access, but may be necessary to pay for the infrastructure that makes access available in the first place. In museums, where collections care and safety is paramount, access must often be filtered through a "gatekeeper" so as not to put the collections at risk. Access, for all its desirability, often cannot be granted on a wholesale basis.

Accuracy of Information

Information integrity is undergoing new assaults in the digital era. Morphing and manipulation of data can be seamlessly achieved and the deceptive use of these techniques is no longer the purview of tabloids or con artists (Wheeler & Gleason, 1994). A more insidious problem, however, is when information integrity is compromised as the result of an unwitting and inadvertent action. This is a particular risk with databases, which allow us to make spontaneous actions on our information that we may not have made upon further reflection, and which often cannot be undone. For example, a seemingly innocuous attempt to portray cultural sensitivity in data records by replacing objectionable terms with culturally appropriate ones may erase information that provides important historical context and insight into perceptions and prejudices of the past.

Ownership of Information

The question of who owns information intersects with the legal areas of intellectual property and contract law. While law will dictate who can legally claim ownership, pragmatics and tradition may hold sway in ownership cases. By tradition, U.S. universities do not claim ownership of their faculty members' lectures and publications, although they have the legal right to do so if such works are created as part of their employment. Similarly, a museum may relinquish its legal claim on information from an indigenous culture out of moral and ethical considerations rather than legal obligations.

INFORMATION POLICIES IN MUSEUMS

The History and Current State of Information Policies in Museums

The doyenne of museum information policies is Elizabeth Orna, who addressed various aspects of information management and policy in museums throughout the 1980s and 1990s (Orna, 2001; Orna & Pettitt, 1980, 1987, 1998). A handful of museum-based associations such as the Museum Documentation Association (now known as *mda*) and governmental organizations such as the Canadian Heritage Information Network also explored the periphery of this topic as part of their broader mandates to improve the creation and management of museum documentation. However, efforts to address information policy in the context of museums have virtually ceased since the late 1990s, with the notable exception of Michael Shapiro's presentations on digital information policy (1999, 2000a, 2000b).

The absence of dialogue about information policies is baffling, given the rising importance attached to information in all aspects of our society. Museums aren't averse to policy development, for they routinely develop policies in areas governing collections, facilities, and employees. Nor can museums be faulted for ignoring the demand for information placed on them by their audiences, since nearly all museums now use digital technologies to access, manage, and deliver their information. Then why the dearth of information policies in museums, and the even more troubling absence of discussion about the need for these policies in the museum community?

The answer may lie with a central point made earlier: information is not perceived as an asset in museums, but is instead treated as a by-product of activities. In the digital era, this perception will ultimately unravel, as the unique demands of digitization and networks force museums to increasingly turn their attention to information issues. Eventually, pressure to develop information policies will be fueled by the museum community's increasing reliance on information technologies and the growing perception of museums as providers of information services. Problems will arise so frequently, and the needs will be so great, that museums will turn to policy to systematically address the collection, use, and dissemination of its information resources.

Why Museums Need Information Policies

The rising importance of information in museums is reason enough to warrant a policy to administer and govern it, but there are other factors that also argue for an information policy.

Accountability

Museums are accountable to governmental authorities and to the public—both legally and ethically—for their assets. They must demonstrate wise and careful management of their funds, their collections, their facilities, and any other entity that is valuable to them in an economic, social, or intellectual sense. Since information is valued by museums in all these ways, it too is an asset, and is therefore subject to the same standards of care and accountability as other museum assets.

Accountability is a thread that runs throughout an institution. Staff are ethically obliged to transfer museum information to their successors and users in an enhanced form by incorporating and applying the latest cataloguing methodologies and information technologies available during their tenure. Senior administration are responsible for seeing that the staff members have the resources and leadership to do so. The Board of Trustees, legally bound by the "duty of care" obligation (i.e., to operate responsibly and demonstrate that they did so (Malaro, 2002, p.72)) must ensure that the museum's information assets are handled responsibly. All parties are accountable to their visitors, to those who use their collections, to those who donate or help procure them, and to society at large, with whom they hold a public trust.

One way of demonstrating responsibility is through policy. By creating and endorsing an information policy, a museum makes a public commitment about how it intends to govern and administer its information assets. In an article on collections management in museums, attorney Ildiko DeAngelis (2002) notes that the absence of a collection management policy is a dereliction of a board's "duty of care" obligation, because collections management is one the core activities of a museum (p. 91). One could argue a similar dereliction of duty exists with the absence of an information policy, since information management is also a core museum activity, as evidenced by the resources and attention devoted to it (e.g., IT departments and staff, technology infrastructure, technology budgets, etc.).

Efficiency and Increased Productivity

It is inefficient and costly to address museum information issues on a case-by-case basis. A policy offers a more effective way to manage information resources by outlining expectations and ensuring *consistent* treatment of issues throughout the museum. An information policy can help control the flow of information resources into and out of an institution, and identify how information needs to be internally managed through all its phases (from creation through disposal). This type of governance ensures the best use of human, fiscal, and time resources when managing museum information.

Another way information policy affects efficiency is by allowing for more informed and improved decision-making. With the clear articulation of principles and values that a policy offers, museum administration and staff have a solid basis for making decisions grounded in a careful and measured consideration of factors instead of impromptu or ad hoc "judgment calls."

And finally, information policy can increase intellectual access to the collections for both staff and the public. Guidelines that govern and promote responsible information management, collection, and use also result in improved information resources (such as catalogues, indexes, thesauri) that enhance search and retrieval in networked systems.

Integration

The advent of information technologies allows (physically) separated information to be (virtually) integrated, making it broadly available and accessible in ways not previously possible. In museums that still segregate their information by departments, an information policy can help guide the process of integration by addressing museum-wide information needs, the methods and technologies to be used to foster the integration, and the responsibilities of staff to support the process of making information accessible.

Credibility

Policy bestows instant credibility on its subject. Developing an information policy focuses attention on the topic, and helps promote it among staff, board members, and others who might not otherwise be attentive to the issues surrounding this arena. It also helps strengthen institutional identity by articulating the role information plays in the mission and goals of the institution.

THE COMPONENTS OF AN INFORMATION POLICY

Philosophy and Purpose

An information policy is a statement of principles, values, and intent about an institution's information resources. It is derived from a museum's mission statement, goals, and priorities, and defines the following:

- The objectives of information use
- The technology used for managing information
- The persons responsible for managing information (and their duties)
- The resources available for information activities
- The criteria that will be used for monitoring the policy (Orna & Pettitt, 1987, p. 8)

The purpose of an information policy is to facilitate governance of information created and used by a museum. Unlike procedures, which outline detailed methodologies for undertaking particular tasks (i.e., the "how to"), information policies are a broad-based statement of general principles (i.e., the "what and why"). A well-developed information policy is a progressive document that addresses the potential information has for furthering the museum's mission and relationships with its audiences and enables that potential to be reached. It should promote change and new information strategies, and not lock in traditions and old ways of doing business.

As a governance tool, an information policy allows a museum to acknowledge its respect of, and responsibility for, its information resources. It holds the museum accountable for a certain standard of behavior, helps pre-empt problems and accusations of wrongdoing before they arise, and provides staff with guidelines on how to administer and manage information resources.

Although information policies must conform to all local, state, and federal laws, they are not legal documents or instruments of law. A good information policy will, in fact, aspire to a higher standard than the law requires, taking into account ethics, traditions, and community values.

An information policy also addresses more than just technology issues. Many museums have IT policies that are mistakenly proffered as an information policy. Certainly IT, as the distribution and storage mechanism for information, is a critical component of any information policy. But governance of IT alone does not encompass the myriad other issues that arise with museum information such as records management, information ethics, access and use, privacy, intellectual property, etc. Information issues transcend technology: thus an information policy must address more than infrastructure.

Specific Components

Policies by their very nature are institution-specific documents created from local need and rooted in an individual organization's specific mission, activities, and organizational structure. Thus each institution must define what "information" means in the context of their organization. Orna and Pettitt (1998) provide a structured way to do this, and suggest that institutions begin by asking themselves, "What do we need to know to survive and prosper?" (pp. 23–24).

All policies, whether they address gender equity or information management, comprise the following basic elements (Zorich, 2003):

- Statements about the institution's mission and activities. Policies usually summarize mission and activities to remind staff of their larger institutional purpose, and inform users (who may not know) what the institution's *raison d'être* is.
- Statements about the institution's particular philosophy, values, and principles vis-à-vis the policy topic. Institutions usually enter into policy development with a clear ideology that guides their decision-making and governance about the policy topic. For a museum information policy, a statement of principle might be an acknowledgment that the museum "collects, manages, and disseminates information in the service of promoting knowledge," or that it "views information as a valuable organizational resource."
- Statements of the policy's intent and purpose. Policies include statements that give some insight into why the policy is necessary for the institution. Orna and Pettitt (1998) offer an example from the U.K.'s National Maritime Museum that could serve as a statement of purpose for that organization's information policy: "To make essential information available . . . within a planned structure (Information Technology) and to provide information which is accurate, credible, comprehensive, relevant, easy to use and timely" (pp. 204–205).
- Administrative statements. For a policy to be an effective governance document, it must (briefly) outline some administrative details such as who approved the policy, who will administer it, how it will be monitored, and when it should be revised.

In addition to the generalized statements outlined above, policies have specific statements that address various aspects of the policy topic. These statements vary greatly from one organization to another depending on the specific circumstances of an institution. Following is a list of some areas that museums *may* wish to address in their information policy if they are relevant to their institutional situation[5]:

- Access and security
- Archiving and retention

- Adoption of standards and best practices
- Building information resources
- Information integrity
- Information infrastructure
- Ownership and management
- Permissible and non-permissible uses (by type, by constituency, by technology)
- Preservation of information
- Relationship to other policies (e.g., IT, intellectual property, collections management, etc.)
- Respect for rights and principles (intellectual property, privacy, First Amendment, academic freedom, indigenous knowledge, etc.)
- Respect for the rule of law
- Risk management
- Roles and responsibilities of users

The mechanics of policy development are covered in numerous sources (California State University, Monterey Bay, n.d.; University of Minnesota, 2003; Weber, 2002; Zorich, 2003) and important insights can be gleaned from colleagues who have participated in policy development in other areas. The process can be time-consuming and requires institutional commitment, but the result will offer the museum guidance and insight for years to come.

CONCLUSION

Today's technology resources have propelled us into a world of information that both challenges old assumptions (e.g., intellectual property) and brings forth new issues for consideration (e.g., preservation of digital information). Museums, as social institutions, are affected by the circumstances of their time and must respond to the realities of the Information Age in ways that are consistent with their mission and role in society.

As museums expand their traditional roles to incorporate newer aspects as providers of information and experiences, they need to rethink the role of information in the context of their mission. They increasingly serve global audiences who will never physically visit their collections or facilities, but can still experience much of what they have to offer via virtual visits. More than ever, museum information will need to be digitized, contextualized, authoritative, and persistent.[6] The tradition of tolerance for the vagaries and idiosyncrasies of past information practices will need to be abandoned in favor of more rigorous information management and adherence to policies that govern the information arena.

In general, policy development tends to occur when an issue has reached a critical mass in terms of causing disruptions, confusion, or uncertainty within an institution. Perhaps information policies have still not gained currency in museums because things have not reached this point. But the issues are looming around the corner: intellectual property, digital preservation, persistence of Web resources, privacy, use/abuse of resources, information integrity, and ownership are just some of the challenges museums are now confronting and must address on a systematic and consistent basis.

Ultimately, museums collect information to promote and foster knowledge. Information policies will help guide and govern this process.

ENDNOTES

1. Consider how many databases and files exist on old storage media that have never been migrated over to new systems because time constraints have relegated them to the "we'll-get-to-them-later" category.
2. For an excellent discussion of the complexities in descriptive data for museum collections, see the "Discussion" sections of the various chapters in Cataloguing Cultural Objects: A Guide to Describing Cultural Works and Their Images (Visual Resources Association, 2006).
3. Tragically, a conflation of both these scenarios occurred when the National Library of Bosnia was bombed by Serbian Nationalist forces in 1992. International efforts to "reconstruct" the collections were partially possible because the library had sent microfiche copies of many of its holding to other libraries around the world. In addition, many scholars who had studied and used collections at the Library often had information, and in some instances, copies of works from the Library. (See David P. Johnson, Sr., "Scholars Help Bosnia Rebuild Destroyed Libraries," Northeast News, Dec. 1998: pp. 64–65. http://www.wrmea.com/backissues/1298/9812064.html)
4. Recall the problems caused by the Brooklyn Museum of Art's show Sensation, which offended the sensibilities of then New York City mayor Rudolph Giuliani, or the earlier First Amendment issues that arose when the Cincinnati Contemporary Arts Center hosted an exhibit of photographer Robert Mapplethorpe's work. See ArtsJournal.com at http://www.artsjournal.com/issues/Brooklyn.htm for a synopsis of the almost daily dramas that surrounded the Sensation exhibit, and The Cincinnati Post's Online Edition at http://www.cincypost.com/news/2000/mapple040800.html for a summary of events and the latest controversy stirred up by the 10-year anniversary of the Mapplethorpe/Cincinnati Contemporary Arts Center obscenity trial.
5. This list is not comprehensive. Museums may well identify other issues that have arisen in their institutions that they wish to address in their policy.
6. The use of the word persistent in this context means consistently and reliably available. This definition is increasingly applied to digital resources (e.g., persistent URLs, persistent knowledge bases, persistent databases) in an effort to promote the availability of digital resources whose location identifiers often change.

8 Metadata and Museum Information

Murtha Baca, Erin Coburn, and Sally Hubbard

J. Paul Getty Trust

INTRODUCTION

The word *metadata* emerged in the late 1960s from the information technology sector,[1] and has become ever more inescapable with the advance of digital technology. It is often defined, not particularly helpfully, as "data about data." This means that metadata provides information about content—in the museum context probably a document or work, or group of the same—that may exist in analog or digital form. Metadata may be defined as information by means of which we hope to not only identify and describe, but also to control and continue to exploit our collections, both analog and digital (see Baca, 1998b; National Information Standards Organization, 2004b). This is a very broad definition, and indeed some would argue that the term *metadata* has been stretched to apply to things for which perfectly good and more precise words already exist, such as *catalogue* or *registry*. Cataloging and metadata are in fact terms used by different communities to describe identical, or at least overlapping, activities that serve the same purpose: the management of collections. Because of its expansive brief it has proved necessary to split metadata out into various categories—descriptive, administrative, technical, and so forth—depending on its origin and use. Metadata can exist in almost any form, and again need not be digital, but it is most likely to be useful to a wide number of people over a long period of time when it is structured, semantically-controlled, and machine-readable.

What Does "Structured, Semantically-Controlled, and Machine-Readable" Mean?

Broadly speaking, structured, semantically-controlled, and machine-readable metadata is simply information that is not generated *ad hoc*, according to the idiosyncratic needs of a given project or individual. Rather, it uses the vocabulary, granularity, syntax, organization, and other elements set out in documented and shared standards. The advent of the Internet and later the World Wide Web made it possible for any appropriately connected computer to access and recognize properly formatted—that is, machine-readable—information, and exponentially increased the incentive to share and implement metadata standards. One obvious benefit of this was that it made possible much wider resource discovery than is possible if users are restricted to local card catalogs. Another and perhaps less obvious advantage of implementing metadata standards (and indeed data and technology infrastructure standards in general) becomes apparent as we grapple with some of the negative side effects of constant technological progress: observance of accepted standards is one of the most certain guarantees of future viability.

In the library world, data standards are often broken down as follows:

1. *Data structure standards.* These are the metadata elements or categories of information that can form a schema or "record" (in the intellectual sense). The Dublin Core Metadata Element Set (http://dublincore. org/documents/dces/) and *Categories for the Description of Works of Art* (http://www.getty.edu/research/conducting_research/standards/ cdwa/) are two examples of data structure standards.

2. *Data value standards.* These are the actual data that "fill" or populate a data structure or metadata element set; data value standards usually take the form of a vocabulary, classification, or thesaurus. LCSH, the *Library of Congress Subject Headings* (http://authorities.loc. gov), and the AAT, *Art & Architecture Thesaurus* (http://www.getty. edu/research/conducting_research/vocabularies/aat/), are examples of data value standards.

3. *Data content standards.* These are the rules or guidelines for how the data values should be entered or formulated within the data structures. AACR, the *Anglo-American Cataloging Rules* (Gorman, 2004), and CCO, *Cataloging Cultural Objects* (Baca et al., 2006), are two examples of data content standards.

A fourth type of data standards is *data format/technical interchange standards*, which are data structure standards expressed in a specific machine-readable form. MARC21 (http://www.loc.gov/marc/bibliographic/), the MARC XML schema (http://www.loc.gov/standards/marcxml/), MODS (http://www.loc.gov/standards/mods/), CDWA Lite (http://www.getty.edu/ research/conducting_research/standards/cdwa/cdwalite/), and the Dublin Core XML schema (http://dublincore.org/documents/dc-xml-guidelines/) are examples of these types of technical expressions of data structure standards.

When we put all of these types of standards together, what do we get? Cataloging—aka, descriptive metadata.

DESCRIPTIVE METADATA NO LONGER
A FOREIGN CONCEPT FOR MUSEUMS

Descriptive metadata is "used to describe or identify information resources" (Gilliland-Swetland, 2000, Categorizing metadata section, ¶1). It is the type of metadata that most closely resembles traditional cataloging, and it is made up of a combination of metadata elements and the data values—the terms, numbers, etc. used to populate those elements—that deal with what an object is, and what it depicts or means. Descriptive metadata is what makes it possible for users to search for, retrieve, and understand museum objects and their visual/digital surrogates. In comparison to the library and archival sectors, museums have been slow to embrace the practice—and concomitant rewards—of adhering to descriptive data standards. While they initially shunned data standards because of the perceived constraints they would exert upon their scholarship and academic freedom, museums are now beginning to embrace the notion that data standards can accommodate the variety and the occasionally contradictory nature of museum information. Cataloging, or creating and managing good metadata, is an extremely important part of a museum's work: in addition to creating efficient access to users both inside and outside the institution it supports registration and a host of other activities. It is resource-intensive (in both human and financial terms), but standards-compliant metadata and cataloging is more easily re-purposed, easier to maintain, migrate, and share, and more cost-effective and scalable. If data is already standards-compliant there is no need to re-invent the wheel every time one's institution wants to make its collections accessible to a new audience or in a new environment.

Data Structure Standards: I Use an Off-the-Shelf
Management System with a Built-In Data Structure,
and I'm Not Going to Build My Own System, So Why
Do I Need to Care about Data Structure Standards?

Data structure standards guide the way we structure information about our collections in information systems. Museums may think they are exempt from having to pay attention to data structure standards since many are not in the position to create their own in-house systems for managing and documenting their collections—nor do they necessarily need to do so. However, even in this situation, understanding how data structurally resides in an information system is extremely important for a number of reasons. Firstly, information technology continues to evolve, and as a result, museums are—and will continue for the foreseeable future to be—continuously faced with having to migrate to new systems. Knowledge about the structure and organization of data in a system makes it possible to make informed decisions when selecting a new system, and greatly facilitates the process of migrating data. Secondly, when consortiums such as AMICO, the Art Museum Image Consortium (http://www.amico.org/), were introduced, museums were faced with the need and desire to make their collections available to audiences outside of their own institutions in a "union catalog" environment along with the collections of other museums. Harvesting and federating services are also increasingly including museums' collections. In many cases, the ability to share information in this context involves mapping one's own data structure to a common data structure. There has already been a tremendous amount of research and work done in designing crosswalks that map the relationships of various data structure standards,[2] but utilizing these crosswalks requires that one understand how information lives in its local system.

Data Content Standards: Aren't Things Like Anglo-American Cataloging Rules (AACR) for the Library and Archival Communities?

The library and archival communities have well-established rules for the order, syntax, and format in which they enter data values about their holdings, primarily through data content standards like *Anglo-American Cataloging Rules* (AACR) and *DACS* or *Describing Archives: A Content Standard* (the heir to *Archives, Personal Papers, and Manuscripts*). However, while the museum community can and does sometimes refer to AACR to help with best practices for cataloging objects in their collections, data content standards (aka cataloging rules) better able to address the idiosyncrasies of unique cultural objects, as opposed to bibliographic library holdings, have long been absent. A recently published data content standard, *Cataloging Cultural Objects* (CCO) was developed expressly to fill this gap. CCO has the potential to facilitate enhanced access to and across cultural heritage and visual resources collections, if it is adopted as a shared resource by museums, libraries, archives, and the visual resources community for describing works of art, architecture, and material culture and their images. While there may be some in the museum community who revolt at having to pay attention to yet another standard, CCO should actually provide relief to museums and other heritage institutions. Listservs that traditionally served as forums for asking questions such as, "Do you construct a title for an object where the work type and purpose are unknown?" to "Do you express materials and techniques in the singular or plural?" can now point to CCO as the authoritative voice in such matters.

Data Value Standards: How, Exactly, Will They Liberate Me?

Data value standards have shown remarkable development through the course of varied implementations and interpretations, and as a result have become extremely effective at accommodating the growing needs and expectations of museum audiences when it comes to accessing information on museums' collections.

There is a common misconception in the museum community that having to use a controlled vocabulary means being constricted to using whatever the vocabulary considers to be the "preferred" term, even if a particular curator is not in agreement with that nomenclature. In fact, a vocabulary can liberate a cataloger or curator from having to follow unpalatable pre-scriptions in his or her choice of words. This is because while controlled vocabularies and thesauri are essential for ensuring consistent and accurate use of names and terms for describing the objects in a collection, superim-posing a vocabulary onto the searching mechanisms of a collection manage-ment system allows all relevant or equivalent words for describing an object to be treated as equally important access points.

To help further explain this concept, let's look at how one might cata-logue a *firedog*, used to support logs in a fireplace, by its object name in a decorative arts collection. If the cataloger were instructed by in-house guidelines to reference the AAT in selecting the preferred object name for objects in the collection, the cataloger would select the term "andiron." From an access point of view, a curator might search for this object by the name "andirons"; but a docent in the education department might search for "firedogs," while a scholar visiting from France might search to see if the museum has any "chenets" or "landiers" in the collection. These are all valid names for this type of object, and all valid access points.

In an ideal situation, the museum's collection management system would have direct access to published vocabularies such as the AAT and LCSH, as well as the ability to build and maintain local vocabularies and term lists. If this were the case, then users of the system could utilize controlled vocabular-ies when performing searches. In other words, a vocabulary-assisted search will automatically know that a search on "firedogs" also means a search on the terms "andirons" and "chenets." This may sound revolutionary, but in fact it is already possible with numerous commercially available museum col-lection management systems today. The key is for museums to start utilizing vocabularies which are already available to them in order to facilitate more effective access to their collections. If a collection management system does not yet have the capabilities to link to vocabularies or to perform vocabu-lary-assisted searching, then it is important that catalogers include equivalent terms and even broader (more generic) terms in the object record, in order to accommodate the various ways users may attempt to search for objects.

If You Build It, There's a Better Chance They Will Come: Freeing Oneself from Having to Adopt an Entire Published Vocabulary Such as the *Art & Architecture Thesaurus*

The above example shows the power of utilizing a controlled vocabulary such as the AAT to assist in retrieval and to free museum professionals from having to agree on a single "preferred" term or name for an object or an artist. However, museums are increasingly attempting to accommodate the growing audience of visitors and online users who are not subject experts, and therefore unfamiliar with the nomenclature that museum professionals use to name the objects in their collections. The words used to describe objects in this scenario might involve terms that are not found in published vocabularies such as LCSH, *Nomenclature for Museum Cataloging*, or the AAT; but it is still possible to create access points for these types of non-expert users.

As an example, let's look at *lekythoi*, vessels used to hold oils that were often part of funerary rituals in ancient Greece. A museum visitor who encounters these vessels on display might later go back to the museum's Web site hoping to learn more about them. If the visitor is not knowledgeable in art history or ancient art, it's possible he or she might try to find the vessels by searching for "containers" or "bottles." The AAT classifies *lekythoi* as vessels and jugs, but not as containers or bottles. Despite the fact that the AAT classification is correct from a scholarly point of view, the more generic and less scholarly words that might be used by non-expert users constitute important keywords or access points for *lekythoi* for non-scholarly users.

It is becoming more common for museums to create a vocabulary of terms specific to the materials represented in their collection. This can take the form of a simple, but controlled, alphabetical list of terms that clusters together valid equivalent names. It can also take on the form of a hierarchical structure, more like a true thesaurus, which may be referred to as a "local" thesaurus. Local thesauri are becoming very attractive to museums because they can fulfill the needs of audiences who are unfamiliar with the subject matter represented in a particular collection, and at the same time give data creators the ability to have deep and rich hierarchical structures wherein concepts and classifications are broken down into valid scholarly or scientific terminology. This allows the headings or object names that are meaningful and necessary for enhanced access by curators, researchers, art historians, and other subject experts, to be maintained.

The decision whether to build a local thesaurus, to utilize all or part of an existing published vocabulary, or do some combination of both, is an important one. While the initial labor in building a thesaurus can be burdensome, over the long term terminology applied to collections through a thesaurus will be significantly easier to maintain. For example, if the decision is made to change the preferred form of an object name from the plural to singular, editing only one term within a thesaurus is considerably more efficient than having to recall all the records in a system and change each one. Additionally, a local thesaurus will contain terminology that is entirely specific to the scope of an institution's collection, whereas incorporating an entire vocabulary such as the AAT means a tremendous amount of the terminology and structure will never be applied to an institution's collection. Factors such as time, resources, audience(s), intended use, accessibility of published vocabularies, and available tools for creating thesauri should all be taken into consideration when deciding whether to build a thesaurus or utilize an existing one.

The J. Paul Getty Museum is an example of an institution that has built a local thesaurus to serve the needs of both academic and non-academic users. The Getty Museum decided it would be beneficial to build a single data structure able to serve multiple audiences, as opposed to maintaining multiple controlled lists of terms or thesauri in an attempt to serve the needs of a variety of users. This is an important factor to consider in constructing local thesauri, or when indexing using terms from existing controlled vocabularies: no one published vocabulary is going to meet all of an institution's or collection's needs, just as no single application can serve all of an institution's needs in managing information. This is why there are a variety of vocabularies available to museum documentation professionals, including *Nomenclature for Museum Cataloging*, ICONCLASS, AAT, LCSH, LCNAF, TGN, to name only a few. The Getty Museum chose to build a local thesaurus because it realized that existing published vocabularies such as the AAT alone were not sufficient to meet their objectives for enabling non-expert access to the collection.

The Getty Museum's objective in creating a local, collection-specific, object type thesaurus was to create a way to search the collection by object type for all sorts of users, from a visitor with no art history background and little knowledge about the holdings of the museum's collection, to a scholar wanting to discover if the Getty Museum has any black-figure *volute kraters*. The thesaurus resides in an application that is part of the museum's collection management system: that is, it is maintained and managed in one application. Portions of the thesaurus are then re-purposed, or brought over into different applications depending on the needs of the given audience and environment. The thesaurus was designed to allow the data to be flexibly re-purposed in a variety of different systems or settings, and the choice was made not to structure it according to the museum's six curatorial departments, or even its seven different collecting areas, in order that users should be able to effectively search for an object without knowing the internal structure or collecting policies of the museum.

For example, in the object type thesaurus under the heading *Serving Drinks* is a list of the names of those vessels in the museum's collection that fit this classification (Figure 8.1). This list is not made available for the audience on the Getty's Web site or in the public access kiosk system, located in the galleries, because it was determined that the names were too specific and too unfamiliar for the users browsing the collection in these environments. Instead, the selection of the heading *Serving Drinks* in the museum's public access kiosk system (Figure 8.2) produces a more user-friendly results screen, made up of thumbnail images of relevant objects (Figure 8.3). Similarly, vessels that form part of the Antiquities collection and the vessels that form part of the Decorative Arts collection were combined into user-friendly

Figure 8.1 Screen capture of the J. Paul Getty Museum's local object type thesaurus showing the generic heading *Serving Drinks* located under the *Decorative Objects and Vases* category. © J. Paul Getty Trust

categories such as *Bottles and Pots* and *Dining Vessels* on the Web site and in the kiosk system in order to assist users browsing by object type. A user selecting *Serving Drinks* will see a results list of images that includes *volute kraters* from the Antiquities collection and *tea services* from the Decorative Arts collection (Figure 8.4).

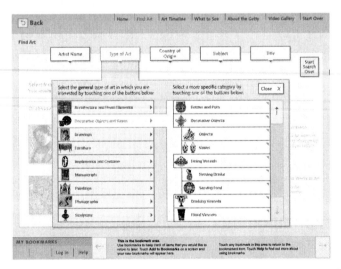

Figure 8.2 Screen capture of the J. Paul Getty Museum's local object type thesaurus presented in the museum's public access kiosk system. © J. Paul Getty Trust

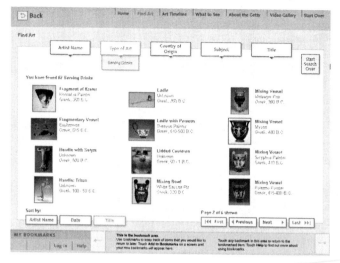

Figure 8.3 Screen capture of results for *Serving Drinks* in the J. Paul Getty Museum's public access kiosk system. © J. Paul Getty Trust

However, the "correct" names of the objects in the collection are of great importance to curatorial staff and other users who know specifically what they are looking for, and internal users of the museum's collection management system are able to reap the benefits of vocabulary-assisted searching: a search for specific vessels such as *volute kraters* within the museum's database will retrieve all relevant objects, including those associated with

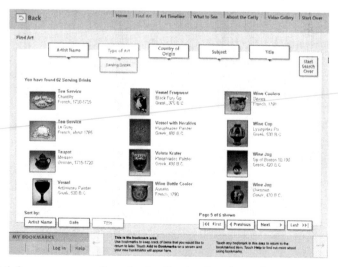

Figure 8.4 Screen capture of results for *Serving Drinks* in the J. Paul Getty Museum's public access kiosk system, showing objects that come from both the museum's Decorative Arts and Antiquities collections. © J. Paul Getty Trust

equivalent terms such as *volute craters*. Additionally, a user could simply search on *kraters* in the collection, and through the power of the thesaurus retrieve all types of *kraters* represented at the Getty Museum, such as *bell kraters, calyx kraters, column kraters,* and *volute kraters.*

The above discussion highlights an important issue: the question of how to express a thesaurus takes on a completely new dimension when data is re-purposed for the Web, and potential users therefore include anyone who has Web access. It is possible to make certain assumptions about visitors to a museum's Web site, for instance that they include people interested in visiting the museum, educators putting together art curriculum, students doing a class assignment on art, researchers furthering their scholarship, individuals looking for works of art that were illegally obtained from their family, and so forth. However, powerful commercial search engines such as Google are now also drawing users into museums' Web sites who may not have been interested in museum collections *per se*. This new audience is one which museums want to reach, and they are attempting to captivate these serendipitous guests. The ability to browse collections by thesaurus-powered headings such as *Serving Drinks* at the Getty Museum, or classifications such as *Arms and Armor* at the Metropolitan Museum of Art is an effective way to introduce collections to a completely new audience with no preconceptions of what those collections might contain.

However, there are also users who have a specific search in mind when they come to a museum's site, and in fact the expectations of these users are very high, largely because of their experience with search engines such as Google. They expect immediate, relevant results, even when searching with a museum's site-specific search engine. For users such as these, the Getty Museum utilizes its data structure and local thesauri to facilitate retrieving relevant and exhaustive results from the collection, by transporting the local thesaurus terms that have been applied to objects in the database over to the keyword tag on an object's Web page. The keyword tag also provides a place to add further descriptive words that Web-crawler-based search engines can read and index, thus enabling additional access points that may not be available through the visible text displayed on the page. (These tags exist in the source code of Web pages, and are invisible to the user in normal Web viewing.)

The same strategy allows the local thesaurus to serve both scholarly and non-scholarly users. For example, a user performing a keyword search on *drinking vessels* will retrieve examples of vessels from both the Antiquities collection and the Decorative Arts collection, such as *oinochoai* or *tea services*, because the thesaurus' hierarchical structure has been embedded in the keyword tags for the Web pages devoted to those objects, enabling them to be read by the search engine. Similarly, an antiquities scholar could perform a search on the Getty's Web site for *denarii*, Roman silver coins from the late third century BCE until the mid-third century CE, and would retrieve a coin from the collection in which nowhere in the visible text of the page does the word *denarii,* nor its singular form *denarius,* appear. This is made possible because the Getty Museum's local thesaurus has a facet dedicated to coins, with the specific names of the coins listed as "children" of the more generic term. The data is then carried over to the keyword tag for the coins on the museum's Web site.

It is important to point out that inclusion of thesaurus terms in keyword tags is just one option for improving searching. If a museum were to superimpose a thesaurus upon their internal search engine, for example, then vocabulary-assisted searching would be possible without having to duplicate terms, equivalents, or the hierarchical structure from a thesaurus in the keyword tag. Moreover, commercial search engines may not support the keyword tag. Google, for instance, relies upon other criteria to retrieve pages. However, the Getty's internal search engine does read the keyword tag. While a lot of traffic is driven into Web sites through commercial search engines, it is extremely difficult to keep up with their ever-changing search algorithms. It is therefore advisable to start with the internal search engine used by an institution's own Web site, whose method of reading and indexing content is presumably known and where changes can be controlled and planned for.

Another potential strategy for improving access to collections for both scholarly and non-scholarly users, besides building a local thesaurus, is to utilize or effectively "localize" a published vocabulary. When users of the Getty Museum's public access kiosk system requested the ability to search for objects according to style and period, such as Impressionism or Baroque, the museum obliged by creating a timeline that displayed this information. When creating the timeline the museum considered building a local vocabulary of style and period headings relevant to the material represented in the kiosk. However, given that their collection management system provided access to the AAT, and that the AAT has a whole hierarchy dedicated to style and period terms, there was no need to build a local thesaurus in this case. The museum therefore indexed the necessary objects from the collection, linking them to their corresponding style and period terms from the AAT (Figure 8.5).

The Getty Museum did, however, encounter the potential dilemma that they wanted to use headings that were not considered scholarly, or that were not the accepted style and period headings included in the AAT. For example, the Department of Photographs wanted to be able to use thematic headings for its collection in the timeline such as *Discovery and Invention, Early America,* and *The Sixties.* These headings are of course not part of the AAT, nor would the Getty Vocabulary Program, which produces and manages the AAT, want the museum to contribute them as possible candidate terms. However, the fact that these are not "scientifically accurate" styles or periods doesn't mean that the museum is prevented from using them to describe material in their collection. The museum simply chose to build their unique thematic headings directly into their licensed copy of the AAT

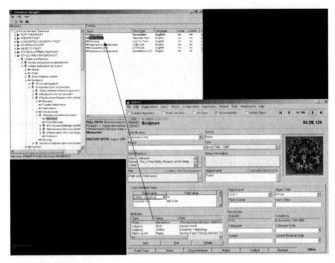

Figure 8.5 Screen captures from the J. Paul Getty Museum's thesaurus and collections management system, showing the connection between linking a term from the *Art & Architecture Thesaurus* with a museum object record. © J. Paul Getty Trust

and designate them as "local terms," citing the public access kiosk system timeline as their source (Figure 8.6).

In addition to the thematic headings created by the museum, the style and period terms from the AAT were also brought over to the timeline, where they serve as access points to the collection (Figure 8.7). Building the timeline with terminology from a controlled vocabulary, linked to object records within the museum's collection management system, opens up many possibilities for re-purposing data. For example, the museum can now choose to bring these style and period headings over to the keyword tag for objects available on their Web site, giving searchers the ability to retrieve tables in the Rococo style even if they don't know the specific object names for those tables used by the Decorative Arts department.

ADMINISTRATIVE, TECHNICAL, AND PRESERVATION METADATA

Descriptive metadata, discussed above, is used to describe or identify the content or "essence" of resources, which is generally only subject to occasional and gradual change (for instance, if a painting is reattributed through some scholarly epiphany). There are other aspects of any resource that are likely, in contrast, to be subject to constant modification. An object may be moved to a new location; it may be bought or sold. Its legal status might change from copyrighted to public domain. An analog object may be digitized either to create a surrogate or as a preservation measure,[3] or a digital

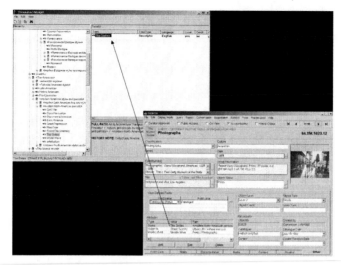

Figure 8.6 Screen captures from the J. Paul Getty Museum's thesaurus and collection management system, showing the connection between linking a local term built directly into the *Art & Architecture Thesaurus* with a museum object record. © J. Paul Getty Trust

object may be migrated to a new format with new technical requirements, and it might be that the new version does not fully replicate all the features of the old. As electronic and digital works enter museum collections, this last factor is becoming increasingly significant, and raising new and sometimes disturbing questions. For example, what does one do when the cathode-ray tube television sets used in a piece by Nam Jun Paik burn out and cannot be replaced because they are no longer manufactured? If one replaces them with new technology, is it still the same work? How does one both preserve a work and guarantee its authenticity when it is subject to built in obsolescence, or is it even possible to do so?[4] Even those museums that do not collect electronic and digital works will almost certainly maintain collections of digital surrogates, such as images of the analog items in their collections.

This section will concentrate upon metadata that documents the progress or evolution of museum artifacts, that records benchmark characteristics and values and the changes they undergo, and that may itself be used as the basis or trigger for future management decisions. Such metadata has usefully been divided into administrative, technical, and preservation categories (Gilliland-Swetland, 2000), and it will inevitably accrue over time and itself need careful management if it is not to become unwieldy. Administrative metadata, as the name implies, is used in activities such as acquisition, registration, auditing, and rights assignment; and it can apply to physical as well as digital assets. Such metadata is already rather well understood in most museums, and collection management systems are generally designed to hold it, but it is likely these same systems will not be so able to manage

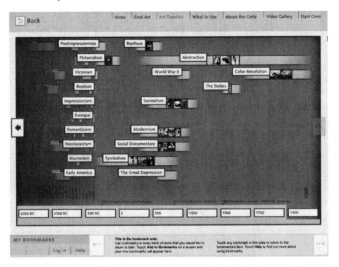

Figure 8.7 Screen capture of the interactive Timeline in the J. Paul Getty Museum's public access kiosk system. The Timeline uses headings from the *Art & Architecture Thesaurus*, in addition to local terms built directly into the *Art & Architecture Thesaurus* by the museum. © J. Paul Getty Trust

technical and preservation metadata, which applies primarily to digital or electronic objects, or objects that are reliant on some form of technology. Technical metadata, as the term implies, documents technical factors such as hardware and software requirements; recording, capture, or scanning settings; formats; or authentication and security measures (which may impede or block access or preservation). Preservation metadata tracks preservation activities such as refreshing, migration, and quality assurance. Note that while digital asset management systems are generally good at holding technical metadata, at the time of writing there is very little commercial support for digital preservation, and very few management systems are robustly able to either hold preservation metadata or handle long-term preservation itself, though this is a situation likely to improve over the next few years.

It should be noted that this division of metadata is not universal, and that the classes themselves are not mutually exclusive. For example, both preservation metadata and technical metadata may be viewed as primarily subsets of administrative metadata, and most technical metadata will be highly relevant to any preservation strategy. Data standards for these kinds of metadata are not as mature as those for descriptive metadata or cataloging, but they are evolving rapidly.

At this point a discussion of the nature of digital information might be useful. (While the discussion will confine itself to digital phenomena for the purposes of simplicity, much of what follows also refers to the wider world of electronic phenomena.) The notion that once something has become digital it has somehow been made permanent lingers, but in fact the strength of digital files is in such things as their transportability, or the ease with which they can be processed by computer or reproduced—and they are actually likely to face an early death. While digital files hold out at least the theoretical possibility of perfect duplication down the ages, they have no absolute, independent existence, and no definitive physical form. Rather, they exist on one of a variety of physical carriers, such as hard disks, floppy disks, or CDs and DVDs, all of which have limited life spans and which may not be accommodated within the technological landscape beyond a few years even if they survive in robust condition, and in specific formats that may similarly cease to be supported. It has been argued that it is not actually possible to save a digital file, but only to save the ability to reproduce it, in the sense that digital entities are so highly mediated that every time a file is opened it may manifest itself differently—depending on the equipment, specifications, calibration, and other factors in play—rather as every performance of a musical or theatrical piece will be different.

In short, digital files are hard to control and maintain, and a preservation and technical metadata strategy is an essential component of any digital management policy. It is helpful to think of metadata and the file content it describes as irretrievably united, and as together constituting a single asset. It is metadata that provides the access points to digital content that is otherwise opaque. (One cannot hold a naked CD, one with no labeling or other contextual information provided, up to the light and know that it contains a particular work of art. To find that work one must first of all know where to look, and secondly being able to read the CD, and for the reasons listed above this may be difficult to achieve.) Metadata document and facilitate both day-to-day management and preservation measures such as migration, the translation of files from old to new formats; emulation, the running of old files in their original format on new systems; or recreation, the attempt to reconstruct or reinterpret file content based on the available information. Metadata can provide documentation of works that have otherwise disappeared entirely, or provide the means by which we know that works are authentic and reliable.

Given their potentially extremely short life span, it is prudent not to separate the issues of long-term preservation and routine daily creation of and access to digital assets. It is in fact as necessary to develop and enforce a digital preservation policy as it is to have a conservation policy in place to protect analog originals. However, because of the nature of digital objects preservation must begin much sooner in their lifecycle than is the case with analog objects, ideally at the point of creation. Not all preservation policies will be identical, as each must be adjusted to the type of collection and institution in question, but it is possible to map out some of the likely rules. Perhaps most important is making the long-term management of digital files a routine, everyday requirement of business, rather than an afterthought in a series of digitization projects. In fact, museums (and other institutions) need to stop thinking about the creation and dissemination of digital resources as discrete *projects*, and start thinking about them as *programmatic activities* (Hamma, 2004a). Beyond this it will be necessary to set—and enforce—appropriate data and metadata standards, and to create or subscribe to a trustworthy archival digital repository.

All content should be created to the highest standard possible, using standard and preferably open-source formats, and all content should be immediately linked with metadata that describes it. Content not created in standard formats likely to be supported for a considerable period of time should be converted to standard formats wherever possible. This may involve a trade-off between functionality and normalization. Unfortunately, the more unusual or avant-garde the technology used, the more quickly a digital asset dependent on it is likely to become inaccessible when, for instance, the sole vendor of the required proprietary software goes out of business. All format and format migration information should be captured in the metadata of every asset. A unique and persistent identifier should be immediately applied, perhaps created by following a documented file-naming protocol. This identifier will itself essentially act as metadata, and ideally will echo or match whatever identification scheme is used within the broader museum. For example, an object's ID number may be reproduced in the file name of the digital image of that object. In purely practical terms it can be difficult to search and retrieve objects when no systematic naming protocol is in use, as anyone with a hard drive full of objects called file1, file2, file3 knows.

All descriptive, administrative, technical, and preservation metadata deemed necessary to ensure future viability should be created and/or gathered and bound to the digital content by being entered into the collection or asset management system in use and associated with the appropriate file name or other identifier. It is always preferable to choose published, open standards, and more and more technical, administrative, and preservation metadata structure, value, and content standards are becoming available. If it proves necessary to develop qualifications or variants of existing schemas, or even completely new schemas, these should be documented and shared. Note that some technical metadata is generated automatically in the "headers" of digital files, which contain information such as file size, format, and dates of creation and modification. (It is this information that is displayed when, for instance, "View/Details" is selected when viewing folders in a Windows system.) It is possible to expand the information that is held in file headers to include other useful information, and conversely is it also possible to extract information from file headers for inclusion in whatever management system is used. In an ideal world all metadata on digital content would be stored both internally within files and externally within management systems, thus minimizing the chances of any asset becoming disassociated from its metadata and unrecognizable, and in fact this is likely to become increasingly practical as the tools available for the management of digital collections develop.

One tool that is available for creating a digital preservation policy is the Open Archival Information System (OAIS) reference model, also known as ISO 14721:2003 (Garrett, 2005). While this model might seem rather arcane, its greatest value is that it provides a simple way of conceiving of—and therefore planning for—the lifecycle of digital assets from ingest into a digital repository to dissemination on the Web. OAIS has been widely shared both within and between communities, and it quite clearly and simply lays out the many different points at which metadata is likely to be required for managing digital assets, and demonstrates that metadata needs will change over time. It postulates "Information Packages"—content and its metadata—moving through archives from producers to consumers. The "Submission Information Package" (SIP) is created by the producer and presented for ingest to the archive, which can place restrictions on the type of formats accepted or the quality or size of files, stipulate which metadata elements must be included, and so forth. The "Archival Information Package" (AIP) is created within the archive, where the additional preservation metadata required for long-term management is added (and tasks such as migration to new and viable formats are performed). The "Dissemination Information Package" (DIP) is distributed from the archive to the consumer. It may include content data in a particular form, and is likely to include only the subset of metadata required for the consumer's needs or that the producer or archive is willing to share.

Nearly all the recent administrative, technical, and preservation metadata developments within the cultural heritage community have been influenced to a greater or lesser extent by OAIS, though it should be noted that is not the only possible digital archival model. These have primarily focused on digital still images, as these were the first aggregations of digital objects with which most cultural heritage institutions had to deal, although some schemas are specific to, for instance, multimedia or audiovisual content, or applicable to a wide range of media.

The digital library community has, not surprisingly given their strong background in cataloging, been particularly active in developing technical and preservation metadata schemas and initiatives. For example, the late 1990s saw the development of sixteen core Preservation Metadata Elements for all digital assets by the Research Library Group, and the definition of twenty-five preservation metadata elements, with media-specific sub-elements, in the National Library of Australia's (1999) *Preservation Metadata for Digital Collections*. The UK-based CEDARS project and European NEDLIB also developed preservation metadata element sets. These schemas remain useful, but they have been followed by more developed, and inevitably complex, metadata proposals. The RLG/OCLC PREMIS (Preservation Metadata: Implementation Strategies) initiative is looking at a wide range of issues around the practical implementation of a digital preservation strategy, and delivered a core set of preservation metadata elements and a supporting data dictionary in 2005. METS, the Metadata Encoding and Transmission

Standard, is maintained by the Library of Congress and provides a way of "wrapping" together the different types of metadata—descriptive, administrative, etc.—that applies to particular content.

Museums and similar institutions have also developed some interesting metadata strategies. The Variable Media Network (http://variablemedia. net), which emerged from a collaboration between the Guggenheim Foundation and the Daniel Langlois Foundation for Art, Science, and Technology in Montreal, proposes re-creation, rather than migration onto current technology, as a preservation strategy. It uses a questionnaire (in the form of an interactive database) to capture information on works of art that is independent of their current medium. It relies on being able to work with living artists, and asks them to imagine their work appearing in a different form, once the original has become obsolete. This is a powerful approach for some works, particularly conceptual art whose central ideas it may be possible to express in various ways, although other works may be less easily divorced from their medium or physical form. The Rhizome ArtBase (http://www. rhizome.org/artbase) is an associated initiative that aims to present and preserve new media art, and also uses a questionnaire to capture descriptive and technical information from the artist at the point of submission, with the aim of guiding future preservation measures. The information gathered through either of these initiatives can at the very least serve as documentation of works that fail to survive in any other form, and in the best-case scenario it can provide the basis for their preservation.

The InterPARES 2 research project (http://www.interpares.org) is an interdisciplinary project that examined how the requirements of accuracy, reliability, and authenticity developed by the archival community can be achieved in the digital realm. Metadata and description were included in its remit, and its final report was delivered in 2006. Industry groups have also developed some interesting metadata initiatives. The Joint Photographic Experts Group (http://www.jpeg.org) JPEG 2000 image standard includes an enhanced ability to contain metadata (Joint Photographic Experts Group, 2004). The Motion Picture Experts Group (http://www. mpeg.org) has developed MPEG-7, designed to describe multimedia content, and MPEG-21, a broader "multimedia framework" that is intended to describe complex and composite digital objects—that is digital content, its metadata, and the structural relationship between content resources—from creation to delivery. Metadata schemas of different origin can be used in combination. For instance, CIDOC-CRM, the object-oriented Conceptual Reference Model developed by the International Committee for Documentation of the International Council of Museums, is well able to describe the identity and physical attributes and location of traditional museum objects, but is less adept at describing their digital surrogates in any detail. However, if it is used in combination with MPEG-7 it is possible to achieve a comprehensive description of both analog object and digital surrogate (Hunter, 2002).

At this point it may be useful to mention the format of metadata, as opposed to its content. While the quality of metadata is always more important than the format in which it is captured or the system in which it is kept,[5] XML (Extensible Markup Language) is asserting itself as the format of choice, because it offers a way to provide structured information in a platform-independent form that is based on international standards and enjoys a broad knowledge base (http://www.w3.org/XML). "Markup" languages essentially provide a way of codifying simple text files—by placing tags around certain content—to make them machine-readable and therefore potentially extremely powerful. Additionally, plaintext files are extremely easy to access by any number of programs with no licensing or copyright restrictions, and are consequently likely to remain viable for a long time. This allows metadata to be captured in a form that will itself be robust and lasting. Many metadata element sets have already been developed as XML schemas. However, as long as metadata is well structured it should be possible to convert it to XML, or indeed other formats, when this becomes desirable or necessary, and it may be more expedient to use other formats while support for XML data develops.

One of the big advantages that we may eventually hope to reap from the use of XML is the ability to automate more and more tasks associated with the capture and use of non-descriptive metadata. At the time of writing, it is still feasible, if cumbersome, to undertake many digital preservation tasks manually. However, over time digital collections will become so large that this will no longer be practical, and much of their management will need to become automated if it is to be possible at all. Fortunately, technical and some other forms of metadata are much more amenable to such automation than descriptive metadata, which is likely to always be heavily dependent on human intervention (i.e., cataloging), even if automatic indexing becomes more sophisticated. A general trend towards seeking ways to automate and facilitate the creation, extraction, and manipulation of metadata where this is practical is discernible among vendors and manufacturers as well as within the cultural heritage and other user communities.

To give just some examples of this trend, the Research Libraries Group (2004) initiative *Automatic Exposure — Capturing Technical Metadata for Digital Still Images* examined the steps that might be taken to facilitate the automatic capture of technical metadata for digital still images that might be essential for their preservation. These included engaging scanner and camera manufacturers in a dialogue on the technical information necessary for archiving, an examination of existing industry metadata standards, and an evaluation of existing and emerging technical metadata harvesting tools. Adobe Systems Incorporated has developed XMP (Extensible Metadata Platform), an XML-based labeling technology that allows metadata to be embedded into files themselves, thus creating "smart assets" that are self-identifying and theoretically capable of retaining context information even if passed across multiple formats and devices (http://www.adobe.

com/products/xmp). Standard computer operating systems are also moving towards enhanced support for metadata creation and retention in response to the proliferation of digital data and the consequent difficulty in managing it: Windows, Macintosh, and Linux are all reported at the time of writing to be shifting away from traditional ways of locating information, for instance via file folders, in favor of indexing and searching file metadata, probably through some combination of database technology and XML-formatted data. Developments such as these are extremely interesting, but they should not mislead institutions into assuming that technological progress will provide a metadata fix without requiring any active or immediate engagement on their part.

Other tools in the metadata management arsenal include asset, media, and collection management systems. These are available in a broad range of costs and levels of complexity, and over time more of them are becoming capable of handling XML-formatted data. Such management systems can perform useful tasks: aid in consistent metadata capture; provide a centralized and shareable metadata repository; bind digital content and metadata together; access the technical metadata hidden in file headers. Additionally, they may allow batch processing of files, provide security barriers, and schedule migrations or other tasks. As with all things technical, it is advisable to choose a management system based on open rather than proprietary standards, and remember that these tools cannot create a management or preservation policy, however useful they may be in implementing one, and generally do not incorporate an archival digital repository.

CONCLUSION

No single metadata schema or controlled vocabulary is likely to answer all the needs of any institution. For the foreseeable future, it is likely that every institution will be required to piece together its own metadata and cataloging strategy from the available options, not all of which have been listed here and any of which might be helpful in a specific instance, based upon the resources and needs of the institution. However, it is becoming clear that carefully crafted, standards-based descriptive, administrative, technical, and preservation metadata are a crucial part of any strategy aimed at creating interoperable, coherent, intelligible, and long-lived information sets.

ENDNOTES

1. Entrepreneur Jack E. Myers coined the conjoined term metadata in 1969, and registered the capitalized form, Metadata, as a trademark in 1986. See the Oxford English Dictionary Online New Edition, draft entry Dec. 2001: "1969 *Proceedings of the International Federation for Information Processing Congress 1968* I. 113/2: There are categories of information about each data set as

a unit in a data set of data sets, which must be handled as a special meta data set."

2. See, for example, the crosswalk of cultural heritage metadata standards at http://www.getty.edu/research/conducting_research/standards/intrometa data/3_crosswalks/

3. Digitization is gradually gaining acceptance as a reformatting option, although the preservation of digital assets is itself a new and uncertain field and it remains vital to retain originals whenever this is in any way feasible (Arthur et al., 2004).

4. Pham (2004) documents that Nam June Paik has come up with the solution of offering certificates that allow new technology to be used to maintain a work without destroying the authenticity of the work. Museums may consider making sure that they have the legal right to maintain electronic and digital art at the point of its acquisition, a more straightforward process—perhaps—when there is a living artist available.

5. The NISO Framework Advisory Group (2004) identifies the following principles of good metadata: "Good metadata should be appropriate to the materials in the collection, users of the collection, and intended, current, and likely use of the digital object. [. . .] Good metadata supports interoperability. [. . .] Good metadata uses standard controlled vocabularies to reflect the what, where, when, and who of the content. [. . .] Good metadata includes a clear statement on the conditions and terms of use for the digital object. [. . .] Good metadata records are objects themselves and therefore should have the qualities of good objects, including archivability, persistence, unique identification, etc. [. . .] Good metadata should be authoritative and verifiable. [. . .] Good metadata supports the long-term management of objects in collections."

Section 4

Information Interactions in Museums

9 Interactive Technologies

Paul F. Marty

Florida State University

The modern museum offers visitors many ways of interacting with exhibits, from hands-on interactives that help visitors learn basic science principles in children's museums, to touch screen computer displays that encourage visitors to delve more deeply into the background and context of important works in art museums. To be effective educational or informative tools, museum interactives need not involve advanced computer technologies; indeed, one finds very few museums today that are without some form of interactive technology. With the exception, perhaps, of museums that take an extremely minimalist approach to exhibition (going so far as removing all exhibit labels, for instance, from their galleries), it is rare for someone to encounter a completely passive experience when visiting a museum.

Interactivity is very important for museums; museum visitors frequently report being more engaged with the museum's exhibit when they have opportunities for interaction (Falk & Dierking, 2000). These interactions can take a variety of forms, such as conversing with docents or other visitors about the exhibits, manipulating the museum's artifacts in some way (by touching or turning, for instance), paging through a flipbook or gallery guide while touring the museum, or learning from multimedia kiosks and other information stations available in the galleries. The appropriate use of interactive technologies of all types in museums involves many issues (e.g., interactives should be integrated within exhibits so that they augment but do not detract from the visitor's experience); and many have written about the issues of constructing interactives, incorporating them into exhibits, and measuring their impact on the museum visitors (Boehner, Gay, & Larking, 2005; Economou, 1998; Evans & Sterry, 1999; Milekic, 2000; Paterno & Mancini, 2000; Wakkary & Evernden, 2005).

This essay focuses on the sociotechnical implications that arise when interactive technologies are used in museums to enhance the visitor experience. Advances in museum interactivity have had two major impacts on museum professionals and visitors. First, these advances have removed many of the pre-existing barriers that placed artificial restrictions on the abilities of museum visitors to interact with museum artifacts. Second, they have offered new ways for individuals and groups to conduct interpersonal social interactions centered around museum collections and exhibits.

In tearing down old barriers between visitors and artifacts, museum interactives have done much more than allow the users of museum resources to manipulate artifacts (or their surrogates) in ways previously impossible. They have removed certain restrictions of space and time that historically have constrained artifacts and collections to individual museums or galleries; using digital surrogates, for instance, museum visitors can interact with artifacts from diverse collections regardless of the actual physical location of any particular artifact. Advances in interactivity have also removed barriers between public access and behind the scenes activities, allowing visitors to "perform" the tasks of the conservator, for instance. They have even helped remove the physical boundaries that separate "inside the museum" from "outside the museum," with the result that the line between online and in-house offerings blurs to create the "museum without walls."

The theme driving much of modern museum interactivity is that of constant integration, where access to all types of resources (behind the scenes, in the galleries, online, etc.) becomes uniform, seamless, and transparent. In the case of online museums, for instance, the removal of these barriers can result in interactives so transparent that visitors may not be aware of a separate physical museum identity, or that certain physical barriers between artifact and access even exist. As these barriers are blurred or otherwise removed, many new ways of reaching museum users become available.

The virtual museum, for instance, has the ability to offer many new experiences to new audiences. Visitors physically unable to come to a particular museum are frequently able to experience some of the museum's offerings online, no matter where in the world they may be, as long as they have access to an Internet connection. Visitors planning visits to a particular museum can prepare for their trip by downloading information resources such as highlights of the museum's collections or driving and parking directions. Online access to museum resources can even be targeted for specific users, attracting a wider variety of visitors by offering resources specifically tailored for their interests (Bowen & Filippini-Fantoni, 2004). As museums explore more interactive ways of making collections available electronically, museum professionals are better able to meet the needs of different users more effectively. By integrating online resources with in-house experiences, museums can offer visitors the ability to create personalized museum visits, tailored to each person's individual needs.

The ability to personalize the museum visit through interactives can be potentially problematic; as interactives in museums become ever more personal, there is the risk that individual users will explore only issues of interest to them. In breaking down the traditional barriers of information access in museums (removing many concerns of time and space, as well as certain physical limitations of artifacts), visitors have the opportunity to build new walls, concentrating only on their own particular needs and interests. For these reasons, some have argued that interactive technologies in museums can be inherently asocial, and many have taken steps to limit this problem (Galani & Chalmers, 2002; Hsi, 2003; Woodruff et al., 2002). Such concerns make it all the more important to carefully analyze the impact of new interactive technologies in exhibits, and in particular their impact on how museum visitors interact with the exhibit and with each other.

Much recent research in this area has focused on aspects of personalization and pervasive (or ubiquitous) computing experiences in museums (Arts & Schoonhoven, 2005; Hsi & Fait, 2005; Jaen, Bosch, Esteve, & Mocholí, 2005; Manning & Sims, 2004; Parry & Arbach, 2005; Silveira et al., 2005). As information resources from diverse sources become merged with information technologies allowing access to data independent from the users' physical location, it has become more common to offer different museum visitors different experiences. Underlying many of these systems is the (certainly debatable) notion that the best way to visit a museum is to tour the galleries with an expert who knows not only information about the objects on display, but also how these objects relate to things important in the life of the visitor. While this may or may not be true, museum professionals have for centuries taken steps to help people learn more about the museum's collections from a personal perspective: grouping items by subject or time period, providing interpretive labels, offering multiple versions of gallery guides, and so on; pervasive computing technologies are simply the latest step in the process of personalizing the museum visit.

It is not uncommon for museum visitors to employ a variety of interactive devices in museums; technologies such as audio guides or handheld computers provide visitors with the ability to access detailed descriptions of individual works of art or create their own personal tours complete with online access of their favorite collections. Like all new technologies, of course, the development and use of these devices can be problematic, and the history of interactive devices in museums can be read as a history of failure (Schwarzer, 2001; cf. Taxén & Frécon, 2005). While some may fail for purely technical reasons (e.g., limitations in the physical capabilities of the device), the majority will fail for sociotechnical reasons that generally result in unanticipated negative consequences for museum visitors. The constructivist perspective that lies behind many of today's interactive technologies (especially those where visitors are able to develop and follow their own experiences in the museum) raises many questions about their relative advantages and disadvantages.

First, there are many questions about whether personalized, interactive technologies actually improve the process of visiting the museum. Studies demonstrating that visitors spend more time in galleries while using interactive technologies raise important questions about whether additional time means that visitors are spending their time constructively (Evans & Sterry, 1999). Visitors with interactive devices may spend more time in galleries trying to figure out how to use the device than actually learning from or interacting with it. Visiting a museum with interactive technologies is clearly attractive to many museum visitors, as recent examples such as the growing popularity of pod-casting museum audio tours, designed and developed by the general public for museums such as the Museum of Modern Art in New York (cf. http://mod.blogs.com/art_mobs). Nevertheless, not everyone prefers the technological to the more "traditional" approaches to museum visitation. As more museum professionals consider incorporating interactive technologies in their galleries (especially portable, personal devices), it is important to study how, when, for whom, and under what conditions these devices improve the museum visit.

Second, there are many concerns about the social implications behind interactive, personalized technologies in museums. Does the widespread use of audio guides in galleries, for instance, enhance or detract from the social experience of visiting a museum with friends or family? Recent studies have explored methods of making audio guides less socially isolating, allowing visitors to eavesdrop, for instance, on their fellow visitors (Woodruff et al., 2002). Researchers have also explored ways of extending the social experience outside of the museum itself, allowing visitors within physical galleries to interact and converse with online visitors to those same galleries using the museum's Web site (Galani & Chalmers, 2002). Other researchers have studied the social interactions among museum visitors who co-visit museum Web sites in groups while online, taking virtual tours led by virtual docents over the Internet (Paolini, et al. 2000). Projects such as these have redefined the social aspects of museum visitation for all visitors, whether they are located online, in the galleries, or in some combination of physical and virtual visits.

Third, the availability of interfaces capable of adapting information resources dynamically to meet the needs of different users (Paterno & Mancini, 2000) raises questions about whether this type of personalization is beneficial for either the museum visitor or the museum professional. Online interfaces that allow visitors to play the role of a curator, e.g., building their own personal digital collections by book-marking their favorite artifacts (Adams, Cole, DePaolo, & Edwards, 2001; Bowen & Filippini-Fantoni, 2004), can be an excellent way to attract visitors to the museum, but can also limit the abilities of museum curators to develop exhibits that cut across universal lines. Similarly, it can be argued that personalization technologies that allow individual museum visitors to draw upon their likes and dislikes to create their own personalized set of museum artifacts place artificial restrictions on visitors, limiting the role of serendipitous discovery and their ability to develop new, previously unknown interests. Interactive technologies that adapt to the personal needs of individual visitors, therefore, should be examined very carefully for their benefits and drawbacks to both museum visitors and museum professionals.

Fourth, as interactives in museums become more popular and more personal, concerns about information usage, user profiling, and privacy policies have become more common (Hsi & Fait, 2005). Interactive technologies offer museum visitors greater opportunities to engage with museum collections; they also increase the museum professional's ability to document and track the interests and needs of museum visitors. In most museums, visitor studies are certainly not uncommon; many museum professionals learn what they can about their visitors in order to improve their experiences. What is new, however, is the extent to which museums can learn details (possibly very private details) about the visitors' interests, desires, needs, and so on. Should museums have the right to track, for instance, how long each visitor spends standing in front of each exhibit or visiting each exhibit online? While one can argue that the more information museums collect, the better (i.e., more tailored and more personal) experience they can offer visitors, there are many difficult questions here which museum professionals will need to address.

The bottom line is that new interactive technologies are changing the experience of visiting a museum, yet technology of any type cannot exist in isolation. The museum professional or researcher planning interactive technologies must consider those technologies in many different contexts. Interactive technologies are a form of mediation between museum and visitor, and one cannot simply implement new technologies without considering all the social, personal, or physical ramifications of bringing those technologies into the museum environment. To succeed, it is important to understand how interactives will be used by different users, how they will be integrated into existing exhibits, and how they will help the museum fulfill its overall mission of meeting the needs of visitors.

None of this can be accomplished if museum professionals do not know what museum visitors want. The reason so many new museum interactives fail is because they are based on incomplete knowledge about the needs, interests, and behaviors of museum visitors. One cannot, or rather one should not, spend a tremendous amount of money bringing new interactives into the museum environment simply because they represent the latest and greatest technologies. If one's visitors have no desire to carry around handheld computers, for instance, then it makes little sense to spend money on technologies no one will use. The first step in designing and implementing new interactive technologies is to evaluate the changing needs and expectations of the museum's visitors, and to design solutions that meet those needs and expectations, providing opportunities visitors will actually use and enjoy.

10 A World of Interactive Exhibits

Maria Economou

University of the Aegean

INTRODUCTION

The use of new technologies has transformed the way museum professionals interpret the collections in their care, as well as the way visitors interact with them. Digital media have changed the way visitors are experiencing exhibitions and cultural content, whether they are in the museum or offsite. This is also affecting the public profile of cultural organizations, their internal procedures, and the way they communicate with their audience.

This chapter examines the way information and communication technology (ICT) has been used in the museum setting for public displays and interpretation and the different types of application we can distinguish. It also explores the potential of ICT, the special characteristics it can offer for enriching the museum visit, which go beyond traditional interpretation media, and looks at the important issues which the use of these tools raise.

Its specific focus on interactive exhibits is set in a wider context which regards the museum as an organization that creates, preserves, and communicates information and knowledge and examines the ways ICT can facilitate these functions. The idea of interactivity is not easy to define, despite the different attempts that have been made by authors from various fields (e.g. Adams & Moussouri, 2002; Lévy, 1995; Roussou, 2004; Sims, 1997; Steuer, 1995). Adopting Adams' and Moussouri's (2002) museological perspective, in this chapter we refer to interactive experiences as those which actively involve visitors physically, intellectually, emotionally, and socially, but limit our focus to computer-based interactive exhibits (as this definition can also include manual ones). In this sense, the level and type of visitors' engagement can vary considerably as is illustrated by the examples discussed below.

TYPES OF APPLICATION

Information Kiosks

One of the most popular and early applications of new technologies in exhibitions is the information kiosk. This can play different roles and take many forms, from a stand-alone computer with a touch screen or other interface device, to a setting where a whole section of the exhibition floor is devoted to information stations.

In one of the possible scenarios, information kiosks can act as electronic labeling systems, installed next to the display cases, allowing exhibition organizers to keep minimal or no information next to the objects inside the case. The computer application usually includes a digital photograph or a graphic of the case with the objects numbered or as active areas in systems with touch-sensitive monitors. When touching on an object or typing a number, the user can see the relevant label. This way, the cases are not loaded with interpretive material which can interfere with viewing the exhibits. The visitor is given the choice whether to access the supplementary information, which can be offered at different layers and levels, starting at the basic object label and offering the possibility to browse deeper in different directions according to the user's particular interests. This can be adapted to very different museological approaches: in the traditional one, where the objects retain their central place and are left to "speak for themselves," the electronic labeling system can contain all relevant information, thus freeing the displays from the interpretative layer, allowing the visitor to focus on the experience of encountering the original artifacts. Following a different approach, where great emphasis is placed on the interpretative information, an electronic labeling system can extend to provide in-depth multimedia information (incorporating, for example, sound, moving and still images), which can be adjusted to different visitor profiles.

When properly designed, systems of this type offer museum staff the flexibility to easily update the information, accommodating more frequent and easy changes of the permanent displays. Of course, in order to work effectively in the gallery, they require careful positioning of the electronic labeling stations, intuitive design of their interface and navigation, and consideration of visitors' flow in the gallery. In general, when new technology applications are closely linked to a specific exhibit, object or case, the design of the layout and the physical relationship with the other objects and components of the exhibition are of vital importance and require systematic planning and evaluation. A point to consider with systems of this kind is that when there is a queue to use the application at busy times or when any of the kiosks are not operational, the visitors' experience might be disrupted.

An early example of the electronic labeling system is the Music Room, an exhibition of musical instruments at the Horniman Museum in London, where information kiosks with touch screens were designed to provide an alternative to text and graphic display panels, enabling visitors to find out more about and to listen to the sounds made by the more than one thousand musical instruments on display.

In other cases, information kiosks are used in the gallery for offering background information about the exhibition themes, not necessarily specifically tied to the objects on display. For example, they can focus on the artist or artists whose works are on show, refer to an artistic movement or historic period, or present aspects of the culture which produced the ethnographic or archaeological artifacts in an exhibition. Well-designed applications bring the objects to life and show aspects of their original context in interesting and engaging ways. To some extent, they can bridge the gap created by the "museum effect," the isolation of the objects and the new role they take as exhibits when they are taken from their original environment and placed behind museum cases. For example, interactive multimedia applications can offer a good idea of what life was like for different members of the crew aboard the historic vessels displayed at a maritime museum, show the particular celebrations where the objects exhibited at an anthropological museum were originally used, or include interviews of artists talking about their work and videos showing them during the creative process.

Taking the example of the Horniman's electronic labeling system further, a museum in New Zealand is using an information kiosk to contextualize and bring to life the collection of traditional musical instruments (Puoro) and other artifacts displayed. At Puke Ariki, a combined library, museum, and visitor information centre in New Plymouth, in the North Island of New Zealand, the touch screen interactive kiosk installed in the Taonga Maori Gallery (a gallery displaying the material culture of the local indigenous people) is used to tell the story of traditional musical instruments[1] (Figure 10.1). These are featured on display in glass cases beside the interactive installation in the gallery, while the program aims to convey aspects of their significance, the material they are made of, their history, how they are used, and most importantly what they sound like (Figure 10.2). In this case,

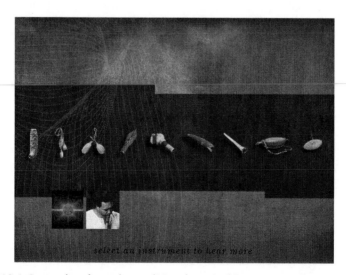

Figure 10.1 Screenshot from the traditional musical instruments (Puoro) information kiosk application showing the main menu. © Click Suite

Figure 10.2 Screenshot from the traditional musical instruments (Puoro) information kiosk application showing a demonstration of playing an instrument. © Click Suite

due to the positioning of the information kiosk and considerations for visitors' flow, users' interaction has been planned for a very short period.

Another way that museums are using information kiosks is as "conceptual pre-organizers," placing them before a gallery or exhibition and including introductory information about the display they are about to see in order to help them familiarize themselves with the basic concepts and the way the particular exhibition is organized. In some cases, the computer application includes also practical information to assist orientation and planning of the visit. An example of using multimedia stations as both conceptual pre-organizers and orientation tools is the exhibition on ancient mathematics organized in 2004 by the Foundation of the Hellenic World in Athens (Moussouri, Nikiforidou, & Gazi, 2003). The computer stations that have been installed before the entrance to the exhibition include, among others, information about the different paths visitors can follow depending on their level of interest and available time, as well as suggestions for visitors with children.

In other cases, the information kiosks are placed after an exhibition gallery, usually in an adjacent space, offering more in-depth information to satisfy the curiosity that the exhibits might have roused. They can act as a reference system allowing more concentrated and longer interactions without disrupting the visitors' flow in the gallery or assist with comparisons with objects from the collections not on display, thus improving access to all the objects in the collection. At the British Galleries of the Victoria and Albert Museum in London, a comfortable space with armchairs, books on British art and crafts, and computer stations with information on the objects, creators, periods, and styles is used as a reference point after visiting part of the galleries and before continuing the visit (Figure 10.3).

Information kiosks have also been used as museum directories, usually positioned near the entrance to the building and at key locations, offering information about what is on display at the museum, the special activities and events, and the location of exhibits and facilities. The Minneapolis Institute of Arts has cleverly used a Web browser platform for its museum directory "Today at the Museums" (http://www.artsmia.org/directories/) which is built on three different databases (for events, exhibitions, and collections management) and has incorporated visitor research findings in its latest redesigned form.[2]

Moving beyond the examples of information kiosks providing supplementary information about the exhibits on display or visitor orientation, are the computer applications which act as primary exhibits themselves, usually expanding beyond the two-dimensional monitor to the space in the gallery. Artists frequently explore the possibilities of new media in this direction, as is the case with the interactive environment "Beyond Pages" (exhibited in 1995 at ZKM, Karlsruhe), an influential early example created by the Japanese artist Masaki Fujihata (see http://on1.zkm.de/zkm/werke/BeyondPages).

In this three-dimensional space, a data projector loads on a table images of the pages of a leather-bound volume, which can be activated with a stylus, bringing to life the objects referred to in the book (stone, apple, door, light, etc.).[3] In a game between the traditional, linear way of organizing and

Figure 10.3 Computer stations at the specially designed reference and consultation space at the British Galleries, Victoria and Albert Museum, London

storing information represented by the book, and the new digital means of representation, the artist experiments and analyzes the properties and the limits of the two worlds. He uses the idea of leafing through the pages of a traditional book, but goes beyond its two-dimensional limits, introducing elements of surprise with three-dimensional and moving objects, while maintaining a very simple and intuitive interface for the viewers of the work.

Another aspect of the interaction with digital technologies that is frequently explored by artists is the way presence and interaction of virtual and real visitors can shape and influence the work of art. An example from the work of Jeffrey Shaw, one of the best known artists and theoreticians in the field, is the Web of Life (http://www.weboflife.de/), a network project produced in 2002 at the ZKM Institute for Visual Media with Michael Gleich, Lawrence Wallen, Bernd Lintermann, and Torsten Belschner. This is part of a larger project which includes a book and a Web site. The distributed nature of the artwork allows it to be manifested at numerous interconnected locations worldwide (one large-scale environment situated permanently at ZKM, Karlsruhe and four others designed to travel around the world, e.g. exhibited in Bratislava, Bonn, Rotterdam, and Tokyo in 2003). The Web of Life allows visitors from all the locations where the exhibit is on show to interactively influence the performance of an audio-visual environment (projected three-dimensional computer graphic and video images and spatialized acoustics) by imparting to it the unique patterns of their individual hand lines through a hand-scanning user interface (see http://www.jeffrey-shaw. net/html_main/show_work.php3?record_id=117).

Visualizing Three-Dimensional Information and Virtual Reality Exhibits

New technology applications have also been used in other ways that move beyond the two-dimensionality of the information kiosk, experimenting with virtual reality exhibits and immersive environments. These give visitors the feeling that they are actually in the recreated space themselves. The ability to visualize three-dimensional information and recreate whole environments can have a strong emotional and sensory impact (which in turn can influence the learning process).

Public understanding of archaeology was one of the first areas of application of virtual reality in the cultural sector (such as the reconstructed model of the Roman bath complex at Bath in the United Kingdom) as it offered the ability to visualize complex phenomena and put together the disparate pieces of the surviving evidence from the past, making clear how things might have looked.

One example of an immersive virtual reality display is the system installed at the Foundation of the Hellenic World in Athens, which is used to take its users for a "virtual tour" to ancient Olympia (among other places) showing details of the recreated buildings and giving users the impression that they participate in the atmosphere during the ancient Olympic Games or the ability to wander through the city of ancient Miletus (Gaitatzes, Papaioannou, & Christopoulos, 2004; Roussou, 2002). This resembles the CAVE® system, a ten-foot cube where the interior three walls and the floor are projection surfaces, thus creating a three-dimensional virtual reality (VR) experience that can be shared by several (usually up to ten) people who need to wear special goggles.[4]

Virtual reality is also being used at the British Museum in London for a "virtual unwrapping" of the Egyptian mummy of the priest Nesperennub (dating to c. 800 BC). Cross-sectional images running the full length of the mummy's body obtained through a CT scanner at a London hospital and 3D laser images scanned in Scotland were used by Silicon Graphics Inc. to create a virtual model of the mummy, which enables scientists and visitors to examine what lies within the wrappings, and even to make a tour inside the body. The model was the basis of a 20-minute virtual tour shown at an immersive theatre featuring a curved screen and requiring visitors to wear polarized 3D glasses. Integrating successfully the virtual and the real, this is part of the temporary exhibition "Mummy: The inside story," which opened in 2004 and has also on display the real-life mummy (Taylor, 2004).

Simpler and more affordable VR environments can be created for the Web, using, for example, QTVR,[5] VRML, and its successor X3D.[6] These have been used for virtual walks through real galleries (such as the virtual tours of the Louvre, http://www.louvre.or.jp/louvre/QTVR/anglais/index. htm), virtual ones (such as Inuit 3D, http://www.civilization.ca/aborig/ inuit3d/inuit3d.html, one of six inaugural Virtual Museum of Canada exhibitions launched in April 2001 (Corcoran, Demaine, Picard, Dicaire, & Taylor, 2002)), or museum facilities which the public cannot usually visit (such as the Research and Collections Center of the Illinois State Museum, http://www.museum.state.il.us/ismsites/rcc/rcc_tour/tour_start.html), or to provide three-dimensional models of artifacts (such as the QTVR movies taken at the cockpit of aircrafts at the Smithsonian National Air and Space Museum, http://www.nasm.si.edu/interact/qtvr/uhc/qtvr.htm).

The level of interactivity in these virtual environments varies. For example, in the Foundation of the Hellenic World examples, there is always a trained museum educator who acts as an animator, controlling the navigation and interaction with the system. In other cases, such as the VR applications on the Internet, users are interacting with the exhibit with greater autonomy, but even then there is usually a limit to the type of exploration a user can have.

Irrespective of the quality of the graphics and the realism of the simulated environment, user experiences of 3D virtual worlds have often turned out to be unsatisfactory. Users' interest usually wanes off after a short exploration, when no particular task is set or no activity to encourage their engagement (Di Blas, Gobbo, & Paolini, 2005). The last few years the focus has shifted from creating immersive to creating also participatory environments, shared

3D worlds accessible by several users over the Internet (often referred to also as multi-user or collaborative virtual environments). The experience of working with several 3D education projects shows that in order to create a 3D place where people can have a meaningful experience, a fundamental role is played by "virtual presence," the ability to meet and interact with other people in that world and engage in common activities, rather than by the faithful reproduction of a real place (Di Blas, Gobbo, & Paolini, 2005; Roussou et al., 1999).

The cultural world has slowly started showing an interest in this direction, experimenting with the use of agents, avatars, or virtual representations of human or other beings to populate the often isolated and impersonal environments of virtual worlds. In cultural settings, avatars have been used in a traditional way as guides to the virtual exhibition or world, such as the early example of Virtual Leonardo developed by the Politecnico di Milano in collaboration with the Museum of Science and Technology in Milan for an exhibit of Leonardo's machines (Paolini et al., 2000).

Other applications move away from the model of interaction with an authoritative, knowledgeable figure to explore the creation of virtual communities, interacting around cultural digital objects or the specific exhibition theme presented by the museum. Susan Hazan (2004) suggests that the crossing of national and cultural borders of online communities can give voice to new interpretations of cultural objects. Although virtual communities and 3D virtual reality environments supported by private companies are very popular on the Internet,[7] cultural institutions have hesitantly started exploring links with existing virtual communities which have an interest, for example, in genealogy, cultural history, performing arts, monuments, literature, or digital art. Although not all organizations in the cultural sector would embrace with enthusiasm the idea of collaborating with a non-professional online body, work in this direction is already broadening "the reach, value, and relevance of cultural heritage" (Geser, 2004, p. 5).[8]

Mobile Computing and Handheld Devices

Following a wider societal trend for customizable and individual interactions and services, recent research is exploring the use of mobile computing applications in museums which can be personalized, to a degree, depending on the profile and interests of the specific user. Personal digital assistants or similar devices can offer information about exhibits or sites relevant to the location (especially when used with wireless networks and infrared sensors). For example, the personal digital assistants (PDAs) used at the exhibition "Points of Departure", organized by the San Francisco Museum of Modern Art, allowed visitors to access video clips of the artists and archival material related to the works on show. In some cases, visitors were allowed to see videos of Robert Rauschenberg, Louise Bourgeois, and others talking about the specific work of art that they were actually viewing at the time (Samis, 2001).

Applications of this type can have the form of a guided tour, often with audio commentary, suggesting a predetermined path through the museum or exhibition to the user. In other cases, they can provide higher levels of interaction, acting more as a resource to be consulted when the users choose to, but also offering them the opportunity to record their own impressions, participate in opinion polls, and to even take with them snapshots of their experience, or bookmark and email home interesting information (as is the case, for example, with the Multimedia Tour of the Tate Modern Gallery in London).[9] In the case of the Handheld Education Project developed at the Renwick Gallery, part of the Smithsonian Museum of American Art focusing on applied arts, the interpretative approach followed by the staff was to use the PDAs to move beyond the traditional museum label, trying to make the objects "come alive," show the life of the artists behind the objects, or use them as catalysts for enriching the visitor experience, aiming to engage the visitors who often feel at a loss for participating in a museum (Boehner, Gay, & Larkin, 2005). It is interesting that in this approach there is a shift of emphasis from "assets to experience," which follows a wider museological trend. In this case, the PDAs are used to encourage a conversation, a drawing in and engagement with the objects and the related information and not a simple transfer of knowledge.

The use of mobile and handheld devices in museums raises several important issues. First of all an institution examining a project of this kind needs to examine whether it fits with the institutional mission, goals and objectives, and the return on the considerable investment (Exploratorium, 2001). Others have to do with the user interface and the type of interaction, as a key concern with handheld devices is the ability to offer not only a customizable individual interface, but also a transparent, ubiquitous one that does not disrupt the experience of the visit. This needs also to take into account visitors with different abilities and needs. Another issue to be examined is whether mobile devices support or even act as "catalyzers" for social interaction or isolate users from their environment. These devices offer the potential (rarely realized so far) to transfer multiple voices and perspectives, rather than the unique, authoritative interpretation. We also need to explore further their potential for increasing engagement and interaction with the objects on display, moving beyond the initial attraction of the technology to further exploration and active participation. Despite the growing popularity of these systems, the few existing systematic evaluation studies (e.g. Boehner, Gay, & Larking, 2005; Galani & Chalmers, 2003; Grinter et al., 2002; Sherman, 2002) need to be built up to a greater body of research before we can fully assess how these devices are used in practice and their effect on visitors.[10]

On the Web

Apart from the various examples of stand-alone and handheld applications which can enhance the experience of the visitor in the gallery, the use of the World Wide Web by museums drastically improved access to their collections beyond the four walls of the gallery. The arrival of the Web in the 1990s offered cultural institutions the possibility to create online applications that could be viewed by an international audience. The ease of programming in HTML and the ubiquitous and platform-independent nature of the Web browser led some museums to use the Web also for providing applications in the gallery or for publishing in electronic form.

Several museums initially published on the Internet only textual records of their databases, a significant step for improving access to their holdings, particularly for researchers and specialists. Since the 1990s, an increasing number of museums are providing Web access to collections information in multimedia form, thus creating a rich resource that can also be used for educational purposes. The use of color images and in some cases other media, accompanying the textual records made the collections more attractive to a lay audience of different ages and interests around the world. This provided the basic resource, the backbone of core data, on which interpretative applications could be built, offering information at different depths according to users' various interests.

When Web access to the collection's database is offered without any additional layer of interpretation, great care needs to be placed to the design of the user interface and the paths into the collection offered to non-specialists. For users unfamiliar with databases and similar applications and with only a general interest in the subject, the common search box asking them to type in a term can be ineffective and intimidating. In this case, a good strategy is to design options also for those who do not know exactly what they are searching for, such as the possibility to follow a "guided tour" or to browse through the highlights of the museum's collections (see the Getty example noted earlier in this volume).

The rising number of cultural institutions which started offering online access to digital surrogates of their collections, together with the development of new technologies and protocols for searching across different databases, made the possibility for cross-collection and cross-institutional searching and linking a reality.[11] This worldwide sharing of museums' internal data and archives brought greater transparency to their internal documentation procedures and highlighted the need for adopting widely recognized standards and controlled vocabulary.[12] The use of standardized terminology in the documentation of museum collections allows better retrieval of information, successful communication with other museums, and interoperability between the different institutions' catalogues. This has been assisted by the development of widely recognized structured vocabularies, such as the Art and Architecture Thesaurus (http://www.getty.edu/research/conducting_research/vocabularies/aat/), the Getty Thesaurus of Geographical Names (http://www.getty.edu/research/conducting_research/vocabularies/tgn/) and the Union List of Artist Names (http://www.getty.edu/research/conducting_research/vocabularies/ulan/) developed by the Getty Research Institute and the MDA Archaeological Objects Thesaurus (http://www.english-heritage. org.uk/thesaurus/obj_types/default.htm/), which help to record consistently and retrieve accurately information about art, architecture, and material culture.[13]

Another important parameter of interoperability is the use of appropriate metadata for describing the collections. The development of the Dublin Core (http://www.dublincore.org/index.shtml) had a serious impact in this direction, since it is an internationally recognized metadata scheme which defined a minimum set of 15 metadata elements which should be recorded about each object in order to assist resource discovery. As an extensible system, it offers cultural institutions the flexibility to add more layers of information about their collections. The recording of metadata has considerable implications on workflow and resources, which should be taken into account when planning a project, but they are the key both for the staff of cultural institutions to record their expertise and manage the collections, as well as for users to use them.

Apart from providing access to collections information and digital surrogates, the Web has been used as a medium for showing online exhibitions (often complimenting exhibitions onsite, but in some cases, also wholly digital ones).[14] In some cases, this would include all the material of the onsite exhibition repurposed for the Web, with the added benefit of additional material that could not be included in the onsite version due to limitations of space and the particularities of the medium. Apart from richer information, the online version of the exhibition could also include video clips showing how the objects were originally used, comments from the curators and creators of the exhibition, interviews with artists, and other features offering another layer of interpretation.

One relatively early example of an online exhibition accompanying an onsite one includes the one organized in 1999 by the Minneapolis Institute of Arts (MIA) on the restoration of a 17th-century painting (Castiglione's *The Immaculate Conception with Saint Francis of Assisi and Anthony of Padua,* http://www.artsmia.org/restoration-online/). The MIA decided to bring to the foreground the process of restoring the painting, placing the conservation lab in the public gallery for the duration of the exhibition, and improving access to this important aspect of museum work which usually takes places behind the scenes. The online exhibit included video clips showing the conservators at work (live while the exhibition was open) (Sayre, 2000). In this case, the online exhibit provided additional material accompanying the exhibition, but also acted as a medium to record it, as despite the transient nature of the Internet, the online version is still available, long after the onsite exhibition ended.

In some cases, cultural organizations have taken the opportunity to show online exhibitions further, combining them with outreach and a new way of communicating with traditional and new audiences. An interesting example is that of "Virtual Transfer" (http://www.musee-suisse.ch/vtms/), a project of the Musée Suisse Group (http://www.musee-suisse.com/), a network of eight national museums located in various regions of Switzerland, which is the new incarnation of the Swiss National Museum founded in 1898 (Jaggi & Kraemer, 2004). While a new building is being constructed for the National Museum in Zurich, the Group seized the opportunity to use the Web as a virtual platform to develop a new communication strategy with visitors. "Virtual Transfer" is complimenting "Web-Collection" (http://www.musee-suisse.ch/webcollection/), a database-driven and object-oriented display of the digital collection, based on curatorial data and research (similar to the approach of several cultural institutions providing Web access to digital surrogates of their collections) (Jaggi & Kraemer, 2003). What is innovative is the way "Web-Collection" is supplemented by "Virtual Transfer," which presents objects selectively, interrelating them with diverse fields of knowledge, making them come to life, and showing how they can be relevant today to different types of visitors. Digital storytelling (in English, German, French, Italian, and Romansh), multimedia, and personalized forms of address are used to create a range of interpretation approaches. These include a chamber of marvels with favorite items, curiosities, and masterpieces (where specific objects are used to tackle wider themes: for example, the Roman statuette of Hermes-Mercury and a story on trading, traveling, and mediating; the Medieval Chur Madonna and devotional pictures; and the 17th-century Blackamoor automaton and identities as constructs), picture albums, fictional and historical witnesses, audio adventures (such as audio from Mark Twain's text on a trip to the Jodel and its native wilds, accompanied by archival photographs), learning courses (e.g. on historicism), games, and quizzes.

Museums are increasingly interested in using the Web for communication and interpretation approaches of this type, which compliment their fundamental work of providing basic collections information.

Post-Visit Resource or Souvenir

Another application of digital technologies in museums which goes beyond the "four walls of the gallery" takes the form of a "take away" experience with the production of a deliverable product, such as a CD-ROM or a DVD, which can be taken to the classroom or home for further viewing. This allows temporary exhibitions and their accompanying new media interpretative applications to be recorded and viewed again, acting as a reference tool. Additionally, when appropriately incorporated in the curriculum, it offers considerable educational potential.

We referred to all these different types of applications to paint a picture of the various possibilities. In practice, the boundaries between the different applications are often blurred. For example, the technology is frequently used to connect on-gallery with offsite visitors, Web applications can be used not only by remote visitors, but also in the museum, or the handheld program designed to accompany a special exhibition can also produce a post-visit resource which is emailed home.

THE POTENTIAL: ADVANTAGES AND
SPECIAL CHARACTERISTICS

When cultural information is digitized following appropriate standards and procedures, it creates a rich resource that offers significant advantages complimenting the analog media. One of the first that comes to mind is the fact that high-quality digital surrogates can be displayed and examined by a large number of visitors without any wear and tear to the original. This is particularly valuable in cases where conservation reasons prevent the display in galleries where lighting and temperature levels might damage irreplaceable items from a museum's collection, such as medieval textiles or illuminated manuscripts. Although viewing the digital version will never replace the experience of examining the original, in certain cases this is the only way to provide access to important objects that would have otherwise remained known only to a few scholars. Furthermore, the display of digital collections offers some unique advantages, such as the ability to zoom in and examine minute details of the original that would not have been clearly visible with a naked eye. Some of the numerous possibilities for presentation and interpretation include the ability to combine on one screen objects normally displayed in different galleries, buildings, cultural organizations, or countries. In this way, new technologies offer a medium which circumvents often-arbitrary limitations and boundaries imposed by the history of the collections, the division of academic disciplines, practical considerations of space, or just chance.

One of the most attractive features of the digital environment is the ability to reuse information, although it is important to remember that in most cases, this needs to be properly adapted and often re-designed to suit the medium. The display of digital collections can be adjusted according to different educational or research needs, offering the possibility to even intervene with the digital surrogate in ways that would not be possible with the original. For example, you can draw a line on a painting's surface to indicate the use of perspective in Renaissance art or join together the different pieces from the frieze of the Parthenon currently dispersed in different museums in Europe and place them back on the building.

Another attractive characteristic of digital media is the ability to personalize and adapt the educational or interpretative material according to different learning styles or visitor characteristics. This can include the use of different languages, but also depth and complexity of information, or style of navigation. When appropriately designed, these applications can also offer a more active type of learning or general visitor experience compared to more traditional means of interpretation, and in some cases, they invite users' input. The applications of new technology in cultural organizations we have seen until today incorporate differing levels of interactivity. They range from rather passive presentations, where the number of choices available to users is quite limited and lead to specific routes predetermined by the programmers, to fully participatory exhibits where users' input is encouraged and can shape the final application. For example, art applications for children can include an exploration of known artworks in the manner of a game, which can also include a part where children create their own artworks experimenting with color, shape, and lighting.[15] These can then be stored in the system and be viewed by future users.

The flexibility of reusing the digital information according to different situations offers also advantages for adjusting interpretative applications for users with disabilities and special needs. When designers of these programs pay particular attention to Web and interface design, the preparation of content, the selection of interface devices, and the incorporation of evaluation and consultation procedures, as well as to following the increasing number of relevant guidelines for accessible Web and multimedia content,[16] they can create systems which are better suited for users with physical disabilities, visual and hearing impairments, and learning difficulties (Mattes, 2001; Museums, Libraries, and Archives Council [MLA], 2005).

Another attractive feature of digitization for cultural institutions is the ability to offer worldwide access to the whole of the collection, large parts of which are often not on display. This way, a significant number of items which are held in storerooms, are on loan, or are in conservation labs, can be viewed publicly, an important consideration for museums with large collections which can only display a small percentage (in some cases, less than five percent) of their holdings.

The labor-intensive task of digitizing cultural collections at appropriate standards creates a rich resource which can be used for multiple purposes. A very important one is the support of teaching and learning in formal educational institutions, but also at informal and life-long learning environments. Cultural organizations are particularly appropriate environments for encouraging creativity, self-exploration, and independent learning. Digital resources can support all these in multiple ways, either onsite or from a distance, when for example, the interest triggered by a visit can be cultivated further using a Web-based activity or a CD-ROM using the museum's collections.

Museums can be even more effective in their support of learning when they collaborate with other cultural organizations. The digital revolution has led to an increase in the number of partnerships and joint projects. Funding schemes have often helped form such collaborations, as has the focus on specific themes which can foster an interdisciplinary approach that crosses institutional boundaries and can, for instance, encourage museums, libraries, archives, and schools to work together.

The digital museum also offers considerable potential for combating social exclusion and the marginalization of various groups. Although the potential of ICT in general has often been exaggerated by several governments, computer companies, and technology evangelists, new technologies can indeed be valuable tools for fighting isolation and improving access. In order to achieve this, however, we should "bear in mind the developments of the technology, the patterns of usage, and the culture of use—i.e. *what it can do, who is using it, and how they are using it*" (Parry, 2001, p. 112).

By providing public access to their digitized holdings and services, cultural institutions can play an important role in combating the "digital divide" between those who have and those who do not have access to the technology.[17] Offering multiple ways of exploring the digital collections (e.g. by themes) can enable users to find what is relevant to their own background and circumstances. Other socially inclusive approaches for cultural organizations include outreach projects which can take ICT products and services to the home of older people and children in deprived neighborhoods, to those with poor health at hospitals, and to prisons; creative projects which lend computer equipment to disadvantaged groups for the creation of new content; and, the provision of information in the language of minority groups and immigrants and in form accessible to disabled users (Cultural Applications: Local Institutions Mediating Electronic Resources [Calimera], 2005).

The lessons learned from successful projects (working, for example, with local community groups or schools) indicate that cultural institutions which use ICT in order to combat social exclusion can make a difference when they work towards removing barriers to the use of cultural assets. Apart from institutional barriers (e.g. charges for access) and environmental ones (such as the physical access to and within the building), there are important personal and social ones which include the lack of relevant skills or the lack of confidence to ask what you need, while another problem is people's perception and awareness of cultural organizations (e.g. "museums are not for me"). The lack of personal contact and the freedom to browse around as you want offered by the online environment is attractive for users who might feel intimidated visiting a real museum. Successful social inclusion projects have helped to familiarize marginalized groups with the use of museum Web sites and online material, have given them the basic skills and confidence to explore what is of interest to them, have increased their trust of public services, and have offered them a sense of involvement and of being part of history (Parker, Waterson, Michaluk, & Rickard, 2002).

PROBLEMS, ISSUES AND CHALLENGES

Information and communication technologies have now been integrated into every aspect of our lives. Cultural organizations feel increasing pressure to use ICT in all their activities and keep up with technological changes. This push from society together with the strive of museums to remain relevant and popular at times when there is increased competition with other leisure-time organizations and activities, has sometimes led to thoughtless use of the technology for its own sake. In some cases, cultural organizations have spent their limited resources for state-of-the-art technological tools, without carefully examining first whether these fulfill the institution's mission and respect its character.

When ICT is used in cultural organizations, there is often emphasis on what is visible and high-profile, sometimes to the detriment of important background work on collections documentation. Although often mundane and repetitive, this behind-the-scenes work is fundamental and should be carried out first, as it forms the backbone on which to build educational resources and interpretation programs. One of the advantages of recording information in digital form is that it allows the linking and integration of all the information relevant to particular objects, such as entry-level object documentation, in-depth curatorial analysis, conservation records, or educational materials.

The creation of digital assets is currently a priority for museums and cultural organizations around the world. This intense activity in digitization, creation of databases, and data entry is resulting in a mass of cultural information in electronic form. However, without further processing, this is simply a mass of data which is not sufficient on its own for the creation of knowledge. In order to become meaningful, it needs to be contextualized, selected, interpreted—activities that fall under the traditional role of curators (Dietz, 1998). So, despite the views expressed about the diminishing role or even the extinction of the curator in the face of rapid technological and societal changes (Besser, 1997), it seems that now, more than ever, his or her role is needed, but is simply adapted and adjusted to follow the wider developments around the museum and the cultural heritage world (Prochaska, 2001). It is important to provide appropriate training to all related museum staff in order to create a multi-skilled workforce able to meet the digital challenges (Smith, 2000). One of the opportunities of the digital revolution is the opening up of this role to groups and specialists outside the museum who can work in partnership to interpret and present resources in new ways.

Since the spread of the digital revolution, worries have also been expressed about the threat this poses to the traditional role of the object. Would the digital surrogates end up replacing the originals? Would people be satisfied with viewing virtual exhibitions and get so used to exploring digital images of the real objects from the comfort of their home that they would stop visiting museums? After the initial upheaval and anxiety created by any technological change, it has become obvious that the traditional object is not dead yet, the same way that analogous fears about the death of the printed book after the advent of the electronic one, did not materialize.

In the 1950s André Malraux (1951) thought of his famous imaginary museum (made possible by the technology of photographic reproduction) as an extension and never a substitute of a real one, with specific functions of artistic appreciation and historical research. In the same way, the virtual museum offers the possibility for extending but not replacing the real one, creating new types of visitors who can interact in the virtual environment in different ways (with regards, for instance, to access, educational opportunities, discovery and interaction with works of art, and shopping of cultural products) (Battro, 1999).

The evaluation of multimedia applications used in the gallery has indicated that visitors spend considerable time using them, but in most cases, this is add-on time, rather than taking away from the time spent looking at the exhibits (Economou, 1998). The experience of virtual exhibitions and online museum material so far indicates that these have not replaced the need and desire to physically visit the related cultural institutions. On the contrary, most museums with an online presence would agree with the view of Victor Rabinovitch and Stephen Alsford (2002) of the Canadian Museum of Civilization Corporation, who pointed out: "Public feedback on our Web site indicates that an online presence draws attention to our physical installation and stimulates a desire to visit. The virtual exhibitions promote an appreciation of the rich knowledge resources that museums hold in trust" (Digital Technology section, ¶2).

It has become clear that the virtual visit is a very different, and in some cases complimentary, experience to the real one, for which it can never be a substitute. Although the technology is constantly improving in this field, the quality of digital representations of three-dimensional space is still not very high. Furthermore, the experiments for providing some form of social interaction online (for example, through the use of e-guides, avatars, chat rooms; e.g. Galani & Chalmers, 2003; Geser & Pereira, 2004; Paolini et al., 2000; Woodruff et al., 2002) have not answered the problem of the lack of the social context, which plays a fundamental role during the visit to a museum. On the other hand, virtual exhibitions and online experiences can appeal to users who would normally never be part of the traditional museum audience, such as younger users who feel comfortable with and are attracted by high-tech applications, but who believe that museums are not for them. Virtual exhibits can also compliment real exhibitions by helping visitors to prepare for their visit or as educational material for further analysis after the visit. When properly designed and carefully planned, online applications can trigger interest and curiosity on the subject of the exhibition, while poor quality implementations can have exactly the opposite effect of disappointing users and creating a negative impression about the organization. As with all media used in the museum, it needs to be employed with care and with an understanding of how it can best support the institution's mission.

Some might argue that the very notion of virtual exhibits goes against the heart of the idea of a museum, which is normally rooted in authentic material evidence and direct encounter with works of art, unlike that offered by the recreated, artificial, simulated environment of virtual reality. Today, though, the traditional model of the museum that "lets objects speak for themselves" has been replaced by a more dynamic and flexible one which focuses more on people and the different ways they can interact with objects. The emphasis has shifted from the objects to the ideas that these represent or to the information they hold (MacDonald & Alsford, 1991) and the way they influence visitors in making meaning (Roberts, 1997) and constructing knowledge (Hein, 1998). The postmodern museum highlighted the different, and sometimes conflicting, stories that can be told about the objects. Many museum professionals advocate the use of ICT to find new ways to tell these stories, rather than recreating in the digital environment old paradigms for the museum's role (Morrissey & Worts, 1998; Teather, 1998).

We mentioned that online cultural environments lack the social context of the real visit. Yet the adequate support of social interaction between visitors or between visitors and staff is a cause of concern also about the use of ICT applications in the galleries (Economou, 2003). Today, it is widely accepted that who is accompanying you at the museum or whom you meet can have a significant impact on your experience and what you will take away from it (Falk & Dierking, 1992; Hooper-Greenhill, 1999; McManus, 1987, 1988; Silverman, 1990). Yet, the designers of these applications rarely succeed in successfully incorporating or encouraging communication between different users or support of groups. It is not an easy task to design programs which allow individuals to contemplate the displays at their own pace with a system which can be personalized according to their preferences, but which at the same time maintains contact with their environment and social group, and does not create little "aliens" isolated in their own world behind their headsets or handhelds as they go around the gallery. Incorporating testing with groups on the exhibit floor and evaluation from the early stages of design is very important in this process. This can help to create applications that incorporate or even take advantage of social interaction in public spaces and leisure-time activities.

What we can conclude from the different manifestations of the digital museum we examined is that new technologies cannot substitute for the direct encounter of our senses with the material culture of the present and the past. But they have the potential (which has not always been realized so far) to compliment, enhance, and extend the cultural experience in new ways, and share it with new audiences. In order to achieve this, the rules which apply to traditional museum activities are also relevant here: it is important for museums to be clear about the aims of these applications and the audience they are trying to reach; to follow widely recognized professional standards; to retain the importance of cultural content and not compromise its essential characteristics; to consult with visitors and users; and to generally be self-reflexive about their work practices. This way, rather than slavishly following the technological wave, cultural organizations can master these powerful tools and use them only where appropriate and in ways which support their primary mission and public.

(ENDNOTES)

1. The application received an Honourable Mention at the 2004 Muse Awards organized by the American Association of Museums Media and Technology Committee (History and Culture category). See http://www.mediaand technology.org/muse/2004muse_history.html.
2. "Today at the Museums" received the Bronze 2004 Muse Award (AAM Museums Media and Technology Committee) in the Collection Database/Reference Resource category. See http://www.mediaandtechnology.org/muse/2004muse_ database.html.
3. The sound effects imitate the movement of the objects on the paper (for example, the stone and the apple roll on the digital page, the light is lit on the desk lamp, the door opens a video screen in front of which you can read apart from the syllables of Japanese script, which can be read by a voice when selected by the user's stylus).
4. The original CAVE® was developed at the Electronic Visualization Lab (http:// www.evl.uic.edu/core.php) at the University of Illinois at Chicago and was first presented to the public in 1992. Few CAVE® systems have been used in the cultural sector. The Ars Electronica Center (http://www.aec.at) in Linz, Austria was the first to install a CAVE system, for which it continually invites artists to design new virtual worlds based upon their own artistic conceptions, similar to the NTT InterCommunication Center (ICC) in Tokyo (http://www. ntticc.or.jp/index_e.html).
5. QTVR stands for QuickTime VR, designed by Apple Computer (http://www. apple.com/quicktime/technologies/qtvr/). By 'stitching together' several digital photographs of three-dimensional spaces, it creates photorealistic panoramas that can have interactive components. The panoramic environment created by QTVR movies can also offer users a certain degree of interactivity in their exploration, but can be displayed only on a small part of the computer screen (e.g. 480×300 or 300×200 pixels).
6. VRML stands for Virtual Reality Modeling Language, which became an ISO open standard in 1997. It was widely used for producing interactive three-dimensional applications on the Internet, allowing users to navigate through a 'virtual world', such as a museum gallery, and to examine three-dimensional objects. In 2001 it was superseded by Extensible 3D (X3D), the next generation of the VRML specification, which uses XML encoding. A special browser plug-in is required in order to view both VRML and X3D 'worlds', but unlike QTVR, these are usually displayed on the whole computer screen. Both VRML and X3D are recognized as Web3D, the general term used to describe all standard forms of Internet 3D technologies endorsed by the Web3D Consortium (http://www.web3d.org).
7. The most popular are Active Worlds (http://www.activeworlds.com) and Second Life (http://www.secondlife.com), which let you visit, chat in, but also create 3D virtual worlds.
8. The idea of an online community in virtual space has also been of great interest to artists, as is the case, for example, with the multimedia art work on the Internet "Conversation with Angels" created by MEET factory and hosted at the Kiasma Museum of Contemporary Art in Finland (http://angels.kiasma. fng.fi). This is a virtual multi-user world in which users can chat with a variety

of different avatars in real-time, creating an online community with the artwork, thus exploring the idea of an artwork as a social process.

9. This was developed in association with Antenna Audio. More information is available at http://www.tate.org.uk/modern/multimediatour/.

10. More information on the use of PDAs in museums, an extensive bibliography, and examples of projects and applications is included in the Canadian Heritage Information Network Tip Sheets on Personal Digital Assistants (http://www.chin.gc.ca/English/Digital_Content/Tip_Sheets/Pda/tip_sheet13.html).

11. For work in this area, see the CIMI (initially, Consortium for the Computer Interchange of Museum Information) Web site (http://www.cimi.org), which was frozen in December 2003 when it ceased operations, but includes useful information and a list of publications.

12. For more information, see the Web site of CIDOC (the International Documentation Committee of ICOM, the International Council of Museums) (http://www.willpowerinfo.myby.co.uk/cidoc/), which includes information on museum information standards. See also MDA (http://www.mda.org.uk/), the UK's leading organization for information management and documentation in museums.

13. More information on various thesauri and terminology control schemes can be found at the *SPECTRUM Terminology* part of the MDA site (previously known as the *wordHoard*), http://www.mda.org.uk/spectrum-terminology/.

14. A variety of examples can be found at the annual Museums and the Web conference site (http://www.archimuse.com/conferences/mw.html/), which organizes a "Best of the Web" competition with a "Best-Online Exhibition" category.

15. As is the case of the "Playground" section of the ArtsConnectEd site of the Walker Art Center and the Minneapolis Institute of Arts (http://www.arts connected.org), and several games of the "Show Me" zone for Kids (http://www.show.me.uk/games/games.html) of the 24Hour Museum site, the UK's "National Virtual Museum" (http://www.24hourmuseum.org.uk), to mention only two of the several available examples.

16. See, for example, the World Wide Web Consortium (W3C) Content Accessibility Guidelines (http://www.w3.org/TR/WAI-WEBCONTENT/ and http://www.w3.org/TR/2004/WD-WCAG20-20041119/) and the UK Royal National Institute of the Blind Accessible Website Guidelines (http://www.rnib.org.uk/digital/hints.htm/).

17. The "digital divide" is of course also experienced by cultural institutions that do not have enough resources and access to technology skills themselves. See, for example, the situation in Africa (Luhila, 2001).

11 Blurring Boundaries for Museum Visitors

Areti Galani

University of Newcastle

Matthew Chalmers

University of Glasgow

INTRODUCTION

A summer day of 2001 in Glasgow, and two friends were wandering around the Mackintosh Interpretation Centre looking at objects and displays and talking about the famous Scottish architect and designer, Charles Rennie Mackintosh. One of them was inside the gallery; the other was a few metres down the street, in an office space but 'visiting' with her friend through her desktop computer. Her computer displayed a three-dimensional model of the gallery with textual information about the artefacts and supported real time audio communication. Building on this experience, this chapter explores some of the issues involved in technology-mediated museum visits, with a focus on group visits. It discusses the sociality of museum visiting and how social conduct among members of a group is organised during a visit. It particularly focuses on how technological applications may enhance the museum visit for local and remote visitors by supporting and encouraging sociality.

For some time, museums have supported the design and implementation of a range of media, analogue and digital, which enhance the visitor experience for diverse audiences. Digital museum applications have particularly capitalised on the informative and interpretive potential of new technologies. Electronic guide books, combined in some cases with ubiquitous technologies, have been implemented in museums and heritage sites to offer information in the form of video, audio and text (Exploratorium, 2001) as well as three-dimensional reconstructions of buildings (Vlahakis et al., 2001) and cities. Furthermore visitors in the Reykjavik Art Museum and the Prestongrange Museum in Scotland (Smith, 2004) may use their own mobile phones to access audio commentaries about the artefacts on display and the artist John Bellany respectively. Museums and research groups have also experimented with personalised adaptive and adaptable information applications for onsite and online visitors, and which may also support pre- and post-visiting. Personalisation of museum information has been explored by a combination of means (Bowen & Filippini-Fantoni, 2004) such as visitor profiling (Paterno & Mancini, 2000), history of use (Oberlander, Mellish, O'Donnell, & Knott, 1997) and so forth. Moreover, projects such as HIPPIE (Oppermann, Specht, & Jaceniak, 1999) and Cooltown (Fleck et al., 2002) progressed from offering location-dependent information to visitors to additionally supporting recording and editing of parts of the experience for later reflection and sharing with friends and family.

Most of these applications, however, are designed to offer additional, diverse and personalised information primarily to individuals who visit the physical premises of a museum and secondarily to Web site visitors. The emphasis on single-user technology, with some exceptions that we will discuss in a following section, and the focus on personalised information reflect, we believe, an assumption of the primacy of the physical experience and the belief that information seeking is the primary motivation of a museum visit. In the museum studies literature, often the discussion about new media focuses on the real–virtual divide (Mintz, 1998) that treats remote visits as secondary or surrogate experiences to the physical ones, prioritising the unmediated experience of the museum object—"the real thing"—over the mediated experience via technology. Museum Web site design, on the other hand, appears divided as to whether to provide genuine online visitor experiences or instead encourage and support physical visiting (Cunliffe, Kritou, & Tudhope, 2001). Local and remote audiences appear segmented, and the connection between local and remote visitors is often overlooked, particularly in the context of a single synchronous visit.

This chapter explores a visitor-centred approach to local and remote museum visiting, and looks at the relation among local and remote visitors from the point of view of visitors' interaction. Instead of focusing on delivery of information in galleries, which is discussed in other chapters in this book, we investigate social interaction among friends in museums and how social conduct may blur the boundaries among local and remote, and may foster shared experiences for combined onsite and online audiences. This approach is inspired by social science, especially the field of ethnomethodology (Garfinkel, 1967) that argues that every activity is effectively a socially accomplished activity, and that the context of an activity is dynamically constituted in and through social conduct. We also believe that current developments in ubiquitous technologies and telecommunications, as well as changes in the mentality that surrounds their use, encourage us to think of digital technologies not only as information tools, but also as experiential processes that fit with one's everyday life and interactions. We do not overlook information—rather, we treat it as a resource for interaction. We are interested in interpretation that is produced in the course of collaborative encounters among participants. In this way we wish to "regard new media, particularly the World Wide Web, as a resource that more closely resembles a museum visit than a museum collection" (Borysewicz, 1998).

The next section elaborates on the notion of the museum visit as a social activity, particularly how social interaction often mediates and shapes personal engagement and vice versa. The section is inspired by observational studies of non-educational groups of visitors in two cultural institutions in Glasgow, UK. We then briefly present museum technologies that support social interaction, to be followed by discussion of an excerpt from a mixed reality museum application that supported simultaneous visiting among local and remote visitors. We then argue that the categorisation of local and remote participants is not a straightforward cut. Instead, boundaries may be blurred with social conduct. Furthermore, we expand this discussion to issues regarding the role of the museum object in mixed reality museum environments, the emergence of a mutually complementing physical and digital museum design, and the practicalities of running and maintaining such environments.

MUSEUM VISITING AND SOCIALITY

Museum visits are social events. Whether treated as educational activity or leisure activity, museum visiting is shaped by social conduct in terms of both visitors' intentions and overall experience. In a pioneering research of visitors and non-visitors in the metropolitan area of Toledo, Hood (1983) identified that 'being with people' was highly valued among occasional visitors and non-visitors, and often a reason for people not to visit museums alone. This finding also persisted in recent studies of visitors' motivation that discussed museum visiting as a 'day out' with the whole family, friends and relatives (Falk, 1998; Falk, Moussouri, & Coulson, 1998).

Furthermore, Baxandall (1987), talking about visits in art museums noted that the bulk of the experience is not about "looking at pictures but about talking about looking at pictures," and the labels are means of constructing the visitors' dialogue about art. Falk and Dierking (1992), following extensive visitor studies, defined social context as one of the three key elements that influence the way visitors experience museums—the other two being the personal and physical—and argued that learning in museums is necessarily socio-culturally mediated (Falk & Dierking, 2000). A series of other visitor studies also looked at how social interaction might affect learning, and how social behaviour is expressed in museums, especially among family members, for example (Baillie, 1996; Diamond, 1986; Dierking, Luke, Foat, & Adelman, 2001). Furthermore, McManus (1987a) focused on specific museum behaviours, such as label reading and so forth and examined how they change in relation to size and cohesion of visitor groups. Visitor studies so far have offered useful insights in the ways interpretation and learning are influenced by social interaction in museums. They have also informed the evaluation and design of museum displays but, due to their primary focus to cognitive aspects of the visit and their often evaluative motivation, they have often overlooked how social conduct is organised and realised throughout the visit.

In a recent extensive study of visitors' conduct in museums and galleries, vom Lehn (2002) adopted a sociological approach to examine how museum objects are constantly constituted in and through social interaction among visitors. Vom Lehn's research was informed by Garfinkel's ethnomethodology and Sacks' conversational analysis (Sacks, 1998), and utilised video recordings to capture social activity around displays. Vom Lehn's study was particularly concerned with social interactions that took place among friends and strangers alike at the object-face. His analytical approach consisted of exhaustive discussion of brief fragments of conduct in front of museum artefacts. Looking at social conduct as it unfolds *in situ* is an approach that stems from the ethnographic tradition in social sciences. In recent years, it has also become increasingly popular among technology designers, among them researchers with an interest in tourist and leisure applications, for example (Brown & Chalmers, 2003; Cheverst, Davies, Mitchell, Friday, & Efstratiou, 2000; Flintham et al., 2003).

Vom Lehn's study looked beyond visitors' conversations to examine other resources that visitors use to organise their conduct with fellow visitors and the exhibition—for instance, the importance of gestures and body posture in the museum experience had been pointed out by Falk and Dierking, but detailed understanding of them had not been pursued before. It also made obvious that the experience of artefacts in a museum is constantly negotiated and re-shaped by social conduct, and that detailed inspection of social interaction with and around museum exhibits may offer insights in the design of novel displays that enable and encourage social interaction. Consequently, it informed the design of such displays in the form of art installations, for example the Ghost Ship exhibit (Hindmarsh, Heath, vom Lehn, & Cleverly, 2002). Furthermore, his work reinforced the potential of sociological approaches in the design and evaluation of museum displays (Heath & vom Lehn, 2002).

However, the majority of visitor studies vom Lehn's investigation included focused on social interaction that happens with and around specific displays, ignoring effectively social conduct that happens in between displays or on the fringes of visiting activity. An ethnographic study of non-educational groups of visitors in The Lighthouse and the House for an Art Lover in Glasgow (Galani & Chalmers, 2002), under the auspices of the *City* project, indicated that other aspects of the visit, such as the pace of the visit (Galani & Chalmers, 2004) and the way friends connect and combine displays, media, and routes throughout the museum environment are also informed and influenced by social conduct. Verbal and gestural activity informs the time people spend with the exhibits, their orientation and exploration of the exhibition content. Verbal and visual cues facilitate both direct interaction and peripheral awareness among members of the same group, as visitors balance personal engagement with the exhibition and social exchanges with their friends.

Moreover, the majority of visitor studies also seem to be preoccupied with social interaction among collocated participants who visit the physical premises of a museum or gallery. Social conduct that takes place during online visiting, which might inform or be informed by a trip to a museum, is often overlooked. However, a small number of studies of remote visitors has indicated that, on the one hand, the number of online visitors is continuously increasing (Rabinovitch & Alsford, 2002) and, on the other hand, that a portion of remote visitors do not visit alone, but instead explore the museum material in the company of others such as members of their family, fellow students and friends (Chadwick, 1999; Goldman & Schaller, 2004; Loomis, Elias, & Wells, 2003; Semper, Wanner, Jackson, & Bazley, 2000).

The study of social conduct in museum visits, i.e. the study of resources and methods on which visitors rely in the production and recognition of social actions and activities, emerges as an overall promising approach to better understand the visiting activity and design for it. Visitors in museums, individually and in groups, engage with and enjoy displays in the presence of others. Mundane, often unremarkable, details of social action such as gestures, posture and so forth support the overall museum experience and inform one's own engagement with artefacts. The rest of this chapter explores how a sociological approach may also lead to a better understanding and design of museum technologies for both local and remote visitors. The next section initiates the discussion by examining existing applications that address issues of social conduct in museums.

TECHNOLOGY-MEDIATED SOCIALITY IN MUSEUMS

Along with museum applications that concentrate on information delivery, there is a series of projects that aim at supporting sociality and communication among visitors in both asynchronous and synchronous activities. Asynchronous social awareness in museums has been traditionally encouraged by visitors' books, and exhibitions that invited visitors' comments as intrinsic part of their displays. New media increased the opportunities for such awareness with online mailing lists and boards, and opinion kiosks, for example the *Tell us what you think* exhibit in the Wellcome wing of the Science Museum in London. For instance, Walker Art Center invites its Web visitors to associate exhibits with ideas and emotions online, and this information subsequently becomes part of the object's profile for other Web visitors to see. Furthermore, in the History Browser (http://www.nma. gov.au/collections/about_history_browser/) on the Web site of the National Museum of Australia one can see other people's pathways through the collections or chat with other Web visitors about artefacts (Peacock, Ellis, & Doolan, 2004).

Synchronous social interaction with the mediation of technology is a relatively new concept for museums. Furthermore, it is treated with concern—and suspicion—among museum professionals. This is partially due to reports from studies of audio guides that have pointed out that the use of such devices in museums often inhibits social interaction (Martin, 2000), and decreases conversation especially among members of the same group (Walter, 1996). These studies, however, report on technologies that are specifically designed for single users. The *Sotto Voce* guidebook project by Xerox PARC studied further the role of electronic guidebooks during group visiting. On the basis of conversation analysis, it defined the conversational role of such devices in visitor's discussions (Woodruff, Szymanski, Aoki, & Hurst, 2001). It also pointed out the often collaborative use of electronic guides in the course of a visit. This research resulted in an electronic guidebook that afforded eavesdropping among the participants: pairs of visitors explored a historic house at their own pace while they remained aware of each other's choices of commentaries. Awareness mediated by the technology created a sense of connection among the visiting parties. Furthermore, the awareness of content that was achieved through eavesdropping informed natural and rewarding forms of conversation, facilitated group cohesion and supported increased awareness of the rooms and their content (Aoki et al., 2002).

Technology-mediated sociality among online museum visitors has also been treated with disbelief among museum practitioners as to how effective technology might be without becoming antagonistic with the physical premises of the museum. In an early paper on new media in museums, Shane (1997) asked: "Does it [the Web] lend itself to the kind of group interaction so central to a museum experience? Can simulations provide genuine experiences that compete favourably with experiencing real objects with other 'real' people?" (p. 193).

Sociality among remote-only museum visitors has been explored in the area of collaborative virtual environments (CVE), for example the virtual tour on the Web site of the Van Gogh Museum (www.vangoghmuseum.nl) and the Virtual Leonardo project (www.museoscienza.org). Both applications used 3D graphics and chat technologies to support synchronous remote visits among distributed online visitors. Each visitor was represented inside the environment by an avatar and s/he could navigate around the environment, manipulate objects, access information about artefacts and interact with friends and strangers. Within the intentions of Virtual Leonardo was to look at the social aspects of the remote visit. Therefore the application prompted the visitor to set an appointment with a remote friend (in a separate workstation) before s/he logged on. The evaluation, however, of the project focused on the technical and usability issues of the system (Barbieri & Paolini, 2000). Limited analysis of conversations that took place in the system indicated that guided tours generated more talk about the artefacts as opposed to discussions regarding the users' whereabouts and the system's functionality that reportedly dominated free browsing. On the other hand, the collaborative virtual tour of the Van Gogh museum has not been evaluated yet (M. Verhoeven, personal communication, July 18, 2003) and despite the fact that the opening page of the tour generates a great deal of interest from Web visitors, the effect of the application to a social visit online is unknown.

Social awareness among distributed local and remote audiences has been loosely afforded by the inclusion of real time Webcam views of the galleries on Web sites, e.g. the Museum of Contemporary History, Bonn (www.hdg.de). Awareness and interaction have also been explored through networked art installations. For example, The Difference Engine #3 (Hershman, 2001) in the ZKM media museum in Karlsruhe was an art installation that deliberately explored social awareness among local and remote museum visitors. It supported a two-way communication between physical and digital by offering views of the virtual museum environment to local visitors and views of the physical environment to remote visitors. A chat channel was also used to support message exchange among visitors. Difference Engine #3 offered local and remote visitors the opportunity to simultaneously interact with a single 'digital sculpture,' and also introduced and encouraged social awareness and interaction beyond the physical walls of the museum. However, visitors' perception and use of the installation has not been studied (A. Buddensieg, personal communication, November 21, 2002).

Furthermore, a series of projects have combined the concept of a shared museum tour with robotics to create robot tour guides for both remote and local visitors. In the Tourbot project (Trahanias et al., 2003), a robot guided local and remote visitors around museum exhibitions while offering Web visitors real time views of the galleries. Both audiences collaborated in the selection of tours through a voting process. In that respect the two otherwise isolated audiences were treated as equally important in the shaping of a museum tour. Although awareness among onsite and online visitors was afforded, direct interaction through the system was not supported. Onsite visitors were aware only of the tour preferences of their remote co-visitors whereas online visitors could also get a glimpse of the onsite participants through the eyes of the robot. The evaluation of the application, once more, prioritised the technological and usability aspects of the system while underemphasising the social and interactional effect on the visit.

THE *MACK ROOM* MIXED REALITY ENVIRONMENT

The *Mack Room* mixed reality environment, which will be the focus of the remainder of this chapter, prioritised and supported social conduct among local and remote participants, and actively sought to capture and understand the visiting activity in relation to its sociality. Its design was informed by initial visitor studies—briefly mentioned in a previous section—as well as by technical, theoretical and interaction design goals within the fields of museum studies, computer supported cooperative work and ubiquitous computing. The prototype explored co-visiting among people who know each other and share an interest in museums, but who may not always be able to visit together due to difficulties such as geographical separation. The *Mack Room* system was designed for a specific gallery: the Mackintosh Interpretation Centre (Mack Room) in The Lighthouse. The exhibition combines textual and graphical displays with authentic artefacts, and over twenty screens presenting video and interactive material.

The *Mack Room* system combined virtual environments (VE), hypermedia technology, handheld devices and ultrasound positioning technology. It allowed at least three visitors, one onsite and two remote, to visit the Mack Room simultaneously. An ultrasound positioning system and a wireless communications network was installed in the Mack Room. The onsite visitor carried a PDA that was tracked via the ultrasonics. The handheld displayed the ongoing positions and orientation of all three visitors on a map of the gallery (Figure 11.1). One offsite visitor used a hypermedia-only environment that comprised a standard Web browser which also displayed the gallery map (Figure 11.2). The other offsite visitor used a first person 3D display with avatars representing the other visitors (Figure 11.3). All visitors shared an open audio channel, and wore headphones and microphones. The system also supported multimedia information for the offsite visitors in the form of Web pages that were dynamically presented upon movement in the map or VE. That information was similar but not identical to the information presented on labels and panels in the gallery. This automatic presentation schematically followed the spatial organisation of the exhibition, so that all three visitors could 'look' at the same display when in the corresponding location. In that respect, the system supported interaction around corresponding exhibits in the Mack Room and in digital form, which we tentatively called 'hybrid exhibits' (Brown et al., 2003).

VISITORS' EXPERIENCE

The user trials of the system took place in the Mack Room. Thirty-four people took part in groups of two and three friends. Each visit lasted approximately one hour and comprised an exploratory part and an activity-based part. In the first part, the members of each group were encouraged to familiarise themselves with the technology and explore the gallery according to their own interests. In the second part, they were given a mixture of open-ended and focused questions about Mackintosh's work, and were asked to come up with answers based on evidence from, or their experience of, the exhibition. The group's activity and discussions were recorded, and a semi-structured interview followed each visit. The analytical treatment of the data explored the organisation of the visiting activity through the detailed observation of the participants' verbal and gestural activity.

Green, Blue and Red are friends and colleagues. In the trial Green was onsite, in the Mack Room, while Blue was visiting in the VE and Red in the hypermedia environment. In the following excerpt, Blue and Red had

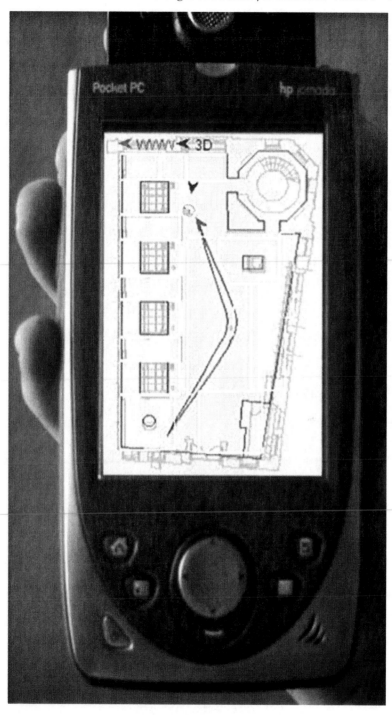

Figure 11.1 The Mack Room Interface (Handheld)

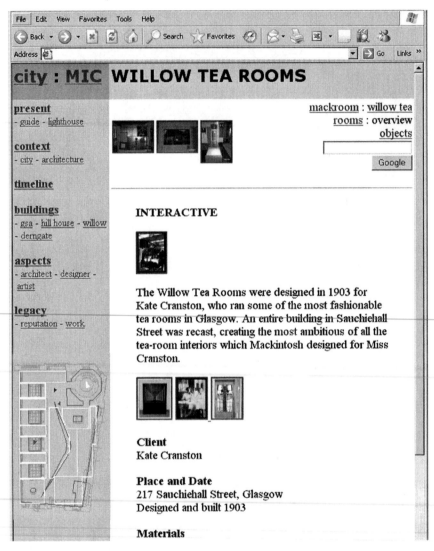

Figure 11.2 The Mack Room Interface (Website)

earlier spotted a display about a bedroom that Mackintosh designed for 78 Derngate Street, and they had a chat about it. Green, who was occupied in the other side of the room, overheard their discussion, checked his map and when he finished looking at the display he was examining, started walking towards the area that his friends were in. We join the action when Green and Blue meet each other halfway:

> Green: (He stops and looks at the handheld.) Do you know where I am passing? (Pauses) Did you see me go by? (Figure 11.4a,b)

Figure 11.3 The Mack Room Interface (3D Display)

Blue: I see, I did, where are you going? (Figure 11.4c) I am gonna follow you again. (She changes orientation) (Figure 11.5c).

Green: (He keeps walking behind the display.) Oh are you? I was going to the bit that you were looking at which was . . . (Figure 11.5)

Blue: (She keeps moving to the direction of Green but she loses sight of him, since he is hidden by the display wall.) I've just walked into oh . . . where did you go again? (Figure 11.6)

Green: Well I was looking . . .

Blue: Who's the hat?

Green: Where?

Blue: (Inaudible)

Green: (Laughs. Pauses.) What (in loud voice) was the exhibition you were looking at before? (Figure 11.6)

Blue: It was the . . .

Red: The Hunterian Art Gallery. (She moves to the '78 Derngate Street' area on the map) (Figure 11.7b).

Blue: The Hunterian Art Gallery, the *guest bedroom.*

(Green checks his handheld and he walks to the other side of the display.)

Red: Yeah.

Blue: A very stripy bedroom.

Green: (He looks at the display.) Ok that's where I am now.

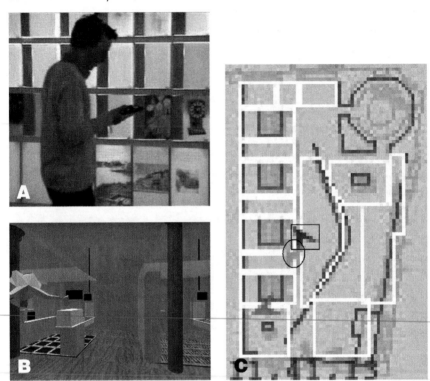

Figure 11.4a-c Interacting in the Mack Room (Part 1)

Blue: Can you see there's like two twin beds and blue and white stripy wall paper?
Green: (He stops and looks at the picture.) Yeah, horrible shape, terrible (Figure 11.7a).
Blue: Oh I think it's bad from here.
Green: Well it probably wouldn't go in your room.
Blue: No it wouldn't, imagine waking up . . .
Green: (Inaudible)
Blue: Imagine waking up with a hang over.
(They both laugh.)

Note: In Figure 11.4 through Figure 11.7, for clarity, the green arrow (local visitor) has been circled and the blue arrow (VE visitor) has been squared. Italic has been used to indicate text taken from display labels.

While Green was moving towards the display in question, Blue moved away from it, and when they met up, Blue decided to follow Green. This decision was verbalised but also acted upon, as shown by the new orientation of the blue arrow on the map. This is not an unusual behaviour in museums. Friends during their visit may attend different displays related

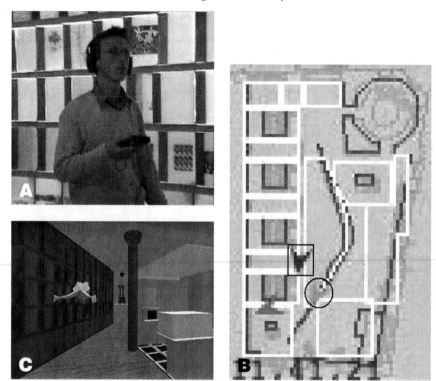

Figure 11.5a-c Interacting in the Mack Room (Part 2)

to their own interest but remain peripherally aware of their friends' activity due to their proximity in the gallery or by retaining visual contact with them. This awareness is facilitated by visual cues and helps the members of the group to keep track of their friends, develop a shared visiting pace and also inform their own exploration. In the mixed reality environment, visual cues were limited in the display of the participants' position, hence movements of arrows or avatars on the map and the 3D model. This limited cue, however, kept the onsite visitor aware of where his friends were and the rough location of the artefact in question; it was further confirmed verbally. Shared orientation towards the display involved several stages: the onsite visitor approached the area based on his; the remote visitor gave a rough description of the artefact, which included its title, as mentioned on the available Web page, and a reference to group-specific knowledge: the location of the original artefact, in the Hunterian Art Gallery. After the orientation stage, the onsite visitor adopted a relaxed viewing position towards the display (Figure 11.7) and the two visitors started talking—while being overheard by their other friend—about the room decoration. Their discussion began with an aesthetic appreciation of the room and concluded with humorous comments about the potential effect of the decoration on one's

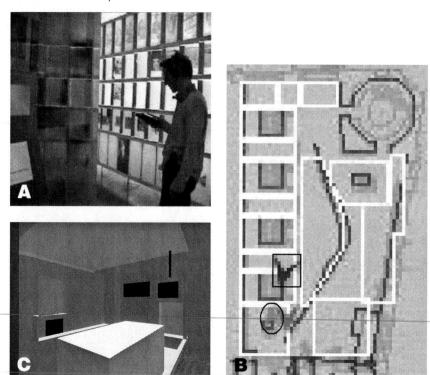

Figure 11.6a-c Interacting in the Mack Room (Part 3)

mood. The latter, appeared to stem from Blue's personal experience but also Green's knowledge of his friend's lifestyle and taste.

DISCUSSION

The richness and topical coherence of visitors' interaction with each other and with the exhibition is the basis of our claim that local and remote museum visitors had a shared visit. In this co-visiting experience, the museum's remote presence was treated not strictly as an information space, used in isolation, but also as a social place to visit, enjoy and relate to others. The latter afforded a set of behaviours that, as we have shown, constitutes a social experience that shares several significant attributes of traditional museum co-visiting. The experience offered plentiful information and afforded rich interaction within a heterogeneous mix of media. This approach moves away from the traditional design focus on single users, toward multi-user interaction that treats the traditional and new media aspects of a museum as equally important elements of the museum experience (Galani, 2003). Furthermore, it broadens design to address both personal and social aspects

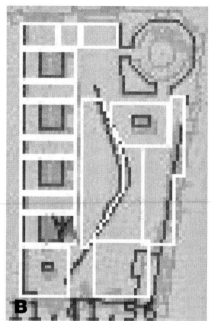

Figure 11.7a-c Interacting in the Mack Room (Part 4)

of the visit, and does not restrict the visitor to either one of these modes. It supports the individual's engagement with displays, which can become a resource for social interaction, and which in turn might inform later individual interpretation.

This work does not attempt to substitute or reproduce a visit to a traditional museum—the same applies for applications such as Virtual Leonardo and the virtual tour in the Van Gogh Museum. It supports, however, a mixed reality museum visit that may cover needs and expectations that are not easily addressed by the traditional museum. Remote visitors, disenfranchised by geographical or other barriers, may interact with the layout and content of an exhibition and become immersed in exploration of and discussions about artefacts. Local visitors may also access information online, with the difference that they can use the contributions, experience and understanding of their remote friends, in a fashion similar to collocated visits and other leisure activities.

This approach to remote access to museum environments creates new opportunities for museum visits and exhibition design. It is not however unproblematic both in terms of technological implementation and museum practice. We would like to explore further the issues that arise from supporting social visits among local and remote participants, such as the shared engagement with museum artefacts, along with the empowerment of the remote visitor and the emergence of a mutually complementing physical and digital museum design. Additionally, we discuss some practical considerations regarding mixed reality environments and how they might fit with a museum's practices and priorities. This section is primarily informed by the Mack Room application, since evaluative studies of other applications of the kind are not available.

Shared Engagement

Recent studies of technology in museums, and especially of use of personal mobile devices, have shown that interaction with technology might inhibit social interaction as well as redirect one's attention from the museum object to the information that is delivered on one's device (vom Lehn & Heath, 2003). Among the most reported disadvantages of such technologies is the decline of talk among visitors. On the contrary, the evaluation of the mixed reality environment in the Mack Room reported a radical increase in talk among participants. For some of the offsite participants the experience was liberating: "I think it is fun though. I quite enjoyed the social engagement in that way, being able to talk about everything more and not feeling that you are disturbing . . . not thinking about other users in the gallery. You know it's kind of liberating," and for others it was a good laugh: "I thought it was actually fun, and I thought it was a laugh; an easy pleasure."

The relaxed manner of the visit brought increased production of funny, unexpected and imaginative comments and reactions by the participants, and the affective rather than scholarly approach to the available content. Co-visitors used and appropriated the available information to suit their shared knowledge and experiences. Although part of the conversation involved giving directions and instructions to one's friends regarding one's where-abouts, well-reported museum behaviours were regularly observed: participants read aloud phrases from the exhibition text, communicated their own knowledge, made connections to their own everyday lives, expressed opinions and verbalised imaginative thoughts.

Unlike the displacement of the object that is reported with mobile devices, in the mixed reality environment the constant focus of the attention was the displays and the environment. The hybrid character of the displays, which meant that the participants interacted with different presentations of the display according to their media, provoked extended discussions around the displays. Initially the discussions aimed to develop a shared understanding of what was available to each participant, to "translate" and "compare" it with each other, as one of the participants said, a process that "gives a different kind of perspective"; then to discuss the content. Asymmetries in the presentation and the amount of content, afforded by the variety of the media, as well as the participants' eagerness to share, often sparked further investigation and exploration of content that was not accessible at the first glance. The attractiveness of displays in the different media was also variable. As a result, people often were prompted by their friends to see objects that they would have skipped otherwise. How asymmetries in the visiting environments influenced interaction is the topic of the next section.

Crossing the Physical–Digital Boundary

The separation between physical museum experiences and digital visits on the basis of the media in use is at the heart of the discussion about new media in museums. The same scepticism seems to inform Mintz's (1998) claim that "a virtual visit to a museum is fundamentally a media experience, not a museum experience" (p. 28). In our opinion, however, this distinction appears to stem from focusing on the individual media and their differences, instead of their use in context, this being dynamically (re)constituted through social conduct. We look at the issue by taking into account the overall interaction among local and remote companions and the exhibition. In the excerpt presented earlier, one of the remote participants pointed out that the decoration of the 'guest bedroom' on display had a similar bad effect on her monitor as it did in the gallery display. In that instance the difference between the media and the distinction between online and onsite did not seem to impose problems in discussing the display and participating in the shared joke that followed. On the contrary, and again using the given dialogue as an example, the two friends used the displays at hand to initiate their discussion, and complemented it with their knowledge of each other's habits and tastes. The distance, the diversity of the environments and media did not seem to inhibit their shared appreciation of the display. We suggest that a more fruitful way of looking at mixed reality environments in museums is to treat all media—new and old—as potentially equal resources in the course of interaction. By that we mean that within their features and possible uses they may embody equally significant opportunities and threats with regard to the engagement with and enjoyment of an exhibition.

This concept is further supported by another point in the excerpt: the moment where the two friends decided to follow each other. Participants in the Mack Room trials often followed each other in the course of their visit. Remote participants followed their local friends around; they also invited them to displays or suggested points of interest to them. Local participants invited their friends to join them where they were standing in the gallery, and shared recommendations on where to go next. Social conduct supported their interaction in and through physical and digital environments, and facilitated the blending of media and environments in one common activity. The participants appeared willing to follow their friends regardless of the media they were using, passing the 'leading role' among them. Although one might expect the onsite exhibition to have primary impact on people's choices, we believe that participants often treated all environments as equal resources for interaction as long as they supported the activity at hand. Additionally, their personal relationship and previous shared experiences appeared to also influence their actions.

Furthermore, by supporting social cues the Mack Room system created a sense of togetherness and engagement for the participants in the visit, which was highly valued in the debriefing interviews, as one of the remote visitors said to her onsite friend: "It would actually be nice to share opinions as you were looking, rather than sit down and have a coffee afterwards to talk about what you've seen. A bit more engaged. . . ." We however feel that, in many cases, social interaction was favoured above individual engagement with the museum displays. In our initial studies of collocated visitors we had established that collaborative exploration of displays is based both on strong personal engagement and social interaction. We believe that mixed reality environments, similar to the Mack Room prototype, would benefit from focusing equally on attracting and sustaining personal engagement with the exhibition along with the support of group collaboration. The design team of the *Sotto Voce* electronic guidebook dealt with a similar concern by awarding strict primacy to the commentary choices made by the user, as opposed to available commentaries through eavesdropping. In mixed reality environments, one way of achieving this is by further exploring and exploiting the individual characteristics and affordances of each medium, for example by introducing complementary asymmetries in the quantity and type of information, e.g. having historical information about a painting presented to one person while another contributes technical information about its production.

We believe that a design approach towards a diverse but mutually complementing physical and digital museum design would also fit with visitors' expectations as an offsite visitor said: "That's what I expected. I expected that I would have more text so I could look up and tell you more things than you would be able to get." Additional information is often available on museum Web sites but there is no provision for the online use of this information either to be reflected on the onsite experience or to be shared among users.

Practical Considerations

We have discussed the social interaction among local and remote friends in a mixed reality museum exhibition, and presented examples of both navigation around the exhibition and lively discussions around displays. In this section, our attention shifts to practical considerations regarding the application and maintenance of mixed reality technology in museum settings. Mixed reality environments may enhance visitor's experience but they also introduce practical challenges. This section explores two aspects of the challenge: the ecology of the museum environment and issues of maintenance and updating.

The remote participants, free of constraints usually imposed by the museum's sheer materiality as well as the corresponding social etiquette, were able to explore the displays and the environment in an unusual manner. Technology enabled them to do things impossible by human standards, for instance passing through walls, as well as things incongruous with museum customs, such as racing each other. In the interviews, most of the remote participants mentioned this kind of freedom as one of the advantages of the experience. They were however aware of the fact that the person in the gallery was accountable for her behaviour not only among the group but also among other visitors. The unexpected navigation choices, e.g. radical changes of direction, *impromptu* disruption of other visitors' field of view and so forth, was the most noticeable change in the visiting manner of the onsite participants. In the interviews, local visitors confirmed that they did not feel intimidated by this freedom; however they expressed concerns that it might be proved impractical in crowded exhibitions—socially disruptive behaviour was also reported by vom Lehn and Heath (2003) in users of audio guides. Based on our experience with technology, we anticipate that subtler behaviours are usually developed as users become familiar with systems over longer or more regular periods of use. A theoretical account of this concept is discussed in (Chalmers & Galani, 2004). Nevertheless, the impact that social interaction among onsite and offsite visitors might have on the navigational ecology of the gallery is worth revisiting.

Furthermore, hybrid exhibits that enable social interaction around and about displays also impose maintenance challenges to museums. Although asymmetries in the content appeared fruitful and often sparked further exploration, the hybrid character of the exhibits effectively means that changes in one environment should be reflected in the others so people can orientate themselves towards the same display. In our studies we found out that asymmetries in content were tolerated better by participants than asymmetries in spatial representations, which almost unerringly lead to confusion, disorientation and distrust of the technology. We emphasise that such asymmetries have to be carefully designed, just as any other exhibition feature would be.

CONCLUSION

This chapter looked at novel museum technologies from the point of view of sociality. It discussed the role of social conduct in the shaping of the visiting experience and it explored in some detail museum applications that explore issues of sociality for both local and remote visitors. It particularly focused on the *Mack Room* mixed reality environment which was presented as an indicative case study of technology that supports social conduct among visitors across a variety of distributed technology-mediated museum environments. Two additional points need to be added here with regard to remote museum visits: the potential of social technologies for the expansion of museum practice and the need for a sociological perspective in museological research.

It is evident, we believe, that social conduct among museum visitors facilitates and often shapes personal engagement with the exhibition and vice versa. Therefore, social conduct might effectively contribute towards wider and deeper access to the collections locally and remotely. This may be achieved through online and onsite guided tours for a mixture of audiences led by docents or visitors alike. Furthermore, the capability of mixed reality systems to support communication around displays from a distance may particularly fit with a museum's educational activities for local and remote school groups and other educational parties. However, the main opportunity of such technologies, we believe, lies in its promise to connect the museum visit, either onsite or online, with visitors' everyday activities, relationships and experiences.

Sharon MacDonald (1993) suggested that "anthropological approaches may have a useful role to play in trying to cope with the enigmas of the visitor sphinx" (p. 80). Research presented in this chapter indicated that sociological methods in the study of museum settings, such as observational studies and ethnography, may be particularly insightful in understanding the museum visit from the point of view of the participants. It also indicated that analytical approaches that are concerned with untangling the dynamic character of the visit, as an activity that shapes and is shaped by social conduct, may be a constructive platform for technological design in museums. Furthermore, the investigation of resources that visitors use to make sense of the visiting activity, these being verbal and visual cues, gestures, objects and so forth, may become particularly useful in the development of inspiring remote visits that take advantage of a variety of technological capabilities. Marty (2004a) anticipated that museum webmasters of the future will be expected to act as users' advocates, to ensure that the information needs of remote users are met. We suggest that understanding the sociality of online visiting needs to be in the forefront of their agenda should they aspire to succeed in their role.

Technology in museums is not only about presenting information but also about supporting social interaction. The advent of wireless communications makes remote communication possible, but in this chapter we have argued that it may also be desirable since it can support social interaction that enriches exploration, appreciation and interpretation of collections. While there are undoubted costs of design and maintenance of new technologies and associated materials for display, we suggest that trends in computing and telephone technology will make such interaction widely accessible among local and remote visitors. Such technology may, therefore, offer a practical means to enhance the accessibility of collections and the educational activities of an institution.

ACKNOWLEDGMENTS

We would like to thank all members of the *City* project, past and present, and the staff and visitors of The Lighthouse and the House for an Art Lover for their help and support. We also thank Holger Schnädelbach from the Mixed Reality Lab at the University of Nottingham. This research was conducted within the Equator IRC, which is funded by EPSRC UK, and was assisted by a donation from HP.

Section 5

Information Behavior in Museums

12 Changing Needs and Expectations

Paul F. Marty

Florida State University

New information technologies have opened a world of new possibilities to museum visitors, bringing with them new ideas about the experiences museums offer and new expectations about the information resources museums provide. The prevalence of digital museum resources and the ease with which digital collections information systems can be made available online has encouraged the functional convergence of digital museums, libraries, and archives (Rayward, 1998). Unaware of the historical barriers to information access that have separated physical information repositories, visitors today expect equal and immediate access to information regardless of the type of organization involved. No longer content with limited access to information resources about museum collections, many now expect that museums will provide them with the information they need, when they need it, no matter where in the world that information might be located (Cameron, 2003).

As the users of digital museum resources become more information-savvy, their information needs and expectations become more sophisticated and more demanding. From the perspective of the museum professional, the expectations of the modern museum visitor can seem rather outlandish, especially when visitors are unfamiliar with the typical organizational schemes and limitations of museum information systems. A student writing a paper on the "Labors of Herakles," for instance, might expect to be able to ask to see all objects in the museum's collections that are related to that topic. A researcher studying the history of glass-making might ask for a selection of glass artifacts that illustrate advances in glass-making technologies over the past two thousand years. Expecting the museum to be able to give them the information they need in the format they need it, individuals making such requests may be surprised to learn that their needs may place unreasonable or extremely difficult demands on museum employees.

Despite the inherent difficulties, most museum professionals support the idea of the museum as an information service organization, where large numbers of people are encouraged to access museum informatics resources. Many museums, therefore, are grappling with the concept of meeting new, demanding, and relatively unfamiliar expectations in order to support the needs of their users. Advances in information technologies have helped museum professionals meet information needs: the move from card catalogs which only curators could access to online databases which all visitors can access, for example, provides more extensive opportunities for information access. These same technological advances, however, serve to highlight any existing limitations of the museum's information systems, and enable the museum's visitors to express more clearly their demands or dissatisfaction with the museum's ability to support their information needs. For each new technology developed to meet current expectations, new needs will arise requiring new technologies, leading to new expectations, and so on in an endless cycle of evolving information needs and technologies.

If museum professionals are serious about keeping up with the evolving needs and expectations of their visitors, it is important that they find ways to assess the information needs of all users of the museum's information resources. If the purpose of the modern museum is to connect visitors with information, museum professionals will need to adapt to the new expectations modern visitors put on modern museums, and be able to cope with the fact that their visitors' needs will continue to change over time. In recent years, these needs have led to extensive research into who the museum's users are, what they want, and whether museums are meeting their information needs, with a particular focus on the online environment.

Studies of the information needs, seeking, and behavior of the typical users of museum information resources explored how museum visitors use information technologies in museums (Economou, 1998; Evans & Sterry, 1999; Galani & Chalmers, 2002; Schwarzer, 2001b) as well as the needs, characteristics, and interests of visitors to museum Web sites (Chadwick & Boverie, 1999; Goldman & Schaller, 2004; Herman, Johnson, & Ockuly, 2004; Ockuly, 2003; Sarraf, 1999). Kravchyna and Hastings (2002) stressed the importance of understanding the information needs of users at all stages of the museum visit, including online access before and after physical museum visits.

Understanding the needs of museum visitors can help museum professionals better serve their clientele (Muller, 2002; Teather & Wilhelm, 1999; Zorich, 1997). Cameron (2003) studied how helping museum visitors conceptualize and use museum information resources can transform the way museum professionals build relationships with their users. Coburn and Baca (2004) looked at how different metadata schemas help or hinder users seeking collections data from museums. Hamma (2004a) examined the changing expectations for online museums engaged in outreach to many different audiences. Bowen and Filipinni-Fantoni (2004) explored different methods of targeting unique user needs through personalization and pervasive computing technologies both inside and outside the museum.

Museum professionals need to evaluate the steps they are taking to meet the information needs of their users. Recent studies explored the value of methods such as usability analysis, user-centered design, or transaction log analysis for museum professionals (Gillard & Cranny-Francis, 2002; Harms

& Schweibenz, 2001; Hertzum, 1998). Dyson and Moran (2000) discussed the importance of creating accessible and usable information resources for online museum projects. Cunliffe, Kritou, and Tudhope (2001) explored different methods for determining whether museum Web sites meet user needs, and discussed the dangers of not evaluating museum Web sites for usability. Peacock (2002) discussed statistics and transaction log analysis for museum professionals interested in evaluating user satisfaction with museum Web sites.

Such research into the "information behaviors" of museum visitors, whether those visitors are in the museum or online or both, is extremely important for current and future museum professionals. If museum professionals cannot learn what users' needs are, they will be unable to determine how users' changing expectations are reshaping the museum experience. Changing needs, perceptions, and expectations of what museums should offer have the potential to transform the role museums play in the information society. If museum professionals do not understand their visitors' changing needs, they cannot predict where those needs will take them.

Thinking about the museum as a user-centered, information organization where user needs are continually assessed, studied, and evaluated, raises many difficult questions for museum researchers and professionals. Should visitor expectations drive what happens in museum exhibits or determine what information is made available? Should visitors be given the ability to construct their own virtual exhibits or maintain collections lists of personal favorites which they can share with other visitors? Should visitors be encouraged to focus on artifacts they know they like, to the exclusion of unfamiliar collections? These questions, many of which arise from advances in digitization and information technologies, raise unfamiliar concerns and fears with many museum professionals.

An excellent example of concerns many have about the changing needs and expectations of museum visitors can be found in the question: Will access to online collections information cause a decline in physical museum visitation? This is a very important question; according to Haley-Goldman and Wadman (2002), "the relationship between virtual museum sites and physical sites has not been extensively researched." In theory, the ability to access digital museum resources online serves as a lure, encouraging potential museum-goers to visit the physical installation, and several studies have provided useful data to back up this theory. As early as 1999, Bowen found that museum visitors, while traveling, were more likely to visit physical museums not in their home town if they had visited that museum's Web site in the past. Kravchyna and Hastings (2002) found that 57% of online museum visitors visit museum Web sites both before and after they visit physical museums. More recently, Thomas and Carey (2005) gathered additional data to support the theory that virtual museums encourage physical museum visitation, and that online museum visitors are also physical museum visitors.

Despite these studies, museum professionals continue to face a seemingly endless debate over this issue. One possible reason that some museum professionals have difficulties with the question of whether access to virtual museums negatively affects physical museum visits is that finding appropriate answers requires taking a different approach to assessing the information needs and expectations of museum visitors. Instead of asking whether people will stop going to museums if they can access digital collections online, it is necessary to evaluate the role of the digital museum in the life of the museum visitor. That is, to understand the relationship between online and physical museum visits, one should focus less on what happens with visitors when they are physically in the museum, and more on the role the museum plays in the visitors' lives, inside and outside of the museum itself.

Keeping up with the changing expectations of museum visitors requires shifting perspectives to see how museum visitors are using museum resources in their everyday lives, not just in the museum. This shift in perspective parallels a shift in the Library and Information Science community, where researchers have realized that studying only those resources people use while they are in the library does not provide an accurate picture of what their information needs actually are (Augst & Wiegand, 2003; Zweizig & Dervin, 1977). Focusing on only what the museum visitor does while in the museum provides a similar skewed perspective. In order to address difficult questions about the use of online museum resources, museum professionals need to know more about the role those resources play in our visitors' lives outside of the museum. Shifting to the "museum in the life of the user" perspective is difficult, but it will likely be the best way to get the clearest picture of our users' needs as well as whether their needs are being met.

As this shift begins to occur in different areas of museum informatics, the overall results have been extremely positive. An excellent example can be found in the area of educational outreach, where museum professionals frequently evaluate the role of the museum in the life of educational organizations, students, and teachers, rather than the other way around. Current educational outreach initiatives, in particular those projects that reach schools and students over the Internet, illustrate how this shift in perspective can help better illustrate the integration of museum information resources into educational activities.

Many excellent projects document methods of integrating museum resources into educational curricula (Parry & Arbach, 2005; Schaller & Allison-Bunnell, 2005; Teather & Wilhelm, 1999). Educators at the Seattle Art Museum have developed an educational outreach project that allows middle-school students to play the role of museum curators, selecting artifacts and developing their own exhibits (Adams, Cole, DePaolo, & Edwards, 2001). Researchers at the University of Michigan have worked with different museums, schools, and content experts to develop online educational activities connecting students, teachers, and experts with museum information resources (Frost, 2001). Similar projects at other museums and other universities demonstrate the positive influence museums can exert on schools at all levels when museum resources are properly leveraged into educational activities (Bennett & Sandore, 2002).

The end result of these and similar projects is that, for many schools, the museum no longer needs to be merely a once-a-year field trip destination. The availability of online museum resources organized into viable educational lessons meeting state and national standards enable the museum to become an integral part of the educational experience. While budget cuts may make such activities difficult, advances in information technology have changed people's expectations of what museums can offer from an educational point of view. By shifting one's perspective of what museums can or should offer in terms of information resources or educational activities, one can provide educators the opportunity to meet the changing needs and expectations of parents and students with respect to museums, schools, and educational outreach.

Continuing to make positive advances in educational outreach as well as in other areas will require changing our current mindset to match the museum visitor's changing expectations, and keeping up with these changing expectations will mean new ways of working for museum professionals. It will be necessary to learn ever more about the users of museum informatics resources, inside and outside the museum: Who are they? What do they want? Are their needs being met? By continuing to ask and answer these questions, hopefully it will be possible to leverage an increased understanding of museum visitors to design new applications capable of evolving to meet visitors' information needs, now and in the future.

13 Understanding the Motivations of Museum Audiences

Kirsten Ellenbogen

Science Museum of Minnesota

John Falk

Institute for Learning Innovation

Kate Haley Goldman

Institute for Learning Innovation

INTRODUCTION

There has been a dramatic growth in the size of the museum audience over the last century (Lusaka & Strand, 1998). Consider the following data: fifty years ago, two in ten people went to museums regularly. By the end of the twentieth century, three out of every five people visited some kind of museum at least once a year (Falk, 1998). These statistics do not include visitors to museum Web sites, which major museums report outnumber visitors to physical museums at least three to one. This dramatic increase in visitorship can be explained, in part, by the growth of new museums and the changes in the quality of exhibits and programming in existing museums. The creation of new museums and improved exhibits and programming alone, however, does not fully account for this increase in visitorship. Research indicates that a large part of this shift can be attributed to changes in the larger society, where free-choice learning experiences, such as visiting museums, are becoming important parts of everyday life (Falk & Dierking, 2000).

The emphasis on learning experiences and growth in museum visitorship comes at a time when the mission of museums has shifted from a focus on collecting and preserving to one of educating the public. Stephen Weil (1999) described this as a movement from being about something to being for somebody. This ideal confluence—increased museum-going, increased attention to education in museums, and increased interest in learning experiences in the general public—is promising for museums hoping to grow their audience.

This chapter discusses what we know about museum audiences and their motivations, and then compares the motivations and experiences of people who experience physical museums and people who experience virtual museums or other technology-supported activities.

UNDERSTANDING MUSEUM-GOING AUDIENCES

Research has found that museum-going correlates with a number of demographic variables, such as education, income, occupation, race, and age (Falk, 1998; Hood, 1983, 1989). One important demographic variable to emerge from the data is age. Even when school trips are excluded, children between the ages of five and twelve still represent a significant percentage of all people who visit museums. Most children visit museums with their parents, making the elementary-school-age group and adults between the ages of twenty-five and forty-four the largest part of the museum audience (Falk, 1998).

Although this information is evidence that families represent the largest segment of the museum audience, it does not mean that demographic variables such as age are the best predictors of who will be most likely to visit a museum. For example, research has shown that race is not an accurate predictor of museum-going: Falk (1993) found that African-American and White science museum visitors were similar in terms of interest in science, occupation, annual income, and frequency of museum-going. These findings suggest that psychographic variables, which describe people's psychological and motivational characteristics, are far more predictive than demographic ones. Despite varying demographics, museum-goers tend to fit a specific psychographic profile (Hood, 1983); they value learning, seek the challenge of exploring new things, and place a high value on doing something worthwhile during their leisure time.

This psychographic profile correlates with the demographic data on museum-goers. For example, psychographic data indicate that museum-goers value education; demographic data indicate that museum-goers have a higher than average level of education. Falk (1998) cautions that this does not mean that a higher than average level of education is needed to value the museum experience. It is more likely that this correlation is because people who value learning are more likely to pursue higher education, such as college or graduate school. Psychographic data about museum audiences' values and beliefs may help predict museum-going behavior, but like demographic data, they do not provide a complete explanation for why people go to museums.

Audience Segments

Psychographic data have led to some important new generalizations about the museum-going audience. Traditionally, the museum audience has been divided into two segments: museum-goers and non-museum-goers. Marilyn Hood's (1983, 1989) psychographic research led her to regroup the museum audience into three segments: frequent museum-goers (three or more visits per year), occasional museum-goers (one or two visits per year), and non-museum-goers.

Hood found that frequent museum-goers tended to prioritize three characteristics of leisure experiences unique to their group: the opportunity to learn, the challenge of new experiences, and the experience of doing something worthwhile in their leisure time. In contrast, non-museum-goers tended to give priority to experiences that are less important to frequent visitors: socializing, participating actively, and feeling comfortable. The characteristics of leisure experiences that frequent visitors rank high are ranked low by non-museum-goers. Additionally, non-museum-goers typically did not go to museums with their families when they were children.

Hood further found that occasional museum-goers, those that visit museums once or twice a year, more closely resembled non-museum-goers than they did frequent museum-goers. Subsequent research by Falk (1993, 1998) showed that as society moved towards a greater appreciation of learning, all museum-goers, frequent and occasional, began to look more alike with regard to these leisure values. Other factors seemed to be at work resulting in some museum-visitors to become frequent visitors and others to be casual visitors.

One major factor appears to be childhood experience. As children, occasional museum-goers are typically socialized into leisure activities that emphasize active participation and socialization. As adults, they continue to be involved in these sorts of leisure activities, including sports and visiting amusement parks. While as children, frequent museum-goers were more likely to be raised in households that emphasized learning and also regularly partook in museum-like experiences. However, more than anything, motivation emerged as the variable that distinguishes who does and who does not visit.

For example, both museum-goers and non-museum-goers place great emphasis on family-centered activities and value feeling at ease in leisure environments. The characteristic that distinguishes museum-goers from non-museum-goers is that museum-goers believe that at least some of the characteristics they value in leisure experiences are available in museums; non-museum-goers do not have this perception. Often occasional museum-goers visit museums for special events, believing that activities planned during special events are more geared towards family interests and best fulfill their needs.

Audience Motivations

In order to better understand the needs of museum visitors, researchers (Anderson & Roe, 1993; Falk, Moussouri, & Coulson, 1998; Moussouri,

1997; Packer & Ballantyne, 2002; Rosenfeld, 1980) have used in-depth observation and interview techniques to gather data on expectations and motivations. This research reveals that there are specific patterns in visitors' motivations for visiting museums. Although there is some diversity in the various patterns of visitor motivation that researchers have proposed, seven categories are consistently included in the findings:

1. Entertainment: "We always have a good time here."
2. Social: "I wanted to spend some quality time with my brother."
3. Learning: "We always learn so much going to museums like this."
4. Life Cycle: "I went to museums as a child and now I am bringing my children."
5. Place: "You have to see this museum if it is your first time in town."
6. Practical: "It was getting too hot so we decided to duck in."
7. Context or Content: "I love butterfly gardens."

Museum visitors tend to mention more than one of these motivations when talking about why they go to museums, since many are overlapping and interconnected. Researchers argue that visitors not only have specific motivations for their museum visits, but that these motivations directly influence what is learned during the experience. For example, Falk, Moussouri, and Coulson (1998) found that in a content knowledge test administered before and after visitors went through an exhibition, people with a high learning motivation scored significantly higher than people with a low learning motivation. More interestingly, people with a high learning motivation, regardless of their entertainment motivation, showed significant conceptual learning. Likewise, people with a high entertainment motivation, regardless of their learning motivation, showed significant vocabulary development and overall mastery of the topic. This result reinforces the belief that learning and entertainment are not two ends of a single continuum. It suggests that visitors do not distinguish between the value of learning and entertainment and that both are effective, but different motivations in learning.

A recent multifactor investigation (Falk & Storksdieck, 2005) reinforced the important connection between what individuals learn within a museum and factors related to why the visitor was visiting in the first place. This study showed that although a visitor's learning is significantly influenced by what they saw and did and who they interacted with, these variables were far less important than the visitor's entering motivations, interests and prior knowledge. The most important variable appears to be the situated identity visitors enact prior to their visit. Most museum visitors use museums in order to satisfy a finite number of identity-related motivations—for example, satisfying curiosity, supporting the experience of others, or the desire to broaden their horizons by seeing new and interesting things. These identity-based motivations are in turn shaped by prior experiences, knowledge, stage of intellectual, physical and emotional development, cultural and social history, and personal interests. All of these coalesce into a relatively small subset of museum-specific identity-based motivations. In other words, although there could be an infinite number of reasons for visiting museums, this does not appear to be the case. Each variable is delimited by physical and socio-cultural realities into a relatively small set of possibilities (Falk, 2006). Thus motivation, interests and prior knowledge converge into a relatively straight-forward way for better understanding the lenses through which the public experiences physical museums (Heimlich, et al., 2005).

Identity and Audience

It remains difficult, however, to fully understand the museum audience based on the above frameworks for visitor motivations. In most instances, if you walk up to visitors and ask why they are at the museum, they will have some difficulty giving an answer that thoroughly accounts for the many variables that have influenced their decisions to come to the museum. More problematic is that identity-building, one of the strongest reasons for visiting a museum, cuts across the motivation categories that have typically resulted from the motivation research. Identity-building is extremely difficult to articulate, and tends to result in motivations that reflect people's general beliefs about who they are or whom they want to be, rather than an immediate assessment of the reasons behind that day's museum visit. This is arguably an indication that we must allow identity to take center stage in our efforts to understand and meet the needs of our audiences.

There is a great potential for people's existing and desired identities to influence their museum experience (Ellenbogen, 2002, 2003). Parents who want to develop a particular family identity are able to quickly adapt the general museum experience, as well as specific content, to reinforce the desired identity. Everything from behavior modification ("We don't bang on the computer screen like that.") to personal narrative episodes ("Do you remember the last time we saw something like that?") can quickly be used to reinforce identity. Critically, this effort to reinforce family identity sometimes comes at the expense of the museum's intended experience or content. Museums intentionally choose content and create experiences to support a specific theme, learning objective, or general mission. Visitors' efforts to develop their identity may be so strong that they appropriate the museum messages and reshape them to support individual or group identity building (Ellenbogen, 2003).

D.D. Hilke (1987) voiced a concern when she suggested that we focus not just on the museum as a resource for learning, but also the visitor as a resource. This inverted point of view highlights the reality that visitors bring a great deal of resources with them, including previous experiences, concerns, and desires to act or be perceived in a certain way. All of these impact the way visitors interpret the museum experience. Visitors' motivations consequently have a strong impact on the learning that takes place in museums.

Hilke (1987), studying the museum-going practices of families, used the term "family agenda" (in contrast to "museum agenda") to emphasize that visitors' resources should be recognized and accommodated in order to create a museum experience that is successful from the museum perspective.

VIRTUAL MUSEUM VISITORS

Undoubtedly, many of the same distinctions that separate physical museum-goers and non-museum-goers also apply to those that make use of virtual museum resources. The distinction between the two, for the moment, lies in those words. Physical museum-goers are seeking experiences—learning experiences perhaps—but experiences nonetheless. In contrast, the Internet was created for resource sharing and communication. This distinction shapes the current differences in motivation in the two venues.

Motivation is not the only significant difference between physical and virtual visits. Despite extensive collections and interactive elements online, the physicality of the experience changes the entire context. Physical museum visits have high opportunity costs such as investments of time, effort, and money (Haley Goldman & Dierking, 2005). Visitors must know the hours of operation, find their way to the institution (and probably find parking), generally pay to enter, and then figure out how to accommodate food and restroom breaks. Museum visits are often upwards of one and a half hours, not including the time spent planning the visit and traveling to the institution. While these opportunity costs influence the visit in a variety of ways, research has rarely accounted for the influence these costs likely have on the typical physical museum visit. The time spent inside an institution is partially a function of an informal cost–benefit analysis carried out by one or more of the group members, calculating "how much time should we spend here at this museum, based on the effort and money it took to get us here?" In contrast, a virtual visitor would typically only invest a small fraction of time and effort in their visit. A virtual visitor can explore a museum in their pajamas, at any time of day or night, as long as they have a computer and an Internet connection. This is not to imply that the motivations in the online world are necessarily any less strong or distinct than physical visits.

The quantity and quality of research on virtual museum visitation, unfortunately, do not yet support the same depth of discussion as research on physical museums. Older data and anecdotal information suggest that virtual museum-goers are very similar to both physical museum-goers and Internet users in general. There are slightly more males than females, and users tend to be younger and of higher socio-economic status than physical museum visitors (Chadwick, 1998; Chadwick, Falk, & O'Ryan, 2000). Like everything else on the Internet, the demographics are in rapid transition. For instance, as of April 2006, the gender gap is very small: 71% of adult women use the Internet versus 74% of adult men. Seventy-one percent of those over 50 years old use the Internet, and 32% of those over 65. While those who are highly educated do use the Internet (91% of college graduates), this does not mean that others do not: 40% of adults that did not complete high school are Internet users (Pew Trust, 2006). Participation in the information society is not just for the wealthy urbanities; 63% of rural households use the Internet and 53% of households with income under $30,000. The Internet is a part of the lives of most Americans, and as mentioned above, physical audiences have also grown enormously. Thus while virtual museum visitation is high, there is even greater potential for increased visitorship in the future.

What we know about physical and virtual museum-goers suggests differences in several key areas. Preliminary research suggests that the motivations of visitors to museum Web sites differ significantly from the motivations of visitors to physical museums. A review of the literature on virtual museum visitors (Haley Goldman & Schaller, 2004) characterized the most common motivations from Web site visits as:

- Gathering information for an upcoming visit to the physical site;
- Engaging in very casual browsing;
- Self-motivated research for specific content information; and
- Assigned research (such as a school or job assignment) for specific content information.

We suspect that these motivational categories will evolve significantly in the near future. Museums on the Internet are like every other technology that develops and changes form over time. Institutions that were primarily "about something" are now "for somebody"; the Internet was invented for resources and communication and is now evolving into an entirely different way of experiencing life, indeed an entirely different way of living. While there is little or no research on virtual museums comparable to the identity-related motivation and psychographic research within physical museums, there is growing research into identity in virtual space, both on the Internet in general, and specifically in the increasingly complex world of online gaming. Given the evidence that identity-related motivations strongly influence the learning and behavior of physical museum visitors, we would suspect that an analogous situation applies to virtual museum use.

Ray Oldenburg, in his book *The Great Good Place* (1989), emphasizes the need for "the third place," spaces that are neither home nor work, for the development of identity. One of the primary "third-places" is now found in massively-multiplayer online games (Steinkuehler, 2005). The third place of the chat room or game, rather than the local community center or bar, can become a primary vehicle for identity construction. While this research is still emergent, one indicator of this trend is the enormous user populations inhabiting such virtual social worlds as the Sims Online and Second Life; studies of virtual worlds suggest that these sites attract hundreds of thousands of users (Squire & Steinkuehler, 2006). This research suggests that users come for the experiences and have powerful, identity-shaping moments. While there is little evidence yet that virtual museums can drive powerful identity-building experiences, one can envision the enormous possibilities for an entirely new type of virtual museum, and very different reasons for visiting these spaces.

CONCLUSIONS

Museums have always employed technology, whether that technology was an exhibition case, a diorama, a video or an interactive. The new technologies that are flooding into the museum environment such as mixed and augmented-reality museum experiences, museum blogs, and podcasting art installations are just the latest wave. Historically, these technologies have been introduced with good intentions and implicit beliefs about how users would utilize and benefit from them. What we argue is that implicit beliefs are no longer justifiable. Recent advances in our understanding of the museum visitor experience provide a growing understanding of how and why visitors utilize museums. Visitor motivations have emerged as particularly important.

As museums become further embedded with technology in both physical and virtual contexts, it is critical that we align those technologies with an understanding of why visitors actually choose to utilize those resources. Too often, technologies are embraced because of an interest in advancing the museum's missions and agendas, without questioning how well those technologies actually serve the needs and interests of the museum's audiences. In other words, the success of new technologies within the museum context will partially be a consequence of its physical attributes such as adaptability, availability and usability, but equally if not more important will be whether or not the technology satisfies the situated identity-related motivations of users. One clear need in the field is to conduct research on museum user's identity-related motivations within the virtual museum environment. Such research in the physical realm has begun to produce some very useful and provocative findings; parallel research in the virtual realm promises to do likewise.

Ultimately, the goal of introducing new technologies into the museum world is not to merely make the museum experience more "up-to-date", rather it is designed to significantly improve the quality of the visitor experience and in most cases significantly enhance learning outcomes. Research on motivations has shown that both the quality of the visitor experience and the extent and breadth of learning outcomes are directly related to entering motivations. The better we understand the relationship between motivations and learning, and how technology-based experiences support these motivations, the more successful we will be.

14 Partnerships for Progress
Electronic Access and Museum Resources in the Classroom

Jim Devine

Hunterian Museum and Art Gallery, University of Glasgow

INTRODUCTION

The Multimedia Unit at the Hunterian Museum and Art Gallery, University of Glasgow, Scotland, recently undertook a major research project entitled "What Clicks?" (Devine, Gibson, & Kane, 2004) to review the existing and potential capability, human and technological, within the museum sector in Scotland in the use of ICT to increase public access and resulting learning opportunities to museum collections. The World Wide Web and other electronic means (e.g., CD-ROM) offer museums the opportunity to increase access for a wide range of new audiences and to promote new learning styles. In particular, opportunities are created for study by those physically remote from museum collections—from primary and secondary schools through to community and special needs groups, to self-directed, life-long learners.

The What Clicks? Project Team worked with a number of partners in Scotland and overseas, both within the museum sector and in education. Small independent museums also played an important part including the Scottish Fisheries Museum in Anstruther. School and community partners were drawn from local and remote areas including Dumbarton Academy, Cumbernauld Primary School, Castlebay School, Barra; Soroba Young Families Centre, Oban; and T.C. Williams High School, Alexandria, Virginia, USA.

The overall aims of the project were to:

1. Work in partnership with others within the museum sector and in education;
2. Review Scotland's current electronic museum resources, available to the public;
3. Identify examples of good practice that could be employed in the future to aid development of electronic resources;
4. Survey potential user groups in situ, to understand their requirements in an educational context; and

5. Produce recommendations for use by Scottish Museums for the preparation and delivery of electronic heritage resources.

METHODOLOGY

For the purpose of understanding the requirements of potential user groups, visits were made to schools, museums, community groups and life-long learners. In situ visits provided the opportunity for flexible, in-depth discussions on a range of topics and allowed the Project Team to gain an insight into how resources are used and managed in practice and also to develop a thorough understanding of the issues raised.

In all cases the interviews varied in time, depending on the level of participation and interest shown by the respondent and the majority were recorded, either on Dictaphone or Digital Video Recorder. A standardized interview schedule was designed to take account of a wide range of issues relating to electronic museum resources. The design was altered to accommodate each group of respondents, with all the questions being asked in a similar wording between interviewees.

The decision was made to use a mixture of quantitative and qualitative research. In very broad terms, quantitative research entails the collection of numerical data to exhibit a view between theory and research whereas qualitative research places emphasis on the words rather than quantification of the data and seeks to understand behavior, values and beliefs (Bryman, 2001). Using quantitative research methods allowed the project team to take into account the opinions and suggestions from a wider proportion of the population than visits would permit and using qualitative research allowed in-depth interviews that yielded detailed opinions from the respondents, which would not be possible using quantitative research alone.

Visits were decided upon depending on a number of criteria, including geographic location and type of organization. In order to get a representative geographic spread across Scotland, visits to museums and educational establishments were conducted in the following areas: Aberdeen and Grampian; Argyll, The Isles, Loch Lomond, Stirling and the Trossachs; Dumfries and Galloway; Edinburgh and Lothians; Greater Glasgow and Clyde Valley; Eilean Siar (the Western Isles); Fife and Scottish Borders.

Museums across Scotland have been producing a wealth of electronic materials and resources and making them available to the public. There are excellent examples of educational resources being developed in many museums, as well as lots of Scottish museums' resources being made accessible through SCRAN (Scottish Cultural Resource Access Network, http://www.scran.ac.uk/). The wealth of material ranges from simple images and descriptions of objects, through to fully interactive virtual museum exhibitions and lesson materials that can be used by teachers in the classroom and in preparation for museum visits.

PARTNERSHIP PROGRAMS

Partnerships with schools are one way in which museums can seek to further develop their educational online resources. Museums traditionally have partnerships with schools for educational visits. Education is a responsibility of museums and it follows naturally that strong partnerships with schools are formed and maintained over the years. There are examples of excellent education programs in museums across the country, with small and large museums making great provision for schools and educational visits to the museum.

Teachers are keen to point out that they should be involved in the development of museum resources for pupils. They can bring a depth of knowledge on the school's curriculum that is far beyond that of any curator and can provide an excellent source of education end-user savvy which is particularly valuable to those museums operating without the benefit of large education departments.

"THE GLASGOW MODEL"

In 1994, the Hunterian Museum and Art Gallery initiated a collaborative program between the Hunterian and the University of Glasgow's Department of Computing Science, in partnership with Scottish schools, which has led to a wide variety of innovative multimedia projects focused around the Hunterian collections and those of other cultural heritage organizations (Devine & Welland, 2000). This has included an award winning Web site incorporating audio, video and QuickTime Virtual Reality (QTVR) (http://www.hunterian.gla.ac.uk). The projects have involved museum staff and students experimenting with a wide range of multi-media presentation techniques. The students have benefited greatly from dealing with so-called "real world" clients and the museum has gained a leading edge in the exploitation of emerging technologies. The term "The Glasgow Model" was struck by Leonard Steinbach, Chief Technology Officer at the Guggenheim Museum in New York, when at the International Cultural Heritage Informatics Meeting (ICHIM) in Washington, D.C. in 1999, he described the Hunterian's approach as an excellent example of collaborative effort between a museum and an academic institution to efficiently further the aims of both.

Under the aegis of this collaboration there have been projects involving video clips to animate and explain aspects of Roman armor, using image maps to illustrate Captain Cook's voyages of exploration, adding audio for Latin inscriptions to aid teachers and pupils. We have used QTVR to provide panoramas of the Mackintosh House and other areas of the museum and to display objects from the Museum in 3-D. This led directly to the major field project undertaken at the Minoan Palace Knossos on Crete (http://www.bsa.gla.ac.uk) as a collaborative venture between the Hunterian, the

Department of Computing Science, the British School at Athens and the Greek Archaeological Service, which has provided an online tour of the entire archaeological site at Knossos for classroom access from anywhere in the world with an Internet connection (Figure 14.1).

Since 1997, the Smithsonian Institution in Washington, D.C. and the Hunterian Museum and Art Gallery at the University of Glasgow, on their respective sides of the Atlantic, have established a collaboration entitled "SHADE" (Smithsonian–Hunterian Advanced Digital Experiments) (Devine & Hansen, 2001). This partnership encourages leading edge practices in the field of digital imaging for the scientific and cultural heritage sector. This collaborative project has developed examples of best practice in skills-sharing between the Multimedia Unit at the Hunterian and the Center for Scientific Imaging and Photography at the Smithsonian. These projects have actively engaged school partners in Scotland and in the Washington, D.C. area (Figure 14.2).

The focus for digital experimentation in SHADE has been chosen from a range of the partners' respective collection strengths to provide a digital educational resource for school/college classrooms in the UK and the USA. Typically, topics are chosen where each partner has complementary collections material to digitize, thus augmenting the digital collections of both and providing an even greater wealth of educational resources to their respective end-users. The development of these digital resources fits with the Hunterian Museum and Art Gallery's ongoing policy of widening access to resources. It also serves a curatorial/conservation requirement to create high

Figure 14.1 Production of the virtual tour of the Temple Tomb at Knossos, with the author in the foreground

Figure 14.2 The author discussing a Hunterian Roman Frontiers online project with students and staff at T.C. Williams High School, Alexandria, Virginia

quality digital images of artifacts for remote access research, while limiting wear and tear on fragile objects and providing a useful income-stream from reproduction fees.

Collections areas, which have proved of mutual interest and benefit to both SHADE partners, have included (1) Hominid Evolution, which resulted in the Hunterian's Hominid Evolution CD-ROM project, incorporating digitally captured skull specimens in QTVR; (2) Flora and Fauna/Ethnography/Anthropology, based around the production of "Virtual Voyages of Discovery and Exploration" with Captain Cook and Lewis and Clark; (3) History of Science and Technology, incorporating the Hunterian's Scientific Instrument collection including Lord Kelvin's involvement in the first successful transatlantic telegraph cable; and (4) Archaeology, including digitization projects to develop learning resources based on Roman and Egyptian collections.

ONLINE RESOURCES REVIEW

Having a presence online is becoming more and more important to museums across Scotland as the benefits of such exposure and access become apparent.

Increasingly, the Internet is an obvious choice for making museum resources and information available to the public. Web sites vary from a single page of basic textual information about a museum to extensive, fully interactive virtual exhibitions and targeted educational resources. To assess the current Web sites of Scotland's museums, the What Clicks? team carried out a Web site review survey. This review used criteria from "Evaluating the Features of Museum Web Sites: The Bologna Report," a paper from the Museums and the Web Conference 2002 (Di Blas, Guermandi, Orsini, & Paolini, 2002). This paper researched the quality and usability of cultural heritage applications and compiled a Contents Schema for Museum Web Sites. This schema was used as a basis for the Web site reviews to assess what types of Web site exist at the moment and the content and quality of those sites.

There are many examples of good Web sites illustrating electronic resources at their best. For example, the Hunterian has several interactive QTVR panoramic and object movies to show users all angles of an object. The Hunterian has developed the use of this technology to add another dimension to put an object back in its original context. The example shown in Figure 14.3 demonstrates a skull that can be seen from 360 degrees; laid on top of the skull is the *Australopithecus Boisei* head, complete with hair and skin.

The Web review conducted by the project team, on museum Web sites across Scotland revealed that 51% of museum Web sites in Scotland currently exist as part of another Web site, such as a local authority or as part of related heritage site.

Chart 14.1 illustrates the most common features of museum Web sites across Scotland. The most common feature is the provision of an "e-mail contact address" which was given by 83% of Web sites. A "mission statement" was the next most common feature (40%), closely followed by "new updates to the site" (38%). Very few Web sites provide "a sitemap" (24%), "search facility" (21%), "help" (18%), "basic technical information" (7%), "other languages" (7%) and "frequently asked questions" section (4%). It is worth mentioning here that a search facility is one of the features most commonly requested by end users.

Chart 14.2 looks at what information each Web site provides to enable the end users to organize a trip to the museum:

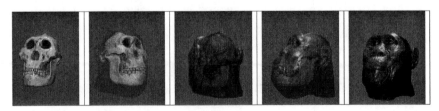

Figure 14.3 Australopithecus Boisei rotating through 360° and morphing from skull to full head. © Hunterian Museum and Art Gallery, University of Glasgow

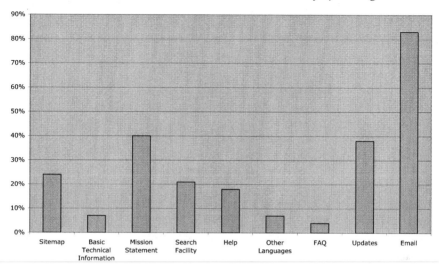

Chart 14.1 Common features of museum websites

- The criteria provided most often was "location" information (91%). This was closely followed by "opening hours" (84%) and then "ticket information" (70%).
- 24% of museum Web sites surveyed in the project contained a description of their collections.
- 21% have a specific collection section on their Web sites and 8% contained virtual tours.

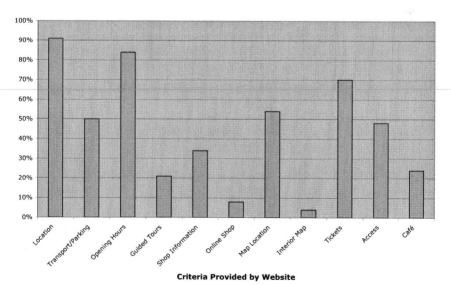

Chart 14.2 Information provided by museum websites

- 36% of museum Web sites contained an education section.
- 21% of sites had a section directed at teachers and pupils.
- 15% of sites had both resources catered to the national educational curriculum and educational packs available to order.
- 10% of sites contained downloadable resources for teachers and 7% had downloadable resources for pupils.
- 92% of museum Web sites had the general public as their main target audience, with 31% also targeting teachers and 24% targeting pupils.
- 71% of Web sites contained material of local significance and 32% of sites emphasized their wider appeal.

Considering the resources available online, the purpose of the Web sites surveyed is outlined in Chart 14.3. The main purpose stated of museum Web sites is to promote the museum (93%). The next most popular purpose is tourism (71%), followed by visitor services (46%), education (40%) and genealogical information (14%).

INTERNET ACCESS IN THE CLASSROOM

The Internet provides a significant source of information for both pupils and teachers. Teachers who took part in this research use the Internet for a variety of reasons and to varying degrees. Teachers use the Internet for lesson planning and researching topics for use in the classroom; 73% of teachers surveyed use the Internet for this purpose.

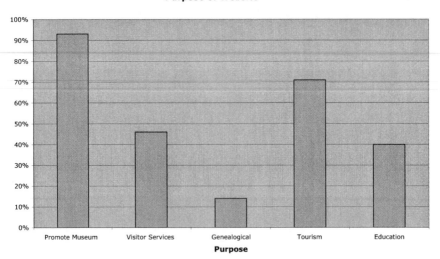

Chart 14.3 Stated purposes of museum websites

The Scottish Schools Digital Network (SSDN) recently carried out a comprehensive survey into connectivity in Scottish schools, contacting all 32 Local Education Authorities as well as Education Advisors and Corporate ICT Support. The purpose of the survey was to determine the quality of connections between schools and their local authorities, identify the technologies used, establish software used and to assess local authority expectations.

The findings revealed that 70% of secondary schools and 30% of primary schools across Scotland now have Broadband (i.e. connection speed in excess of 2Mbps). The operating systems in use in schools vary from Microsoft Windows 95 to Microsoft Windows XP and Macintosh Operating Systems 9-X. The aim for the provision of computers by local authorities was 5 pupils per computer for secondary schools and 7.5 pupils per computer for primary schools; which has been mostly achieved. However, the organization of computing facilities into labs that are used by any class make this more difficult to assess. Local authorities are responsible for the connections between schools and their Internet provider. The authorities have connection speeds varying from 45 Mbps to 1000 Mbps (1 Gbps) available to them. Under-performing local authorities have also been identified and noted for further action.

The current provision of hardware in the schools that took part in What Clicks? research varied widely, from a few basic computers to sophisticated wireless and dedicated computing facilities. The remainder of this chapter describes the provision of the specific schools that were visited, to show specific examples of the varying provision that exists in schools across Scotland today.

Many schools have organized their computing facilities into dedicated computer labs where most of the equipment is kept. Often, the labs are equipped with enough computers to cater for a full class with at least 20 computers and the necessary peripherals such as printers, scanners, digital cameras, etc. Depending on the school, there may be one or more dedicated labs that teachers of any subject can book in advance.

Other schools such as Dumbarton Academy and Castlebay School have organized their computing facilities differently, distributing computers throughout the departments which require them. For example, at Dumbarton Academy, every department has three computers, with other departments such as Computing, Business Studies and Technical necessarily having more. There are laptops, projectors and other equipment available for teachers to borrow as required and there are banks of computers in the library that can be booked by teachers for use with a class.

Another example of the organization of computing equipment is through the use of laptops. Some schools are housed in old buildings, making the physical initial installation or the retrofit of further network access points and cabling difficult. Increasing computer access in schools has led to more and more computers coming into the classroom, however, not necessarily with the adequate networking capability. Therefore other solutions have been employed, one being the introduction of laptops and wireless networking. Dumbarton Academy uses a wireless network to connect its science department to the network, which has proved successful and popular with staff and pupils alike.

Cumbernauld Primary School also employs the use of laptops and wireless networking in a different way. The school opened on January 6, 2003, and is a new integrated community primary school. It is non-denominational and teaches boys and girls of all faiths and cultures from primaries one to seven. In addition it is the first shared-campus school to be built in North Lanarkshire. Its sister campus is St. Andrew's Primary and although the two schools are completely separate, they share the playground, library, dining room and gym hall etc.

Distributed over two buildings, each classroom at Cumbernauld Primary School has at least one computer that is directly or wirelessly connected to the network. Computers are not required for every class all the time, so to allow access for classes as required, wireless access points are located throughout the school, with banks of wireless-enabled laptops for use by any teacher whenever required. The laptops are stored away when not in use, are brought out when necessary and are automatically linked to the network. Using the laptops has proved to be popular with the pupils and teachers are using the laptops increasingly as a regular classroom resource (see Figure 14.4).

Figure 14.4 Cumbernauld P3 pupils using laptops as part of a class activity

REMOTE ACCESS AND MUSEUMS

The nature of Scotland's geography is such that the greatest population density is concentrated in the central belt of the country between Glasgow in the west and Edinburgh in the east. The west coast in particular is very sparsely populated, with schools often located in small communities, many of them on the remote islands of the Inner and Outer Hebrides. For many schools in Scotland, particularly those in the Western Isles, a trip to the mainland to visit a museum is not a realistic journey, either in terms of time or cost. Similarly, students from overseas might never discover the wealth of resource material in the Hunterian Collections and elsewhere in Scotland. This situation can be addressed by an intelligent use of digital media to provide curricular resources drawn from museum and gallery collections, selected and presented in a range of formats to fit with a variety of education users needs from pre-school to life-long learning.

Remote access is a particular area of interest for museums, not only in Scotland, but as Katherine Burton Jones has already pointed out in her foreword to the "What Clicks?" Report,

> This major evaluation of what is currently available in Scotland and what e-learners want to see made available, provides a wealth of information and guidance for the museums community both in Scotland and in the United States, to take the next steps forward in using Information Technology to provide access for all to the enormous e-learning potential of the rich resources held in our museums. (Devine, Gibson, & Kane, 2004, p.3)

Much of the emphasis of the work being undertaken in Scotland is being placed on what electronic resources can do for communities physically remote from museum collections. All teachers interviewed universally agreed that Web sites are a good alternative for those who cannot physically visit a museum collection. One teacher from Stromness Academy (on the Orkney Islands) commented that if there were more museum resources available electronically, "this would improve access to pupils who live far away from the main museums of Scotland."

CASE STUDY: ISLE OF BARRA

The Isle of Barra is situated in the Outer Hebrides, off the west coast of Scotland and covers an area approximately 32 miles square. Its population is around 1200–1300, with many of the population aged over 40 or under 16 years old. Castlebay School, which encompasses both a Primary and a Secondary School, was part of the What Clicks? Project Advisory Group, through the Principal Teacher of Art, Mr Bill Blacker. Castlebay

School has been collaborating with the Hunterian Museum for several years and has provided invaluable user-input and evaluation at every stage of museum-university based projects developing multimedia resources targeted at schools audiences. Barra was chosen due to its geographical location, at the southern end of the Outer Hebrides, a long way distant from the physical museum resources available in mainland Scotland (see Figure 14.5). In addition, the presence of the Barra Heritage Centre, next door to the school provided an opportunity to assess the interaction between the school and a local heritage resource. It proved an excellent candidate for an initial review

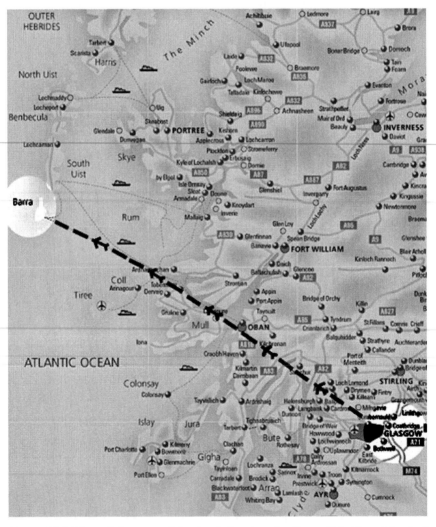

Figure 14.5 Map of the west of Scotland showing the Outer Hebridean Island of Barra

of resources and to be used as a case study illustration of the partnerships and collaborations we were investigating and hope to encourage.

> We will be able to offer education as a community resource and will be able to use IT in order to allow people to share it. (Jim McKeeman, Deputy Head Teacher, Castlebay School, Barra)

The vision of senior education staff like Jim McKeeman and his colleagues at Castlebay has helped to create an environment where the potential of new technology is recognized and widely embraced to enhance learning, both in the School and in the wider community.

The Internet is used by teachers and pupils on a daily basis for classroom-based study. Pupils and teachers will regularly browse the Web to find out a particular piece of information relating to a classroom topic being pursued. However they often found that finding teaching resources can be difficult and time consuming and would like to see some more curriculum focused resources. For example, they suggested that having a database of Scottish literature designed for educational use that was clearly curriculum based would be really useful. Otherwise they tend to use the Internet for specific pieces of information. In the case of pupils, they use "Google" to search for information, then use "Ask Jeeves" if they can't find what they want using Google. They use educational games that develop literacy and numeracy skills in the Learning Base that all pupils have access to. They had heard of SCRAN but don't use it. It would appear that they haven't yet realized the extent of the information available on SCRAN, or that it could be viewed as curriculum based. Other CD-ROMs had been used, for example for studying Macbeth and Tam O'Shanter in English courses produced by Learning and Teaching Scotland. The CD-ROMs had quizzes, audio excerpts, text excerpts, animations and character development information and were described as very good. The pupils enjoy using these resources; however they felt that few such resources as effective as this were available. They found many of the CD-ROMs they purchased to be "disappointing" and this put them off using limited department funds to purchase any more, considering the often significant cost of each educational CD-ROMs.

Specific Web sites are recommended by teachers to pupils for research and assignments, for example the BBC News site is recommended, as well as other tabloid and broadsheet newspaper sites, for comparisons of the type and style of information available and as part of research for argumentative or discursive essays and presentations. Web worksheets are used to supplement the books they use. Many books have related Internet sites with downloadable worksheets and quizzes directly related to the books they already use.

The Learning Support Department at the school also runs a scheme to allow pupils to borrow simple word processing equipment, if they don't have a computer at home, or if their handwriting skills are poor. They are used for typing up reports and other schoolwork and for pupils to use during class if their handwriting is poor. The machines used for this purpose are Alpha Smart 3000s and users are assigned a file number where everything they type is automatically stored; no user saving of work is required so they are simple to use. Usually the document is uploaded to a computer in the Learning Base to allow pupils to format their work before printing it out.

The school uses digital camera equipment a great deal. Following a recent UK Government initiative to allow individual school Head Teachers greater control of locally-devolved school budgets, Castlebay School's Headmaster decided that ICT should be a priority for development within the school and used the devolved funding to buy extra computers over and above what they would be getting through the government's provision to schools through the National Grid for Learning (NGfL) and other organizations. The staff decided that something they really needed in the school was a digital camera and digital video recording equipment—where quality was a big factor. Digital photographic equipment was purchased and was soon in regular use.

The school has been working with the Hunterian to produce local histories and environmental studies using QTVR techniques to create 360-degree panoramic scenes of the locality. QTVR skills were taught to teachers by the Hunterian's Multimedia Unit staff and the teachers in turn taught the pupils who then shot the scenes. Where they encountered difficulties in stitching the resulting panoramas, the Hunterian provided remote technical support and advice. Relevant artefacts from the museum's collections can then be added to these virtual tours, placing the objects back into their original contexts in a form of "virtual repatriation."

Staff and pupils were also able to record school events and to produce higher quality teaching resources. Images are used to support classroom project work and for pupil displays on art and other subject materials. They have even ventured into the area of school business enterprise, taking digital photos of all class groups and other school groups and producing prints that are sold to parents.

Some of the longer term benefits of ICT provision in a remote area like Barra can be seen in the opportunities an early familiarity with technology in school can provide in terms of pupil's preparedness for an ever more technically-demanding job market. Rhona MacDonald, an ex-pupil of Castlebay School and honors graduate in Computing Science from the University of Glasgow worked with the Hunterian on her student academic project work and was the first recipient of the "Hunterian Scholarship" award. Rhona is now employed as a successful Software Engineer with a leading international company based in Glasgow. Rhona commented: "Being in such a remote location obviously pupils can't physically go and visit many places so at least the Internet gives them access to information resources and museum exhibits that they would never otherwise be able to access."

The visit to Barra raised our awareness to a number of important issues for resource providers from the users' point of view, which we have encountered repeatedly on subsequent visits around Scotland and in the USA. In addition to the comments noted above, it also became clear that carefully thought through electronic resources development projects could help meet the requirements and aspirations of a number of areas of museum operations. In addition to the learning-support role that museum Web sites and CD-ROMs can deliver, there is also the potential for significant marketing opportunities for Scotland's museums.

SUPPORT FOR SPECIAL NEEDS

The final stage of the Disability Discrimination Act came into force in the UK on the 1st of October 2004; it is now the legal responsibility of museums to be aware of the issues facing people with disabilities in accessing collections and encourage good practice in the provision of services for people with disabilities. The Disability Rights Task Force encourages the participation of disabled people in all roles of arts and culture including museums, galleries and archives. The Task Force promotes a commitment to best practice in relation to disabled people and ensures that venues demonstrate proactive access policies.

Many museums in Scotland had already been making provision for disabled people to access their collections and displays, for example by lowering cases so that wheelchair users can view the displays and providing alternatives if physical restrictions in buildings restrict access.

This provision of access is not only restricted to the physical building and physical displays. Increasingly museum resources are being made available electronically and therefore consideration must also be taken in making these resources just as accessible to disabled users. The Hunterian Museum and Art Gallery in association with the Department of Computing Science at Glasgow University has undertaken a number of initiatives in this area over the past few years. This has included the use of audio and visual enhancements for Web- and kiosk-based exhibits and preparation of the Hunterian Web site to meet W3C Web accessibility guidelines for users with visual impairments.

Museum resources are already in use by disabled user groups. One such group was seen at Stranraer Academy's Special Needs Unit, in Dumfries and Galloway. Stranraer Academy integrates some of its special needs pupils with mainstream pupils, with additional assistance provided by learning support teachers. One of the pupils there was a fourth year pupil called William. William has cerebral palsy, which is a chronic condition affecting body movement and muscle coordination. William uses a laptop provided by Dumfries and Galloway Council that has special software to support his needs, for example, speech synthesis software so that he can participate in the class activity of giving a presentation. William was very interested in electronic resources and demonstrated the ease of use of a Scottish Ballads

CD-ROM that was created by the Hunterian Museum and Art Gallery and the Department of Scottish History at the University of Glasgow (see Figure 14.6).

William enjoyed using the CD-ROM. He also mentioned using the Internet as part of his studies. His support teacher made the point that museums could do so much to support the needs of pupils like William through resources available on the Internet and CD-ROM. The teacher recognized that more and more electronic resources are being brought into the classroom and emphasized making the resources easy for pupils to access and the importance of targeting materials to different age groups.

CONCLUSION

We have seen how ICT has become an integral part of classroom teaching, therefore, museums need to be more aware of this and play their part in enhancing the classroom delivery of learning resources by working with educators and government agencies to allow us to identify curriculum relevance and to utilize technology to bring the nation's cultural and scientific heritage into the classroom on a daily basis to support the educational curriculum.

The teachers interviewed in our research were unanimous in their opinion that in terms of experience, nothing is better than seeing the real object. However, they were keen to express that in the absence of this possibility,

Figure 14.6 A pupil using his laptop for class work

electronic resources were the next best option. One teacher in the US commented, "The nice thing about technology is they can learn about the museum and actually see different items prior to actually going on the field trip to discover them so they could be a little more aware. I don't want to see it take the place of travel and actual field trips but I think it's a good resource to use side by side."

Although museum trips are the most favored option the reality is that such trips are often difficult to organize and are expensive and therefore do not happen very frequently. In Scotland 18.6% of teachers questioned never take their pupils on museum trips and only a further 18.6% take pupils on such a trip more than once a year. 80% of pupils questioned had "never" visited a museum with their school. Common problems cited for not taking children to museums is lack of money for trips and difficulty of organizing transport and permission to take children out of school. One respondent commented, "The difficulty with a visit to a museum is that visits are time consuming with little guarantee of syllabus relevance."

Questionnaire respondents indicated that in the absence of an actual museum trip, resources online are a good alternative. One teacher stated, "A museum Web site with collections online and a CD-ROM of museum resources are useful because these can be tailored to fit specific lessons in advance." An American respondent commented,

> I think it's the next best thing . . . but there are many schools and many states that do not have the funding or the access and so it's clearly the next best thing because you can get close up pictures, you can get history, what country they're from etc. and so I definitely think that it's a positive resource, not only for us (in the Washington D.C. area) but for very rural areas that do not have the opportunity to venture out but will entice their students later in life to want to see things outside of their comfort zone and to learn more about culture and the rest of the world.

Teachers generally agreed that there is definitely a place for electronic learning alongside more traditional methods of learning such as books and worksheets and expressed that they would like to see them continue to develop side by side.

Interviewed teachers were very positive about any pre-made lesson plans which they could download and alter at will as these resources would have taken them hours to create. One high school biology teacher raised the idea of having pre-made lesson plans that could be made available for other teachers to access and borrow. This is an important issue for museums that are developing electronic resources to bear in mind. Respondents placed emphasis on wanting resources that are interactive and challenging for pupils. One respondent commented that images and animations "often provide graphic presentation which engages the pupil much more effectively than a textbook can ever do . . . pupils usually respond very well to online resources but a fine line must be drawn between material that simply entertains rather than educates."

Many teachers expressed an overwhelming need for "curriculum relevant" material. In the words of one respondent, "It would be a real benefit if museums in Scotland could produce resources relevant to the curriculum in a co-ordinated and easily retrievable way. This would provide pupils with a range of options for investigation and research." It is important, therefore, if museums want to develop resources most useful to teachers that they need to have a sound knowledge of National Curricula and have the relevant staff or partnerships in place for developments towards this.

Several teachers have highlighted the fact that one of the huge benefits of electronic resources is that they allow pupils to work at their own pace. One teacher commented,

> We all learn differently, some of us are visual, some of us can read directions and follow them and I think it is important that we go at our own pace as opposed to trying to keep up or be ahead of everyone else.

Providing that pupils have computer access at home, electronic resources can be used there as well as at school. They also can encourage pupils who are partnered together to perform a task to work cooperatively. Electronic resources can be used effectively and adapted for a range of end users, including lower levels of learners to more advanced learners.

The development of Web sites and electronic resources in museums and in particular small museums has in some cases suffered from poor planning and a lack of focus. Museums typically have had to contend with many pressures and as a consequence other issues have taken higher priority. Traditionally, museums have concentrated on physical displays and are only now slowly beginning to focus on the benefits that electronic resources can bring. Electronic resources often tend to be developed unsystematically within a museum, when the human and financial resources happen to be available, rather than according to a predefined plan. It is understandable that museums end up in this position, for example, when rounds of external funding are available and a museum wishes to capitalize on such funding. Scotland has recently published a national ICT strategy for museums which the author contributed to and which will hopefully make development of internal corporate plans for ICT in Scottish museums more straightforward and encourage individual museums to make such plans and work towards their own goals systematically, rather than the situation that has often been prevalent in many grant-funded scenarios with "the tail wagging the dog." It is important for museums to continue to focus research on the end user in order to understand "what clicks" for them and to rise to the challenge to meet their expectations. Failure to do so will result in disappointment among the users and "virtual museums" with no "virtual visitors."

The examples outlined in this chapter, while they are by no means the only exemplars of best practice which could be drawn upon, illustrate the commitment of the Scottish cultural community to provide access to high quality digital resources for a wide range of education users at all levels. They also serve to demonstrate the considerable expertise in content digitization which already exists in the cultural sector in Scotland and which can serve as a strong foundation on which to build partnerships between the cultural heritage sector and the wider education community to provide access for all, throughout Scotland and the wider world, through digital technology.

The delivery of collections material in a digitized format has allowed the museums community, not just in Scotland, but worldwide, to extend the range of access to these invaluable resources on a truly global scale. Network globalization has broken down barriers of distance, time and language, to bring electronic museum resources into the classrooms of students, from Auckland to Auchtermuchty and from Barra to Boston.

Section 6

Information Collaborations in Museums

15 Collections and Consortia

Paul F. Marty

Florida State University

A growing number of museum visitors, online and in-house, have information needs that require museum professionals to collaborate, sharing data
about their collections to meet changing expectations about information
organization and access. Researchers and scholars (including curators from
other museums) desire the ability to search databases that cross multiple
museum collections to answer research questions or plan future exhibits.
Primary and secondary school teachers want to integrate museum information resources into their curricula and educational materials without
necessarily needing to know which museums contain which collections.
In general, museum visitors are more likely to be interested in finding the
resources they need, and less likely to care where those resources reside, as
long as the information they need can be found easily and with little effort.

The result of these changes is that today's museum professionals work in
an on-demand world, where they face new challenges every day, and where
they are often unable to predict what their visitors' needs and expectations
will be in the future. Meeting even basic information needs will soon require
most museums to work in close collaboration, solving shared problems and
addressing common concerns. Fortunately, museums have a lengthy history
of collaboration, and there is strong evidence that today's museum professionals are ready to face these challenges. In a recent study, Rodger, Jör-
gensen, and D'Elia (2005) found that a) museums are more likely to be
involved in collaborative relationships than any other type of public institution; and b) when organizations other than museums collaborate, they
are more likely to collaborate with museums than with any other type of
public institution. If this encouraging trend is to continue, it is important
that museum professionals understand the relative advantages and disadvantages of participating in collaborative projects.

The most common museum collaborations exist to share data about collections held by multiple museums and to provide online museum visitors
with access to information about those collections (usually including digital
images). Around the world, there are many groups of museums working
collaboratively to achieve these goals. While it is not possible in this short
essay to review every type or instance of museum collaborations, past and

present, it is valuable to consider some of the collaborative projects upon which museums have embarked, examining what they have accomplished and why they were successful. While there have been many attempts in the past to share data using either paper-based or electronic systems (cf. Jones-Garmil, 1997), the following discussion is limited to more recent online museum collaborations.

One of the first consortia of museums to collaborate in order to distribute their collections data over the Internet was the Canadian Heritage Information Network (CHIN), which has been providing online access to the information resources of more than one thousand Canadian heritage institutions since 1995 (Szirtes, 1998). CHIN's Virtual Museum of Canada (http://www.chin.gc.ca) currently provides access to records describing more than two million museum artifacts and nearly half a million digital images. The Australian Museums Online project, now called the Collections Australia Network (http://www.collectionsaustralia.net), was established in 1995 to connect 1500 museums and galleries across Australia (Kenderdine, 1999; Sumption, 2000). In Europe, projects such as Digital Culture (http://www.digicult.info) and Project Minerva (http://www.minervaeurope.org) have been instrumental in documenting and guiding the collaborative digitization activities of European cultural heritage organizations (Digicult, 2002). Major collaborative projects in the United States have included the Colorado Digitization Project (Allen, 2000), Museums and the Online Archive of California (Rinehart, 2001), and North Carolina ECHO, Exploring Cultural Heritage Online (Wisser, 2004).

A number of museum consortia have been formed for the explicit purpose of distributing data from multiple museums to specific audiences through a single source. The Art Museum Image Consortium (AMICO), for instance, was formed in 1997 to distribute digital images from Art Museums to educational organizations. By the time AMICO dissolved in 2005, more than 120,000 images were available online for educational use (Sayre & Wetterlund, 2003; Trant, Bearman, & Richmond, 2000). Museum professionals frequently collaborate to share data with schools, particularly at the K–12 level. Examples of this type of collaboration include the Cultural Heritage Initiative for Community Outreach project at the University of Michigan (Frost, 1999, 2001), the Digital Cultural Heritage Community project at the University of Illinois (Bennett & Sandore, 2001; Bennett & Trofanenko, 2002), and the Smithsonian–Hunterian Advanced Digital Experiments (Devine & Hansen, 2003).

While different collaborations face different specific challenges, museum professionals who wish to share data among their organizations will need to address a variety of common issues (cf. Martin, Rieger, & Gay, 1999). Before beginning any collaborative project, museum professionals need to explore questions about community building, role definition, and needs assessment. They must understand why museum professionals would want to collaborate, who will play what roles during the collaboration, and what audience/user needs will be met through the collaboration. It is extremely important that all partners understand how the project will be used and administered before any collaborative endeavor is initiated. Questions to be answered include: Who does what? Who wants what? And what do they want to do with it?

Second, museum professionals participating in collaborative endeavors must address specific questions about the content they will provide within the consortium. It will often be impossible or undesirable for the collaborators to all provide the same types of content, and therefore it is important for all project partners to agree on the specific content each will contribute to the collaboration. Museum professionals will need to identify and define procedures for the documentation and digitization of collections materials, including the specific metadata standards that will be used for describing the museums' artifacts as well as the technologies that will control the digitization and information access/delivery procedures. Questions to be answered include: What content should be compiled? How will this content be described? And how will users gain access to digitized content?

Finally, the collaborating partners as a group must address important issues concerning digital rights management and access to digital content. Many of these issues are still in flux as information professionals in museums and other organizations grapple with questions about copyright and intellectual property in the digital age as well as the relative benefits of various economic models for distributing digital content (cf. Bearman, 1997; Sherwood, 1997).

While past collaborations have explored several different economic models for delivering distributed digital images (cf. Besser & Stephenson, 1996; Pantalony, 2001), it is likely that future collaborations will require new economic models for administering and accessing intellectual property online (Zorich, 2000). Questions to be answered include: Who controls the rights to museum content? What rights can the museums offer their users? Who will pay for providing access to these information resources?

As part of coping with these challenges, museum professionals within collaborations will face a variety of sociotechnical questions, where the choice of one particular technology solution will have social ramifications for the collaborative partners and the eventual end-users of the museums' data. The following examples provide some perspectives on the types of sociotechnical questions museum professionals may face in three topic areas: metadata standards, digital imaging, and content centralization. While there are a number of organizations exploring these questions, most museum professionals will need some familiarity with these issues, and may end up developing their own answers as they engage in collaborative endeavors.

One of the very first sociotechnical questions to arise in most museum collaborations concerns the metadata standards participating museums will use to share data about their collections. Having everyone within the collaboration use uniform standards can help enhance access to records and objects, provide better consistency in information retrieval, improve the quality and accuracy of data, and make it easier to exchange data between organizations or port data from system to system (Bearman, 1994). Many current content sharing initiatives, however, have been confounded by the difficulty of choosing appropriate standards, and the use of often arbitrary rules for the application of standards across different organizations can result in additional problems. Even the selection of appropriate controlled vocabularies to help users select the preferred search terms can be a challenge for many museum collaborations. In the future, it is possible that museum professionals will develop sophisticated knowledge bases that function as intermediaries between user queries and museum artifact records. As a result, museum professionals may focus less on building records according to particular pre-defined standards and more on created semantically-aware data that can be more easily integrated into different data repositories (Doerr, Hunter, & Lagoze, 2003).

The distribution of digital images presents another example of the socio-technical problems people face in collaborations. Museum professionals face peculiar challenges as they make decisions about providing, and in particular controlling, access to digital images. In an attempt to retain control over who uses their images, they may decide to use any number of technology protection measures, including digital watermarks; some museum professionals have even developed invisible watermarks that can only be seen by the creator of the image (Bissel et al., 2000; Gladney et al., 1997; Mintzer et al., 2001). The problems of controlling access to and protecting the copyright of digital images of museum artifacts, however, are not clear cut. For example, in the landmark case, *Bridgeman Art Library v. Corel Corp. 1999,* it was determined that digital images of works of art in the public domain cannot themselves be copyrighted (Ochoa, 2001). Such developments mean that museum professionals may need to rethink their existing polices about information access, focusing less on copyright for protection and more on retaining control over access to the original artifacts. Given the rapid technological changes that have occurred with digitization, it is likely that the solutions to problems of access to digital images will require novel solutions. In the near future, museum professionals may determine that the best way to protect their intellectual property is to make high resolution images of all works of art in the public domain freely available for use online by their visitors.

Finally, museum professionals building collaborations face questions about whether data shared within collaborations should reside in a centralized location, should remain decentralized across the collaborating partners, or should exist as some combination of the two (e.g. centralized metadata records with distributed digital images). The pros and cons of these approaches have led some to explore the best ways of creating, maintaining, and using collaborative data, researching new methods for sharing collections data (Lagoze & Van de Sompel, 2003; Perkins, 2001; Shreeves,

Kaczmarek, & Cole, 2003). Deciding between vertical, hierarchical models where data is stored in focused, centralized contexts and horizontal, distributed models where data is dispersed at the grass-roots level in multiple contexts raises many unanswered questions: Should museum professionals aim for one universal standard that everyone should use, or should they try to create data sharing standards that allow individual organizations to use different standards internally as long as they can crosswalk their data to a common external standard? Should museum professionals promote centralized repositories that may be easier for collaborating partners to understand or should they rely on decentralized approaches that may be more difficult to master? What is the impact of such choices on metadata standards and information representation in the participating museums? How can the problems of disambiguation and decontextualization that sometimes occur when individual data records are accessed out of context be resolved?

The above issues present no shortage of challenges for collaborating museum professionals. While there are many good reasons to collaborate, ranging from improved access to museum information to strengthened relationships between museums and their partners, these potential difficulties mean that collaborating in museums is not a clear cut process. Solving these problems can be particularly challenging given that many museums still face problems managing their own internal information infrastructures. Hermann (1997), for instance, identified several information management problems museums face, from the challenges of coping with digital media and increased connectivity to the need for museum professionals to work in information-rich environments. Blackaby (1997) explored the challenges museum professionals face when building integrated information systems that connect people, information, and technology in museums. Meeting these challenges will require museum professionals to possess a high level of familiarity with information policy and information management issues. It is these museum information professionals with their strong understanding of the unique sociotechnical issues museums face as information organizations who will lead today's museums into the 21st century.

16 AMOL Ten Years On
A Legacy of Working Beyond Museum Walls

Basil Dewhurst

National Library of Australia

Kevin Sumption

Powerhouse Museum

Australian Museums Online (AMOL) (http://amol.org.au) was an Australian State and Federal Government funded portal that established a reputation, both nationally and internationally, as a highly popular, innovative gateway to Australian museums and galleries. Initially proposed by the Heritage Collection Council (HCC) in 1995 as an electronic register of moveable cultural heritage material, the AMOL project kept pace with changing public and professional needs through a series of redesigns and the development of a diverse range of content and services. In 2005 the AMOL project turned ten, and for seven of these years the Powerhouse Museum has been the project's host. During this time a mix of information products, specially created for both museum workers and the public, gave the site a wonderful eclecticism that set it apart from other metacentres, with contact details of over 1300 Australian museums, sitting alongside half a million item level collection records, discussion forums, an online journal as well as a number of database cultural tourist projects and natural science databases.

In 2005 AMOL was redeveloped as the Collections Australia Network (CAN) with an expanded brief to move beyond galleries and museums, to also facilitate access to the cultural heritage assets in Australia's archives and libraries. CAN used AMOL's information products as a foundation and layered on top of this a self-publishing and administration toolset that facilitated decentralized content management for CAN contributors. As we shall see, like many national cultural portals AMOL's journey over the last ten years was unique as it moved from experimental outreach in the early years of Internet privatization in Australia, through to strategic training and resource dissemination as CAN begins to establish a new place within the complex web of Australian remote, regional and rural professional development providers.

THE AMOL METACENTRE

The AMOL project emerged as a direct response to the information explosion experienced by many Australian museums in the 1990s. In these institutions collection records and interactives were increasingly digitized and the Internet was embraced as a means of communicating with the public and providing rich online information resources. The rapid adoption of information technology tools by some museums was inadvertently "pushed" in Australia by Government moves towards greater fiscal self-sufficiency (Sumption, 1999). At the same time financial pressures to be more cost effective ushered in a period of appraisal of some of the more costly functions of museums such as collection care and management. It is not surprising then that the 1990s witnessed a rapid uptake of computerized Collection Information Management Systems (CIMS). The effects of this were the creation of a wide variety of digital, text, graphic, photographic, and video cultural asset databases. While technological determinism and fiscal expediency "pushed" the process of digitization in museums, the popular explosion of the Internet after 1995 in Australia provided the "pull" to make these resources publicly available.

The net result of these pressures was not only the emergence of AMOL in Australia, but a range of national cultural portal types, across Europe and North America. As cultural institutions recognized the potential of the Internet to make their collections more accessible, governments recognized the benefits of aggregating institutional collection databases into new metacentres. These metacentres were in essence new virtual institutions providing aggregated access to multiple collection databases that were not ordinarily available to cultural heritage professionals, let alone the general public previously. In 2000, Sumption identified three metacentre types (Sumption, 2000): the institutional metacentre, the public metacentre and the professional metacentre.

Institutional metacentres are typically the Web sites of large collecting institutions and, in Australia, include those of the National Library of Australia, the National Archives of Australia and the Australian War Memorial. Characteristically each of these institutions has extensive collection holdings, numbering in the hundreds of thousands, or even millions. Importantly these institutions have over a number of years also developed close working relationships with a large research fraternity and have traditionally provided in-house research facilities. Consequently most of these institutional metacentres have developed online-database driven Web sites that act as surrogates of their corporeal research centers and also provide detailed information about each institution's public offerings.

Public metacentres are Web sites created for, or by, heritage sector advocacy, administrative, coordinating or governance bodies. The aim of these metacentres is to increase public awareness of exhibition and public program offerings from museums by aggregating information about their exhibitions and events and to a lesser extent collections. One of the most advanced of these metacentres is the 24HourMuseum (http://www.24hourmuseum.org. uk), which operates as a public gateway to information about British museums and galleries. To provide for the information needs of their audiences the developers of public metacentres typically use a sophisticated combination of e-zine format, heritage trails, collection and institutional contact databases. They also attempt to target specific user groups through tailoring content and services to them, for example through the provision of children's pages or genealogical resources. Quite deliberately public metacentres then create an online experience that is typically fun and entertaining and thus has broad public usefulness and appeal.

The third metacentre type is the *professional metacentre*. Here the term professional indicates the primary audience for whom these sites were designed. In the case of Australian Museums Online (AMOL) and the Canadian Heritage Information Network (CHIN) these audiences are individuals working in collecting institutions in a paid or unpaid capacity. It is not suggested that professional metacentres are not public-friendly, however there is little doubt that the information architecture, search tools and core content are and in many cases continue to be designed to be primarily useful to those working in the heritage sector.

AMOL's initial information architecture and online tools were very deliberately designed to meet what we then understood were the needs of those working in museums and galleries across Australia. However, intent and outcome don't always align and the effects of technological change and audience diversification, in particular the growth in cultural tourist audiences, had a significant and unanticipated impact on the growth and direction of AMOL. Traditionally, professional metacentres have provided a diverse range of services such as searchable databases of collection objects, databases of professional services, as well as museum directories. However as these metacentres have amassed content they have also attracted, and in some cases deliberately courted, new audiences. AMOL, like CHIN, came to be used by audiences other than cultural heritage workers, and from early 2000 we were aware of and began responding to, the needs of new audiences like natural scientists, cultural tourists, genealogists, teachers, students and even very specialist niche audiences like quilters. Many of these were not significant AMOL audiences in 1998 but by 2005 were major users. By building resources that catered for the needs of more diverse audiences, AMOL ceased to operate as a purely professional metacentre and instead its hybrid form occupied a mid ground between public and professional metacentre.

In 2002–2003 a series of studies were commissioned by the Department of Communications Information Technology and the Arts (DCITA) to investigate the broad needs of the Australian collections sector and, in turn, determine how AMOL should be positioned in future. These studies were to help AMOL both refine its business and operational model and ultimately led to its redevelopment as CAN in 2005. The twin forces of Web-based techno-logical change and audience growth throughout the early 2000s combined to blur the boundaries between public and professional metacentres, and the redesign and relaunch of AMOL as CAN in 2005 marked the final evolution to a new *hybrid metacentre*. This new form resembles that of other contem-porary national cultural portals in the first decade of the 21st century, in particular culture.ca, culture.fr and sweden.net. Similar to these, CAN has emerged in response to a set of unique national and international social, technological and political imperatives. Second generation online Content Management Systems (CMS), growth in online cultural tourism audiences and the political imperatives of better servicing the needs of regional, rural and remote communities all combined to give birth to CAN. However AMOL's transition to CAN was not spontaneous and like the trends it mir-rored, was a slow, steady evolution over a number of years.

OPEN COLLECTIONS: THE DEVELOPMENT
OF A DISTRIBUTED NATIONAL COLLECTION

The HCC originally conceived AMOL as an online service capable of pro-viding a single access point to Australia's distributed national collection, a place where an online aggregated database representing collections from small, medium and large institutions across Australia would make collec-tion records freely available to all working in the cultural sector. This aggre-gation and accessibility was enabled by the broad availability of Internet technologies and also offered the tantalizing opportunity to bring together all collection types, from the natural sciences and humanities, through to the arts and social history. Indeed the rather ambitious mission of conceiving and developing a single space wherein the diverse collection offerings of the Australian distributed national collection could be widely disseminated was the first major challenge and achievement of AMOL and gave rise to argu-ably the project's most significant resource: *Open Collections*.

From 1996, AMOL's *Open Collections* provided users with a basic and advanced search interface for retrieving information held in the collec-tions of museums and galleries across Australia. In response to a search conducted across multiple collections, drawn from a farm of state based regional servers, users were returned a basic catalogue record that was often accompanied by an image. To facilitate the efficient search and retrieval of data across such a diverse set of databases, AMOL used the Dublin Core Metadata Element Set (DCMES). The choice of the DCMES was a reflection of the significant diversity of collection types, object types, as well as the quantity and quality of collection documentation. The use of this metadata standard delivered consistency of records across collections and allowed *Open Collections* to quickly search and retrieve data. In the late 1990s AMOL collaborated with the Consortium for the Interchange of Museum

Information (CIMI) on research into the use of Dublin Core for describing physical objects from cultural collections. From this work emerged the *Best Practice Guide: Dublin Core* (CIMI, 2000). Through mapping information from fields in institutional Collection Information Management Systems (CIMS) to fields in the DCMES, AMOL was able to significantly improve the consistency of records and the performance of *Open Collections*. The use of the DCMES provided for a search across multiple collections using discrete fields such as Title, Description, Date and Coverage thus providing more precise search results for users. In delivering these collection records from *Open Collections*, AMOL not only sought to present records that enabled the discovery of objects in multiple collections, but, critically, also provided a means for users to follow links from these records to view a richer record hosted on the contributing institution's Web site. In this way AMOL was able to persuade contributing cultural institutions that a listing within *Open Collections* would and did increase not only the accessibility of their collections, but also in the early years of the Internet, helped grow their own online audiences.

In the mid- to late 1990s, AMOL was almost entirely reliant on institutions that had the capacity to digitize their collections. At the time it was these, mainly large national and state museums that could contribute material to the *Open Collections* database. Not surprisingly, towards the end of the 1990s, as scanning and photographic technology costs fell and expertise in digitization became more widespread, the number of collections from small- and medium-sized collecting institutions grew dramatically. But technological accessibility alone was not responsible for this upsurge in interest from regional Australia. At the time there were a number of motivating factors attracting collecting institutions to AMOL. One of the most significant was the potential to increase the profile of the museum or gallery to a rapidly growing AMOL user group. In 2004 an average of over 4000 unique users a day accessed AMOL, of these over 20% were using the *Open Collections* resource. As well as an increased profile, there was also the democratizing effect of *Open Collections*, where any public, cultural institution could ultimately place their collection both alongside and across that of any other museum or gallery, be it big or small.

The growing use of *Open Collections* was not just a product of the sector's increasing technological awareness and capacity, but was also deliberately nurtured by AMOL through a series of initiatives that deliberately sought out key collections and provided a number of museums and galleries with collection documentation grants. As well as providing funds, a series of national digitization workshops were developed to increase the participation from small- and medium-sized regional museums and galleries in 1999–2000. The grants generally funded the purchase of digital cameras, scanners and Collection Information Management System (CIMS) software that allowed institutions to undertake collection digitization and documentation projects. Eventually, 30 institutions received AMOL digitization grants and over time numerous local, state and national grants emerged to continue to support digitization projects. Thus, in more recent years, AMOL's emphasis has switched to training support in order to ensure quality control throughout the process of digitization and documentation. One of the most successful of these online training programs, called *Capture Your Collections*, was co-authored with the Canadian Heritage Information Network in 2001.

By 2005 *Open Collections* held over 500,000 collection records from over 100 of Australia's most significant national, state and regional cultural institutions. And as AMOL gives way to CAN, the challenge will be to continue to increase both the quality and quantity of collection documentation. Additionally, CAN's expanded brief, taking in as it does both archives and libraries brings with it challenges to look to new documentation and digitization standards and protocols. With libraries and archives, we quickly recognized that CAN has not only a set of new contributors, but also users whose unique needs and requirements may not be totally fulfilled by *Open Collections*. Thus, alongside *Open Collections*, CAN is now developing a series of specialist search and retrieval tools that very consciously builds on the project's experience successfully working with natural scientists over the last six years. While *Open Collections* catered to the needs of a broad set of professional audiences, such as collection managers, curators, students, and genealogists, the specific needs of natural scientists were met through the development of a specialist natural science portal developed and managed by AMOL, known as *Australia's Fauna*. *Australia's Fauna* is an expression of AMOL's ability to identify opportunities and develop sophisticated collection access solutions.

AUSTRALIA'S FAUNA

Natural history databases within Australian museums have been common since the mid-1970s; however, the development of a single consolidated system capable of searching across all collections was the subject of nearly fifteen years of standards development, technological selection and governance debate. Fortuitously for AMOL, we appeared at a time when the natural science institutions were looking for a technical solution to the searching of their combined databases. Our experience with *Open Collections*, combined with the project's perceived relative independence from any one institution, allowed us to breathe life into the *Australia's Fauna* initiative.

In developing *Australia's Fauna* (http://www.fauna.net.au) in 2003, AMOL's aim was to create a portal to facilitate access to the distributed datasets of Australian faunal collections. All major and a number of minor Australian natural history collecting institutions contribute collection data to the project. Significantly other partners in the project include Australian government departments from the Department of the Environment and Heritage (DEH) through the Australian Biological Resources Study (ABRS), and the National Oceans Office (NOO). *Australia's Fauna* is an example of how cross-institutional collaborations, through a "neutral" but technologically empowered facilitator like AMOL, can provide greater access and meaning to natural history collections on the web.

In Australia, natural history collections have, for a variety of political and professional reasons, tended to sit apart from cultural collections. Where records from cultural collections characteristically emphasize the unique or iconic value and significance of individual objects, natural science collections have tended to reflect faunal and floral diversity. As such, their usefulness and value tends to lie in the aggregation of large amounts of specimen and type specimen data. Consequently, natural science collections tend to hold many duplicate specimens that ordinarily vary only in the time and place they were collected. It follows that the great power of recording specimens in databases lies in the unique spatial and temporal representations that aggregation can deliver. Further, where a single institution's aggregated data is valuable, the aggregation of data from all Australian natural science collections is of greatly increased value to scientists, education and legislative public sector audiences. The resulting representations of these collections provide as complete a picture as possible of specimen distribution and habitat, particularly when they are mapped with other zoogeographic and topographic data.

The *Australia's Fauna* portal provides a user-driven view of biological data. Its search interface allows users to request results be returned as data or as mapped points. If users choose to have their results displayed on a map they are provided with an interface that allows them to pan or zoom in or out of their mapped results, as well as refine their search by drawing a bounding box to restrict their queries further. The map display also allows users to turn on and off various map layers, which can include species, datasets, geographical or habitat features. Each point displayed on the map interface represents one or more specimen records. Points may be individually queried to drill down into the actual data records. Users requesting search results as raw data are presented with a list of specimens matching their query and can browse to individual specimen records. Alternatively, users can obtain all raw data relating to their query in either Comma Separated Variable (CSV), HyperText Markup Language (HTML), eXtensible Markup Language (XML) or Distributed Generic Information Retrieval (DiGIR) format.

AMOL's role in the *Australia's Fauna* project was twofold. Firstly, AMOL hosted the central portal software that allows users to conduct real-time, distributed searches of the faunal collections from around Australia. Secondly, AMOL developed the software hosted at each institution which responds to requests for data from the portal. This software interoperates with the different CIMS software at each institution and returns large datasets in XML format for viewing and manipulation by users on the portal. CAN continues to support *Australia's Fauna* and future developments will focus on the portal providing data to the Global Biodiversity Information Framework (GBIF) project, thereby providing Australian Faunal data to a global repository of the world's biodiversity.

The powerful tools that AMOL developed for *Australia's Fauna* have allowed scientists and researchers to access, for the first time, aggregated specimen records from institutions across Australia. The availability of this information has dramatically changed the ways in which the core audience of scientists view and use information in their daily work. A powerful example of this occurred when a dataset of barnacle specimen distribution in Australian ports was made available on *Australia's Fauna* in 2004. The availability of this dataset has significantly contributed to the tracking of pest barnacle species in Australia's harbors, a task that prior to such aggregation would have been very difficult to achieve. As a portal primarily operating as a professional metacentre, *Australia's Fauna* is indicative of the type and form of tools AMOL typically developed in the late 1990s and early 2000s.

A NEW KIND OF PROFESSIONAL METACENTRE

Where AMOL's focus was traditionally on meeting the needs of professional researchers and academics, as AMOL's understanding of users and their professional needs grew, we realized that the project could deliver other online services to a variety of museum professionals. In retrospect, this shift from the provision of tools for data retrieval and interrogation, to that of real-time information exchange and online market development, marked an important turning point. As AMOL moved to better understand the needs of its audience, it also moved to align itself with the specific needs and changing roles of cultural institutions. In essence, AMOL was moving from a brief to promulgate online research tools to one where the operation and performance of museums themselves was of equal if not greater concern.

In 2000 AMOL launched the National Exhibition Venues Database (NEVD), a database of venue details designed specifically to support the selection and management of temporary exhibitions in archives, galleries, libraries and museums. The NEVD allows exhibition managers to plan their touring programs with relative ease and provides them with quick and easy access to highly detailed records about all aspects of an exhibition space, such as information about lighting, security, fire protection, environmental control information, as well as floor plans. The resource is divided into three main sections that allow exhibition managers to browse for a venue, conduct a search for a venue, and to add or modify their venue's information.

Most importantly, the existence of entries for large national and state institutions alongside those of small regional galleries and libraries, allowed smaller, non-traditional venues to be exposed and thus promoted to mainstream exhibition developers. Following the development of NEVD in 2002, AMOL created the TOUR database, a database of touring exhibitions available for hire in Australia. TOUR provides descriptive records of touring exhibitions currently available for hire in Australia and was a very early example of an e-service for the cultural sector.

The online NEVD had its origins in the paper-based Exhibition Venues Directory developed in the late 1990s. This directory was published by the professional museum association body, Museums Australia. As a paper-based product the original directory was not easy to update, particularly as venue specifications changed relatively frequently and, over time, the directory became out of date. Partnerships with Visions of Australia and the National Association of Museum Exhibitors (NAME) supported the development of the NEVD. Visions of Australia, the Australian Government's national touring exhibitions program that provides grants to support the development and display of touring exhibitions, generously provided funding for the initial development of the NEVD. Most importantly, when launched, Visions of Australia made it mandatory for institutions applying for funding to have an up-to-date entry in the NEVD.

In the conceptualization of the NEVD and TOUR resources, AMOL for the first time provided a strategic, whole-of-sector solution to a broad cultural industry problem: ensuring effective regional distribution of touring exhibitions. This departure from providing purely research based resources and services, to focus on the specific needs of a discrete user community, acted as a catalyst. Its success encouraged us to look to other discrete communities to see how their particular needs could be better understood and met. One of the first of these communities was that of quilt collectors. At a national level Quilters in Australia are very well organized and had for some considerable time administered their own electronic newsgroups. However, in 2001, AMOL was approached by the Pioneer Women's Hut to develop a national register of quilt collectors that could serve to better represent an important and often neglected aspect of women's history. Therefore, in late 2001 AMOL began work on a national repository of Australian quilts: the National Quilt Register (NQR).

CREATING THE NATIONAL QUILT REGISTER (NQR)

The NQR was one of the first high-level database driven resources designed specifically for the general public by AMOL and was developed in collaboration with the Pioneer Women's Hut, a museum located at Tumbarumba in the foothills to the Snowy Mountains in southern New South Wales. As a discrete collection resource, the NQR provides access to some 1,200 significant quilts held in private and public collections across Australia that would otherwise be very difficult or impossible to physically exhibit.

Although the primary intent of the NQR Web site was to provide a virtual showcase for quilts held in private and public collections, the register also functions as a general women's history resource with broad appeal.

As objects, many of the quilts featured in the register carry significant stories about Australian regional history and provide a unique window into women's lives. They also provide invaluable information about the craft of quilting and its history in Australia. While the resource makes available information about Australian quilts, it also has a strong international flavor: many of the quilts were and are part of the lives of migrants to Australia and quite a number were made overseas. As a result the resource continues to attract a strong international audience.

The success of the NQR spurred the development of a range of broad-focused historical microsites designed specifically to contextualize the individual collections and objects held in the *Open Collections* database. However, whilst NEVD, TOUR and NQR were still focused on relatively specialist needs, the new series of non-specialist, regional history focused microsites that emerged catered to the broader needs of students, teachers and most particularly cultural tourists. AMOL's success with the NQR encouraged further experimentation and exploration. With the general public clearly demonstrating an appetite for virtual collections and exhibitions, the AMOL team sought to determine just how far the Web could be used to promote not only virtual collections, but real ones. That is, to develop resources which use objects from a museum's collection to motivate audiences to actually visit the museum and its various real exhibitions.

A BRIEF TO PROMOTE AUSTRALIAN CULTURAL TOURISM

The development of cultural tourism resources by AMOL was a direct response to the user patterns we observed from 2000 onwards. While the AMOL project's professional museum user-base was estimated at a steady 1,000 to 1,300 unique visitors per day in the early part of 2000, the project started to see a significant growth in cultural tourist use, particularly in AMOL's *National Guide* (cf. Scott, 1995). The *National Guide* was a searchable database of Australian museums and galleries. At the time of AMOL's transition to CAN there are approximately 1,300 of Australia's estimated 2,000 museums and galleries listed in the database. Importantly the database was the only authoritative online resource that provided users with searchable descriptions of individual cultural institutions, their facilities, collections, Web addresses and physical locations.

In early 2001 the AMOL team began to develop tools to meet the needs of cultural tourists, the first of which was the *Victorian Art Trails* microsite. Victoria's regional galleries hold nationally and internationally significant works of art and the microsite was designed to not only highlight the regional galleries' unique collections, but, critically, to motivate cultural tourists to visit the individual galleries. This microsite provided a set of six driving trails through the state of Victoria connecting all ten of the states' regional galleries. The trails focused on the main driving routes that tourists were known to use to get to and from the city of Melbourne, Victoria. Having selected a preferred route or trail, the online visitor was encouraged to print off a detailed itinerary, describing the significant sites they would encounter on the route. Importantly the microsite also incorporated high-level motivational content designed to stimulate interest in particular galleries and their collections. This content took the form of a series of high quality interior and exterior panoramas. Research at the time suggested that in order to promote real visitation, it was critical that the Web look to affordance technologies that could provide an engaging and stimulating visual experience (Ryder & Wilson, 1996). And we already appreciated that the look and feel of a gallery was particularly important in helping visitors choose to visit.

Victorian Art Trails was followed in 2002 by *Found and Made,* which focused on craft collections in Tasmania. As Tasmania is a relatively small Australian state and easy to get around by car, we decided there was little need for driving trails. Instead, *Found and Made* placed more emphasis on motivational factors and so sought to illustrate individual public craft galleries in far greater detail, as well as provide compelling personal profiles of artists whose work was featured in individual collections. Three different sub-themes of maritime craft, shell collections and wood design were used to group institutions. Where the distinct trails developed for *Victorian Art Trails* provided a pre-defined sequence of institutions for visitors to systematically traverse, *Found and Made* divided the Tasmanian state into three areas, and visitors were encouraged to choose and design their own customized trail.

For the institutions involved in the development of both these cultural tourism resources, the potential benefit in collaboration was clear: increased visitation through additional tourist attention and with it potential increased general admission revenue, specifically for museums and galleries in regional rural and remote Australia. This potential for AMOL to actively promote the heritage assets of regional and remote Australia has since this time been a strong theme in much of the project's work. This focus on the unique needs of small- and medium-sized collecting institutions (SMCs), often staffed by dedicated groups of volunteers with little formal museological training, was to formally become AMOL's principal concern in 2004, as the Australian Commonwealth government along with the Cultural Ministers Council laid out AMOL's new mission.

CAN: THE EMERGENCE OF A NEW HYBRID METACENTRE

It is estimated that there are 2,500 large, medium and small collecting institutions in Australia that together are custodians of much of the nation's moveable cultural heritage material. Many SMCs are located in rural and remote communities well away from major metropolitan centres and operate with small subsidies and grants from local and state governments. These institutions are run by a diverse range of people from trained full time professional staff, through to the largest group, made up of part-time volunteer workers. All have tended to have a great need and appetite for resources and training that assists with the management, preservation, promotion and display of their collections. Not surprisingly, SMCs had long been strong users of AMOL's many resources; however, within the AMOL project charter, they had not been specifically singled out as requiring particular attention.

In 2002 the Australian Government's Department of Communications Information Technology and the Arts (DCITA) commissioned the first of a number of landmark studies into the state of documentation and digitization in Australia. The first of these was a study into the *Key Needs of Collecting Institutions in the Heritage Sector* (Deakin University, 2002) that investigated the current and future needs of heritage collections in Australia. The study highlighted the need for closer collaboration between the museum, gallery, library and archive sectors and recommended the development of a long-term national strategy to support collections sector growth in Australia. Of particular importance to AMOL was the study's recommendation that documentation and delivery systems for collections, particularly outside of major metropolitan areas, be improved. Not surprisingly, the study singled out small- and medium-sized collecting institutions (SMCs) as those institutions at greatest risk of missing out on the benefits of training, collection digitization, electronic promotion and documentation standardization.

Following on from the Deakin University study in October 2003 AMOL conducted a "grass-roots" study of 400 small- and medium-sized museums and galleries in Australia. This study queried institutions about their access to computers, the Internet, digitization equipment and expertise. The results mirrored the results of the *Key Needs Study* and showed that many institutions severely lacked the support and infrastructure needed to digitize their collections and without this they were not in a position to utilize digital technologies to better manage and promote their collections, let alone engage with AMOL. Thus it became increasingly apparent that in order to better meet this critical sector's needs, AMOL would need to change its focus.

From these studies, and a detailed business analysis review conducted by DCITA in 2004, a vision for the future operation of a revitalized and refocused project emerged. Where in the past AMOL had focused on making accessible predominantly pre-digitized collection assets, a new business model was developed. The new model not only required the project to expand public access to take in libraries and archives, but also to build digitization capacity within SMCs. It was an approach that needed to be cognizant of the fundamental needs and realities faced by the people working in SMCs: a lack of access to regular accredited training, technological infrastructure and financial resources to support adequate collection documentation and digitization. Thus, in late 2004, CAN was conceived and some fourteen months later the first iteration of CAN soft launched.

The newly developed CAN Web site (http://www.collectionsaustralia. net/) offers an array of custom-built tools that significantly enhance both the usability and usefulness of the World Wide Web for SMCs. Most significantly, CAN's new distributed Content Management System had been especially designed to allow users with little technical experience of the Internet to manage their own institution's collection through CAN. In addition, the Web site provides a plethora of new promotional tools to help institutions use electronic means to promote their exhibitions and collections to international, national, and, most importantly, local audiences. Each new CAN subscriber is provided with one or more logins for their staff to access an easy to use administrative interface that allows people to create, edit and archive any image- or text-based data. CAN also provides an online CIMS especially designed to empower SMCs to create, update and manage their own online collection documentation in the form of collection level and item level descriptions. In information architecture terms the transition from AMOL to CAN was akin to moving away from a centralized data repository model, to that of a truly decentralized and distributed model, wherein data storage capacity, tools and training were made available with the specific intent of empowering SMCs to create and manage their own individual institutional metacentres.

The focus on allowing SMCs to use electronic means to better interact and promote their work, particularly amongst local groups like schools, councils, regional newspapers and tourism authorities, is a critical new aspect of CAN. This is largely achieved by allowing SMCs to build and manage a series of discussion list services.

At the level of collection resource management and distribution, CAN has moved away from AMOL in three important ways. Firstly, CAN subscriber institutions are now encouraged, as well as assisted, to develop collection level descriptions. Critically, collection level descriptions are high level finding aids allowing users of CAN's many search facilities to substantially increase searching precision across collections. Secondly, CAN has increased the number of possible record fields that can be made available to users. It has also provided for the association of multiple images, audio or video files with an object record thereby providing the potential for rich collection records to be made available. Thirdly and most significantly, CAN now provides institutions with a free, basic online Collection Information Management System (CIMS). As well as allowing SMCs to manage their own collection data, it allows for a degree of documentation and standardization, which in the future should greatly assist the development and delivery of training in the sector.

Whilst data from AMOL's original *Open Collections* has been migrated to CAN, we took the opportunity to address a significant operational issue. *Open Collections* had always provided access to a plethora of different collection records representing collections covering the performing arts, agricultural implements, natural and physical sciences, decorative arts etc. Invariably, each of these disciplines has over time evolved its own specialist set of thesauri. However, with the advent of CAN, the project is moving to implement controlled vocabulary driven services. In the past, few contributing institutions have used a controlled vocabulary to describe their objects. This is now less of an issue as most institutions are aware of the importance of using thesauri to describe their catalogues and many CIMS products provide thesaurus capability. As we prepared this paper one of the major challenges CAN faces is building and implementing a high level, hybrid thesauri capable of being used by the diverse range of institutions across Australia.

In addition to the technical redevelopment of the project, the new CAN business model also predicated a reappraisal of staff skills and experience needed to successfully roll out CAN to SMCs. The AMOL staff structure was geared towards the technical Web development and administrative skills that allowed development of Web content and resources from within the AMOL team. It was quickly recognized that CAN would require a new mix of skills: one that focused more on being a training provider and facilitator, alongside other heritage training providers in the sector. CAN has a specific mandate to build the resource base and develop skills in SMCs to increase the capacity of collecting institutions to improve the quality, consistency and quantity of collection documentation and digitization. Where the AMOL staff structure was focused on technical development and Web publishing, CAN provides in-the-field training staff, supported by a suite of online resources designed to meet the needs of the people and institutions in the Australian collections sector.

As well as meeting the needs of SMCs, CAN is also now charged with the ambitious brief of providing a consolidated search across a broader scope of collection intuitions, such as libraries and archives. To achieve this we are currently collaborating with projects like the National Library of Australia's *PictureAustralia* to develop the means to interrogate collections in real-time and aggregate search results from multiple sources in a manner similar to *Australia's Fauna*. It is here that the distributed collection search models first developed in *Australia's Fauna* will be used to provide access to the broadest possible array of material to develop a genuinely distributed national collection. In addition to providing a federated search across multiple repositories, CAN will develop Web service interfaces that allow other portals and gateways to access and search CAN content.

To technically achieve this CAN will continue to use the Dublin Core, AGLS and EdNA metadata standards as well as interoperability standards like the Open Archives Initiative Protocol for Metadata Harvesting (OAI-PMH) and OpenSearch. We will also look to use Distributed Generic Information and Retrieval (DiGIR) protocol and Simple Object Access Protocol (SOAP) to create Web services that allow for information to be searched and shared with other online services such as the Education Network of

Australia (EdNA) and PictureAustralia. As CAN moves towards supporting libraries and archives more fully it will also look to support the data and interoperability standards generally in use in libraries and archives, such as Machine Readable Code (MARC), Encoded Archival Descriptions (EAD) and Z39.50.

THE FUTURE OF CAN

Where AMOL started life as a professional metacentre and gradually evolved to take on attributes of the public metacentre model, CAN through its decentralized content management and dissemination model is breaking new ground as it stratifies the metacentre model along operational and marketing functions. At an operational and administrative level, CAN returns to a professional metacentre model with its focus on empowering and building capacity in SMCs. But at a public level, CAN is multi-focused and multi-dimensional, with equal prominence paid to supporting the use of learning objects in schools or encouraging visits to regional centers in an effort to promote local cultural tourism. This multi-dimensional form enables CAN to operate simultaneously as a comprehensive repository of professional resources, and at the same time act as an essential tool for workers in the Australian collections sector. In this way, CAN actively helps promote and sustain the economic and socially cohesive work of regional cultural organizations, which in turn, delivers real and tangible benefits back to the community.

CAN's dual aim of supporting sector development and disseminating content to national and international audiences is important and is part of national strategies to support collection access and sustain and develop the sector. In order to achieve these aims, CAN will need to continue to work closely with state and national bodies like the newly formed Collections Council of Australia (CCA) to support the sector at a grass-roots level. To do so CAN will also in the future look to develop partnerships with Technical and Further Education (TAFE) colleges and universities, particularly those offering cultural heritage education and training. These partnerships could support the development of online courses and training materials accessible by sector workers at all levels.

To ensure CAN's ongoing usefulness to these many constituents, we recognize the project will need to continually evaluate its tools and services to ensure they are meeting the needs of all. As well we will need to embrace a process of continual technical improvement, particularly as we more closely align with the library and archive sectors. In this regard the CAN journey is about to begin and we can only hope that the new projects services will meet the needs of both our museum professional and general public audiences. As we hope this chapter makes clear, the CAN model has been through many iterations. While some of these have been imposed, others have been a consequence of market opportunities and technical innovation seized. Whatever the future might bring, the ongoing success of CAN will be as much about the individual commitment of its small staff and the project's ability to hold onto and foster a culture of innovation and creativity, as it will be about continued growth of online cultural tourism and cultural training and promotional initiatives.

17 Challenges to Museum Collaboration
The MOAC Case Study

Richard Rinehart

UC Berkeley Art Museum/Pacific Film Archive

Layna White

San Francisco Museum of Modern Art

INTRODUCTION

Cultural agents and institutions are utilizing new media to create collaborative networks of knowledge, experiences, and resources that could grow beyond any historical precedent. Such networks may fulfill a vision of seamlessly integrated, globally available, diverse cultural activity, but serious obstacles loom before that vision. Differences in values, practices, abilities, and goals among cultural collaborators and network builders hinder progress already slowed by economic and technical problems. This chapter will identify some of the practical concerns and less tangible obstacles faced by collaborating cultural heritage institutions, and suggest strategies for planning collaboration initiatives. The Museums and Online Archives Collaboration (MOAC) project will be used as a case study to provide examples of problems and solutions to the challenges of museum collaboration.

MUSEUMS AND COLLABORATION

Collaboration is not new to cultural agents or institutions. Scholars, artists, libraries, and museums have been collaborating for decades, in some cases, centuries. Scholars shared research via conferences and professional journals; artists collaborated with others of complementary skill; libraries shared classification systems; and museums created traveling exhibitions that span cities. However, collaborations such as these were usually conducted within the confines of tightly defined professional communities using a highly specialized infrastructure. The public was almost never invited to witness or participate in the collaboration per se, but was often called in to observe or enjoy the finished product. While all of these collaborations continue, several factors compel cultural agents and institutions toward an expanded set of collaborative possibilities.

Within the last few decades, social pressures have forced public cultural institutions to operate more transparently and become inclusive of diversity (American Association of Museums, 1992). These pressures have changed the content of exhibitions, collections, research, and publication, and the methodologies of cultural institutions such as universities, libraries, archives, and museums. New methodologies have almost always resulted in an expanded range of activities. For instance, many art museums have gone from being quiet temples of high culture to bustling community centers with public talks and activities oriented toward different sub-communities of age or ethnicity. However, new social demands and services often are not accommodated within the old infrastructure for collaboration.

Technology itself also has a way of driving collaboration. The Internet has been around since 1969, but it is only since the 1990s that the costs of implementation, broad user demand, and institutional capacity have converged to create the current abundant cultural Internet presence. Libraries that had an interest in sharing cataloging of duplicate items and had mainly textual information to share adopted the Internet at an early technological phase of text-only Gopher and other sites. Museums, relying more heavily on visual information, waited for the advent of the visually capable Web before adopting the Internet in numbers. It is no coincidence that most of the early large-scale museum collaboration efforts involving new media and networks were oriented around material collections rather than the equally typical museum areas of exhibition or education (e.g., Museum Educational Site Licensing Project [MESL, 1995–1999], http://rmc.library. cornell.edu/MESLatCU/Default.htm; Museums and Online Archives Collaboration [MOAC, 1997–], http://www.bampfa.berkeley.edu/moac; Art Museums Image Consortium [AMICO, 1997–2005], http://www.amico. net; Colorado Digitization Project [CDP, 1998–], http://www.cdpheritage. org). Online union databases of aggregated collections take advantage of the low-hanging fruit represented by the rare structured digital information found in most museums early on—the collection management database. While museums have shared collections information before the Internet, there is no catalog raisonné large enough to represent the entire collections of several museums, much less newly expanded educational and scholarly content. Here, new infrastructure enabled, and perhaps even suggested, new services.

Another important factor affecting the future of cultural collaboration is the rise of interdisciplinary research on university campuses and the increasing expectation of integration. The latter stems from the public whose appetite has been whetted by the apparent integration of content from almost every field via the Internet search engine. This integration factor has created the opportunity, if not the need, for cultural agents to collaborate in new ways. Integration necessarily stretches across those previous tightly defined professional communities and beyond the narrow confines of the specialized collaborative infrastructure used by each.

The Way Ahead

New cultural networks strive not only to solve challenges issued by society or technology, but also to identify and seize new opportunities as well. Unlike libraries, museums had no precedent for comprehensive, online catalogs of their material collections. Before 1990, a museum visitor would not expect to find a card catalog in the lobby describing the nature and location of the museum collection in storage. As museums begin to provide collections catalogs online, whether individually or collaboratively, they are not simply finding new technological tools with which to provide an old service. Instead, museums recognize that typically 90 percent of their collection is not on view in galleries, yet there is value in providing access beyond exhibitions. Collections online may provide direct use for researchers, and may be serve as a base resource upon which to build other new educational or commercial services.

Institutional Web sites and online collection catalogs are just the beginning. The expanding reach and pervasiveness of new media in educational and everyday life allows cultural agents to envision cultural information seeping into and spreading across a global network, creating an information resource comprising nothing less than the living record of civilization.

CHALLENGES TO COLLABORATION

While collaboration is not new to cultural institutions, the vision of a cultural information network is—and there are few well-researched models regarding its development, use, or economic and technological sustainability. Every pioneer endeavor is accompanied by risk. Building an open global cultural network is an effort that is as daunting and inspiring as building the Hoover dam, mapping the human genome, or going to the Moon, and with the proper balance of optimism, clarity, and inclusive collaboration this impossible task too will be accomplished.

Most previous projects and reports on cultural network building are glowing with positive results, but all knowledge contributes to further progress, including knowledge gained from failure. In order to build the network, cultural funders and professional peer communities must encourage early pioneers to report on the failures and dead ends as well as the shining successes. The following points of discussion provide signposts with which to navigate this new territory and illustrate the challenges museums face when collaborating.

Traffic Flow

One of the foremost challenges in developing any network (cultural, technical, or otherwise) is designing the shape, configuration, and scalability of the network. Will it be highly centralized, or decentralized? Will traffic flow in one direction, toward a center for consumption? Will traffic be routed back out to nodes? Or will traffic flow in any direction in a peer-to-peer environment? A highly centralized cultural network might be one giant national database to which individual institutions contribute content about collections and activities where it becomes available to the user in aggregate form. Such integration would be useful, but the system's sensitivity to unique material and flexibility for differing regions and users goes down in inverse proportion to integration (think of Google). It is difficult to imagine the impossible logistics of such a monolithic database that could accommodate every type of content and make it accessible, meaningful, and useful to every type of user. The United States in particular does not have the cultural or economic model under which to build such a centralized system. A highly decentralized network is easier to imagine because it already exists as a kind of default; each institution builds its own online presence without much regard to structural integration with other institutions. In fact, this can barely be called a network in all but the loosest sense; it is more a multitude of discrete resources. Here users or participants must seek out each institution separately, searching through hundreds of different interfaces and collating thousands of results. An ideal cultural network would seem to exist somewhere between the two extremes of entirely centralized and decentralized. It should empower individuals and smaller institutions as well as larger hubs of content and traffic.

At the turn of the 21st century, much important work has been done to explore the nature and scale of national-level cultural content aggregation, or gateway projects, sometimes organized by region (Museums and Online Archives Collaboration, or MOAC, and the Colorado Digitization Project) or by subject (ArtStor and the Art Museums Image Consortium, or AMICO). Such projects have encouraged museums, archives, and libraries to agree on common standards for submission of content to a central common database and have tackled related selection, use, workflow, and legal issues. The implicit collaboration or network model of most of these projects has been a centralized hub that draws upon various orbiting institutions. These projects have developed one set of best practices and standardized "languages" that enable institutions to send their content inward toward that center.

However, museums, libraries, and archives have a vital interest in sharing content broadly with multiple content aggregation sites, each of which may add value in different ways. For instance, each may have a different content focus, place the institution's content in a different context, reach different constituent users, or provide different interfaces and functionality on top of the content.

In comparison to "content aggregation" projects, there is a dearth of collaborative projects that focus on the inverse traffic flow of "content dissemination," where institutions disseminate their content outward to a multitude of content aggregation projects. These projects would require a slightly different emphasis that stresses tools and best practices to enable institutions to speak many standardized "languages" as necessary to collaborate and contribute to numerous larger projects. In fact, all but the very largest museums or research libraries cannot afford to actively participate in more than one or two national content aggregation projects. Individual artists, scholars, museums, and libraries do not have cost-effective systematic means of disseminating their content beyond their own Web site. The best practices, tools, and infrastructure to allow traffic to flow in this direction currently await development (the Open Archives Initiative Metadata Harvesting Protocol, http://www.openarchives.org/OAI/openarchivesprotocol. html, may provide a refreshing exception). As a result, content aggregation projects usually reach an artificial limit of 10–50 participants before the most resource-rich or active institutional partners are tapped out and the overall growth of the cultural network is slowed. In addition to the important work already done on how to aggregate content, we need cultural funders and leaders to sponsor projects that investigate and test how individual artists, scholars, museums, and libraries can best disseminate cultural information in cost-effective systematic ways.

It would seem that the ideal cultural network could accommodate traffic flowing in all directions, allowing content to be brought together or disseminated at any level between the individual artist or scholar, to the institution, to the multi-institution super-sites and beyond. So far the emphasis has been on relieving the individual or smaller institution of the burden of that traffic. The emphasis has been on how to create cultural networks that scale up, without as much attention to how they may scale down without fragmenting. Directing traffic at very small scales as well as very large might produce more thriving, diverse, and, ultimately, more robust cultural networks.

Domain Differences

Cultural networks that cross discipline domains or professional community boundaries create additional value, but they also face additional challenges. Museums, libraries, and archives (not to mention individual scholars and artists) have related but distinct developmental histories that have resulted in numerous differences in formalized practice and internal professional culture. Such differences can be easily glossed over when speaking of "cultural networks" or "cultural heritage," but these same differences can rise up as serious obstacles before the vision of integrated cultural information unless they are recognized and confronted.

One significant difference between museums, libraries, and archives is in the area of descriptive practice—how they describe their collections of cultural materials and the kinds of records or documents they produce and preserve. For instance, libraries typically employ cataloging standards to describe mass-produced, published books or journals for public access. Since a key public access question would be what the book is about, libraries have become expert at describing subject—what a thing is about. Museums typically describe unique physical objects for management as much as public access. Museums have become expert at describing the "ding an sich" or the "thing unto itself," including the properties that define unique objects such as their size, material, purpose, maker, unique marks, etc. Archives collect in aggregate, often a collection of papers produced by a person or organization over a lifetime. Archives have become expert at describing relationships between items as well as the history of their ownership, or provenance. So, all of these institutions collect "cultural materials," but they describe them differently. When they aggregate or integrate their respective records of those collections, some descriptions emphasize what the thing is about, others emphasize what the thing is, and yet others emphasize where the thing is from and what it's related to. To be more specific, such a fundamental library descriptive mainstay as subject may be implemented very differently by museums. After all, what is the subject of an abstract painting?

Individual scholars and artists also produce and share cultural content. Scholars produce content that is preserved by libraries, but they also often question methodologies and classification boundaries. Artists question the practice of museums by inventing new art forms like performance, installation, conceptual, and digital art that defy traditional object-centric museum description and bring into question the ideal of systematic descriptive practice. The challenges these art forms present museums are discussed in the project *Archiving the Avant-Garde: Documenting and Preserving Digital/ Media Art* (http://www.bampfa.berkeley.edu/ciao/avant_garde.html).

Cultural institutions also differ in their access models. For instance, libraries usually provide a catalog of their entire collection, along with directions on where to find and use the item. Their access systems are oriented toward self-service and basic description. Museums, however, usually provide a carefully selected or curated group of items in an exhibition. Their access systems, often physical wall labels and brochures, are highly interpretive and rich. Access to items not on exhibition is generally not provided. Would an integrated online resource privilege the unmediated library access or the mediated museum presentation?

Although museums, libraries, and other cultural organizations are typically not-for-profit organizations, the subtler aspects of their respective economic models also differ. For instance, there are many more libraries than museums in the United States, creating differences in scale of funding and collaboration (American Association of Museums, n.d.; American Library Association, 2005). Libraries typically receive a larger portion of their operating budget from the parent university or city than do museums. This latter fact may stem from the perception in the United States that libraries are directly tied to public education and democracy (Benjamin Franklin is the most widely known advocate of free public libraries supporting democracy). Museums, however, are often perceived as optional leisure-time activities and so operate somewhat more in the leisure-time economy, competing with sports and cinema for private dollars. Differences in economic base partially explain differences in operational priorities or incentives to collaboration.

One strategy for mitigating the obstacle of professional community differences is to map between different community practices and find points of commonality from which to build cultural collaborations. For instance, although museum and library records may look different, they share the common concept of an artist, author, or creator. Creator could become a key access point in a shared online database of collections, a way for users to find records from both museums and libraries. However, once found, the museum and library records could still retain their differences "under the hood." A search for a creator could reveal library records that go on to describe what that creator wrote about, while the museum records describe the things that creator made. Although simplistic, this example illustrates a strategy wherein diversity of community practices may be maintained while still allowing integration. Such integration has been the goal of such projects and standards as the Dublin Core (http://www.dublincore.org) and REACH (http://www.rlg.org/reach.html). For a more detailed example, the Online Archive of California (OAC: http://oac.cdlib.org), a content aggregation site, adopted the Encoded Archival Description (EAD: http://www.loc.gov/ead) as their common standard and wrote a best practices document for archives submitting content. The MOAC project produced a second best practices document for museums submitting content to the OAC in EAD format. The two best practices documents overlapped by about 90 percent, thus ensuring mostly consistent implementation of EAD and many common access points, while allowing for differences in descriptive practice between museums and archives. This strategy could be taken much further if careful attention is paid to where the breaking point occurs (the percentage where descriptive practice differs so much that the system is rendered useless to confused users).

Integration on a cultural network need not imply homogeneity that paves over difference, but could instead take the form of compatibility between two systems at certain points if not along the whole spectrum. Creating this kind of compatibility-with-difference requires members of each community to agree on a set of common points that are meaningful to community members and to their intended audiences.

The Promise and Cost of Standards

Building large-scale cultural networks requires the adoption of standards at many levels. Technical standards that include markup (HTML, XML), text encoding (ASCII, Unicode), and media file formats (JPEG, TIFF) allow digital content to be produced and used across heterogeneous computing environments. Community-specific metadata standards such as Encoded Archival Description, Metadata Encoding and Transmission Standard (http://www.loc.gov/standards/mets), and content standards such as the Art and Architecture Thesaurus (http://www.getty.edu/research/conducting_research/vocabularies/aat/) enable some consistency in producing and sharing cultural content. However, many of these standards are too complex for ready adoption by the average cultural institution, not to mention the average scholar or artist. Such standards are necessarily the product of consensus and are often led by the largest cultural institutions or umbrella organizations that are not themselves responsible for implementing the standard. This leads to standards that are elegant and sophisticated and come with a high learning curve and cost of implementation. To illustrate, the museum community had been developing metadata standards for over a decade as of 2004. Yet counting the participating partners in large-scale content aggregation and metadata development projects, it appears that about 100 museums, out of approximately 17,500 total in the United States (American Association of Museums, n.d.), were implementing metadata standards. Simpler technical standards have been adopted by museums in much greater number. It was estimated in 2002 that 62 percent of all U.S. museums had a Web site (Institute of Museum and Library Services, 2002) created using the HTML standard. HTML is much simpler and more flexible than the aforementioned XML-based metadata standards, entailing inexpensive off-the-shelf software, plentiful training opportunities, and a low learning curve.

Clearly, cultural networks are in an early stage of development, and we should not expect too much too soon. It is also true that standards like HTML provide only the most basic level of infrastructure for building cultural networks. However, it is apparent that when the benefit is compelling and the cost low, even cultural organizations of small or average size and capacity will implement the necessary standards.

One way to leverage the promise of standards while lowering the cost of implementation is to develop simpler standards. Many current metadata standards include features that are difficult to implement and are often not even utilized in the end result. Cultural standards development must be better informed from the point of view of the small to average institution or even the individual scholar or artist. Another approach to simplification would be to develop two flavors of every standard, a full version and a light version that is much easier to produce and use, but does not include all of the functionality of the full version. The two versions should be formalized, documented, and tested separately, but an institution could adopt the light version early on, and scale up to the full version in time. The Text Encoding Initiative (TEI) developed a "lite" version of the TEI standard (http://www.tei-c.org/Lite/U5-pref.html), though TEI Lite is still so complex to implement it needs a TEI Lite Lite. One metric for simpler standards might be training time. If one could adequately train a non-technical cultural professional (librarian, museum professional, professor) to use a given standard in one afternoon training session (as with HTML), it might indicate the proper threshold for complexity. If a content organization standard like Dublin Core were as easy to implement as HTML, and a content dissemination tool like OAI or Z39.50 software were as easy to implement as a regular HTTP Web server, we might see much greater participation in building a global cultural network. It is more challenging to make standards simple than to make them complex, and one should by no means gloss over the complexities and trade-offs involved in making the above examples become fact. However, such a feat is feasible and it would help progress building the cultural network by staking the middle ground between high-participation, low-functionality systems like HTML/HTTP/Google and low-participation, high-functionality systems like most current cultural content aggregation portals.

An additional impediment to adoption of cultural standards is the lack of easy-to-use, inexpensive software tools. Although many cultural metadata standards are built upon technical standards like XML, generic XML tools are not sufficient to produce the actual metadata standard. Those tools that are developed by cultural institutions for this purpose are not yet easy to find or share. Two projects attempting to address this problem are MOAC with the Digital Asset Management Database tool (http://www.bampfa.berkeley.edu/moac/damd/DAMD_manual.pdf) and Pachyderm with the Pachyderm tool (http://www.nmc.org/projects/lo/pachyderm.shtml). There is need for a "cultural toolbox" node on the cultural network where institutions that develop specialized products can readily share them with other institutions legally and openly.

Another strategy for simplifying cultural standards development would be to adapt computer industry standards rather than invent new ones. Adaptation of industry standards leverages the work of expert professional communities as well as the availability of tools, training, and documentation. Adaptation of industry standards need not mean slavish devotion to every nuance of the standard as-is, but could include creating a community-specific implementation through the consensus of cultural expertise, including small institutions. Community-specific implementations of standards would support the principle of compatibility-with-difference and could allow diverse practices that include points of convergence between industry and multiple cultural communities including museums, libraries, archives, artists, and scholars.

Standards development always includes a balance of elegance and functionality with simplicity and cost. In order to build thriving and diverse large-scale cultural networks, it may be preferable to incentivize great numbers of cultural agents and institutions to adopt simple standards rather than have relatively few implementing complex standards.

Business Models

The phrase "business model" leads some to connote that cultural networks need necessarily be structured like commercial enterprises. However, "business" here does not mean "commerce"; rather having a business model for cultural networks or projects simply means considering the long-term funding, management, utility, and evolution of the network or project (Bishoff & Allen, 2004). All long-term cultural efforts will need a business model and models may be drawn from the world of commerce or the public sector equally. Standard Oil had a business model, as did the WPA.

Every cultural network must account for and allow diverse business models. Large-scale national cultural networks will necessarily include institutions and projects with different goals and structures, from the public non-profit (i.e., MOAC or CDP) to commercial, and every shade in between (i.e., Art Museum Network, http://www.amn.org; Fathom, http://www.fathom.com/). However, even smaller cultural networks such as might exist between several departments on one university campus (i.e., the UC Berkeley Center for New Media, http://cnm.berkeley.edu) must account for the fact that each participating scholar or department may also participate in other networks under different models.

One way to ensure this diversity is to develop cultural projects that are never exclusive. Cultural projects need not limit a scholar's or institution's content to one purpose or outlet. Instead, projects that hope to create value out of cultural content can do so by adding value in the form of unique content combinations of focus, superior support, or unusual functionality or availability. It is important to design cultural network projects so that they also do not become de facto exclusive because participation is so costly as to prohibit the agent or institution to participate elsewhere.

Cultural institutions and content typically have long-term goals and long-term value. Business models, even those of the semi-commercial flavor, for cultural networks should reflect this long-range vision by addressing sustainability and scalability in ways that may depart from purely commercial models. Cultural networks are simultaneously networks of people, activities, knowledge, content, and values and should be designed with all of these elements in mind.

THE MOAC CASE STUDY

Public museums, libraries, and archives provide access to and disseminate information about their collections for purposes of education, research, and engagement. When information about collections is in digital form (such as digital images and descriptions), cultural institutions can allow instances or copies of the digital content to occur in multiple, co-existing online venues—starting with an intranet and public Web site and extending to one or more content aggregation projects or union databases, some or all of which may be exposed to external search engines. Such saturated access and possibly redundant dissemination may be a boon for people interested (knowingly or unknowingly) in the digital content shared by cultural institutions.

Once found, however, does the digital content selected, described, and presented by cultural institutions support education and research or encourage engagement?

The Museums and Online Archives Collaboration (MOAC) case study describes an ongoing, multi-year collaboration between museums, libraries, and archives to increase and enhance access to cultural content, primarily delivered online but intended to encourage access onsite. Partners in this collaborative project have produced and shared digital content about their collections on the Web as well as examined the value of this content and its presentation for certain audiences. Examples in this case study are drawn mainly from the experiences of museum partners in the collaboration.

MOAC was launched formally as a partnership in 1995 (Rinehart, 2003). MOAC progresses through the steady work of a steering committee comprising individuals representing partner institutions.[1] Steering committee members have diverse job positions—collection managers, digital media professionals, curators, educators, archivists—and represent museums of art, anthropology, and cultural history and special collections within libraries. Many committee members have professional backgrounds in museums and libraries; backgrounds which contribute, in some way, to a shared awareness of how and why to provide online access to cultural content. The objects collected and contributed by partners cross over in types, with museums and libraries, for example, sharing content representing paintings, prints, photographs, artists' books and illustrated books, and functional objects made in various places and time periods.

Several partner institutions, including a digital library, are associated with universities and have lengthy experiences sustaining large-scale public access projects. All MOAC partners bring to the collaboration varying levels of experience in content, digitization, standards, technology, and audiences. All partners draw from this shared experience to a greater or lesser extent, thus increasing their capabilities for content production, management, and delivery. Participating in this project with these partners can help cultural institutions with limited resources share content online and can increase the dissemination and exposure reach of the most active institutions. MOAC is intended to be an open, non-hierarchically structured partnership: new partners are sought, spin-off projects are developed as needs arise, and all partners are encouraged to participate in MOAC to the extent that its activities complement their institutions' goals for access and dissemination. Each partner institution is engaged in collection-sharing projects that exist beyond MOAC and, in fact, those projects may be a higher priority for the institutions.

To date, MOAC partners have collaborated in three formal projects, each of which received funding from the Institute of Museum and Library Services as a National Leadership Grant. In MOAC Classic (1997–ongoing) partners developed and delivered on the Web a testbed of digital content about their collections. In MOAC II User Evaluation (2002–2005) partners examined audience needs and behaviors using this testbed. For MOAC Community Toolbox (2004–2006) partners developed a digital asset management tool for delivering content to multiple, co-existing online venues. The following focuses mainly on MOAC Classic and MOAC II User Evaluation as complementary studies of collaboration as a model for building and sustaining cultural information networks.

MOAC CLASSIC

The purpose of MOAC Classic is to increase and enhance access to museum, library, and archive collections, primarily through one or more online venues. Toward this purpose, MOAC partners produced digital content representing their varied, geographically distributed physical collections and contributed that content to the Online Archive of California (OAC), a considerably sized content aggregation portal operated since 1998 by the California Digital Library (CDL) at the University of California (Chandler, 2002). The CDL is a central partner in MOAC, with the CDL's Robin Chandler serving as the MOAC liaison. The CDL brings to the collaboration an established infrastructure for delivering content online, in exchange for content. MOAC museum and library partners are involved in OAC working groups and other community advisory committees collaborating to develop best practices for producing and delivering digital content.

Partners make digital images and descriptions of collections available for searching in the OAC and other content aggregation projects (such as RLG Cultural Materials and post-AMICO digital libraries) because, among other reasons, exposure may increase the discovery and use of digital and physical collections by audiences such as educators, students, and researchers. Partners assumed people would find access to aggregated cultural content to be useful in that, for example, the potential for discovering content might be greater if digital objects from multiple repositories were pooled easily in one Web site. Disseminating digital content in one or more venues requires management of workflows, copyright clearance, and quality control, to name but a few actions requiring resources (as discussed above). Mindful of resource expenditure, museums (for instance) consider carefully which online venues are appropriate for meeting goals of access and dissemination, just as museums decide which venues are appropriate for receiving loans of objects from permanent collections.

MOAC partners viewed and continue to understand the OAC as an established, sustainable, and appropriate venue for publishing content. The OAC has established guidelines, procedures, and an infrastructure for receiving and publishing content, which means that MOAC partners did not need to start from scratch. Unlike managing local information systems, partners are not immersed in the day-to-day maintenance of the OAC's technical infrastructure for content ingest, indexing, and display. Partners understand the OAC to be sustainable because the CDL is committed to developing and preserving the digital content available in the OAC, which may reassure providers and users that the content will be accessible over time. Finally, the OAC is viewed as an appropriate venue because MOAC partners and the OAC share an interest in having content exposed and used in education, research, and engagement by known and potential users—from the K–12 constituency to experienced researchers—and, as a considerably sized content aggregation portal, the OAC offers one opportunity for users to find and view representations of geographically distributed physical objects.

Standards for Data Sharing

Testing community-based data standards is critical to MOAC's purpose of increasing and enhancing access by, but not limited to, contributing and aggregating content. MOAC partners' choice of standards for providing online access differed dramatically from museum access projects in the 1990s. Partners tested the feasibility of packaging museum content into finding aids using Encoded Archival Description (EAD) as a common data structure standard in order to bring together digital content from museums, libraries, and archives in the OAC.[2] All contributions to the OAC are packaged in EAD thus far and at some level. To date, MOAC partners have contributed a manageable 27 EAD finding aids representing their collections and illustrated with 75,000 digital images. In addition to MOAC contributions the OAC makes over 7,000 finding aids (some with images and some without) from variously sized and specialized California libraries and archives freely available for searching—either singly or collectively—from one Web site (www.oac.cdlib.org).

Using EAD finding aids challenged conventional packaging of museum information by museum professionals because, for one, finding aids function as a tool for archivists to describe archival collections. Furthermore, traditional finding aids have been described as being written for archivists and academics and of limited use to teachers and students (Gilliland-Swetland, 2001). This is a presumed disadvantage for MOAC partners, who see educators and students as well as information professionals and academics as key constituencies.

A finding aid is a model for documenting the context and contents of a collection, such as a collection of stereographs compiled by an individual. Finding aids model the archival practice of organizing items into physical and intellectual arrangements inherited from depositors or previous owners or purposefully imposed by the archivist. MOAC museum partners had little experience packaging content into finding aids. Exhibitions more typically model the museum practice of placing cultural objects into meaningful but often ephemeral physical and intellectual arrangements, with objects selected and placed in exhibitions based on curatorial interests in artists, movements, concepts, and so forth. Past exhibitions, in fact, provided some

MOAC museum partners with a logical source of digital content for repurposing as contributions to the OAC, given that the preliminaries of object selection, research, contextual arrangement, copyright clearance, and photography were completed for the exhibitions. In this case, the concept of a "collection" of objects is understood and used differently by museums and archives.

A finding aid can be an index, inventory, register, or collection guide. MOAC museum partners determined that they would contribute collection guides to the OAC because the concept of a collection guide approximates more closely the museum practice of describing groups of related objects. (The remainder of this case study uses the terms "finding aid" and "collection guide" interchangeably.) For example, a MOAC partner contributed a collection guide for a body of Old Master prints previously exhibited by the partner institution. The collection guide begins with a general description of the scope and arrangement of the "collection," followed by series-level descriptions of logical, imposed groupings of prints within the collection (in this case grouped by culture). Lastly, and more analogous to standard museum practice, this collection guide includes detailed object-level descriptions of each engraving, etching, and mezzotint in the Old Master prints collection, ordered chronologically by artist life dates below the relevant series-level descriptions. Together, the collection-, series-, and object-level descriptions may help communicate context to users about the contents of this "collection." Some, but not all, collection guides and finding aids further describe collections by linking digital images of objects within the collection, as does the Old Master prints collection guide. Legible digital images are highly desirable pieces of information for users trying to determine the usefulness of cultural content for current needs. The MOAC project provided an opportunity to test how well EAD could manage linking to image files.

Using EAD finding aids posed further challenges for partners because EAD is a data structure standard developed by and for the archival community, principally as a means to standardize and move the finding aid model into a Web-accessible format. EAD is an eXtensible Markup Language Document Type Definition (XML DTD) comprising data elements selected by the archival community to represent finding aids, along with a set of technical definitions for structuring and encoding those elements for exchange and integration on the Web (Dooley, 1998; Pitti & Duff, 2001). The Society of American Archivists and the Library of Congress's Network Development and MARC Standards Office are chiefly responsible for EAD, including maintenance of the EAD Tag Library, which names and defines the data elements. Some EAD elements are defined broadly and the definitions do not prescribe precisely how to enter data into the elements. This flexibility has allowed implementers—such as MOAC partners—to use EAD in conjunction with data content and data value standards befitting their professional and institutional practices and needs.[3]

Best Practices: Business Models, Domain Differences, and Traffic Flow

Given the resources required to manage and monitor collection-sharing projects, MOAC partners needed relatively stable standards and stable online venues if they were to participate. In the late 1990s there were few data structure standards commonly used by museums, libraries, and archives in the United States (and few cross-community content aggregation projects). Museum partners in MOAC were interested in testing EAD as a common standard because it was in use by libraries and archives, appeared flexible enough to accommodate museum descriptive practices, and is a non-proprietary, platform-independent standard. In the mid-1990s and immediately prior to the MOAC project, BAM/PFA and the Bancroft Library tested EAD for collections comprised of museum-like objects such as paintings, conceptual art, and functional objects (Rinehart, 2003; Rinehart, Elings, & Garcelon, 1998). MOAC partners subsequently added to this work by having more museum professionals test EAD against a larger group of museum objects.

Early in the MOAC project, partners reached consensus on best practice guidelines for packaging descriptions of collections and objects into EAD finding aids for contribution to the OAC (as discussed above). MOAC guidelines had to concur with museum and library practices (because, for instance, museum partners were not willing to manipulate descriptions for use with EAD) and work within the OAC's expectations for receiving, validating, and publishing thousands of contributions.

MOAC best practices were based on the OAC Best Practice Guidelines for Encoded Archival Description. The OAC guidelines provide a baseline for all contributions to the OAC and are updated as needed by a metadata working group, which includes representation from MOAC, thereby acknowledging a museum perspective. MOAC partners further examined and aligned EAD data elements for object-level descriptions to standards that more adequately represent museum practices: for instance, partners used elements defined in REACH (Record Export for Art and Cultural Heritage), VRA Core Categories, and the Categories for the Description of Works of Art. While partners agreed rather easily on the core set of elements and their definitions, partners were challenged to reach complete agreement on how to fill the elements with data; challenged because partners must retain significant control over how they express data about their collections.

Museum partners, for instance, have local rules for expressing data and were loath to modify data for one project. Partners wanted to export core object descriptions from local collections management systems for use in MOAC, just as they do for other collection-sharing projects. In the end, the MOAC guidelines outline general expectations for using the selected data elements and do not preclude use of local and community standards, such as the more recently published data content standard, Cataloguing Cultural Objects. Implementing local and community standards in a collections management system can promote consistency and predictability across object records in addition to staff departures and arrivals. However, implementing standards consistently across curatorial departments within a single institution, let alone across multiple institutions and projects, can be challenging (and the absolute necessity of reconciling differences might be debated, especially in humanities collections, where opinions and nuances are the norm).

Decisions about descriptions and terminology may depend on who is describing what, in that decisions are influenced by the collecting and sharing goals of an institution, its audiences, and the professional practices of staff. For example, an art museum curator may consider and describe the materials, style, and subjects of a painting differently than an archivist or information professional in a history museum. The "describers" may have similarly named data elements in their local information systems—for instance, Material, Technique, Style, Period, Subject—but each professional may fill those elements differently and with different data: for example, a curator might describe style and period at a very exacting level; an archivist might emphasize an object's function over its style. Differences in expression and usage of data elements may pose challenges for content aggregators (like the OAC) when attempting to aggregate—with assurance—object records based on common data elements. It may be that content aggregators will develop underlying authority files for terms and concepts so that connections can be made transparently between contributions. MOAC continues to provide a testbed for assessing the capabilities and value of aggregating records for multiple object types, collected and described by different cultural institutions.

MOAC partners produced collection guides decentrally, and then delivered text and image files to the OAC or delivered text files with embedded links to image files stored on partner institution servers. The OAC loads and indexes contributed data in a centralized database, from which the MOAC content is published on the OAC Web site as one "virtual collection." Nearly all museum partners created collection guides from scratch; that is, partners did not convert legacy, paper-based collection guides to EAD. Each partner managed independently staffing requirements, budgets, content selection, extent of description, image production, and copyright permissions. For example, most partners used existing equipment and expertise to produce digital images in-house and to host images on local servers. A minority outsourced imaging and delivered access images to the OAC along with text files. Partners retain components of their collection guides—e.g., collection-, series-, and object-level descriptions and master images—in local systems and servers for use in other collection-sharing projects.

Partners control how data is expressed, when pushed out from local systems and into the EAD collection guides. However, in contributing to a content aggregation project with brand recognition like the OAC, partners give up direct control over the look of the product displayed to users in the

OAC Web site (such as the layout of data and the fonts and colors used). Looks matter to museums (and artists/makers and their estates), therefore museums and other content providers may consider interface design when determining the appropriateness of a venue for sharing.

Evolving Access: New Tools for Collaboration

The means by which partners continue working toward the goal of increasing and enhancing access have evolved since the project's inception in 1997. In addition to EAD, MOAC partners explored use of the Making of America 2 (MOA2) standard—a Digital Library Federation project—to structure and present multiple digital images and associated metadata for complex objects, such as artists' books and albums, included in selected collection guides. Partners linked MOA2 encoded files for complex objects to their associated EAD collection guides, with the MOA2 and EAD files displayed to users in separate but linked windows and interfaces. In 2002 MOAC and the OAC switched from MOA2 to the Metadata Encoding and Transmission Standard (METS), a digital object standard that succeeded MOA2 for complex objects and is in wider use. METS can be used to group and organize structural and technical metadata along with media files for digital objects into one record, which can point to other data structure standards for descriptive metadata (e.g., EAD and Dublin Core).

METS changed significantly the way users find, view and perhaps use all MOAC digital content, not just complex objects. In late 2002 CDL borrowed, but did not strip, simple digital images (i.e., a single image representing a "simple" physical object, such as a painting) and associated object-level descriptions from EAD collection guides and packaged those images and metadata into METS digital objects. The lot was then moved into a separate METS digital object repository, allowing the digital objects to be searched independently of collection guides or finding aids. By 2003 the OAC had a collection of several thousand EAD finding aids (some with images, many without) and a collection of many more thousand METS digital objects derived from the finding aids, with both available for searching and browsing in the OAC, in slightly different interfaces. More recently the OAC communicated intentions to receive single METS digital objects (still, with associated finding aids) directly from content contributors.

Prior to this, there was no easy way for users to combine and view images and metadata from different finding aids in the OAC. Digital images were accessible solely by combing through one finding aid at a time. Packaging digital objects separately from finding aids (while maintaining access to the finding aids) meant that OAC users could now view digital objects singly or mixed with objects from other collections or, as in the past, users could view objects within the greater context of a finding aid. Having set out to test the appropriateness of EAD for museums, MOAC partners made the logical progression to using METS for packaging images and associated metadata.

The latter is typical of the museum practice of producing and highlighting access to digital images and object-centric descriptions. The adoption of METS provides an example of how participating in a content aggregation project can help institutions with limited resources see their contributions advance to new data standards (assuming that is desirable) and, as in this example, tested on a large scale by an established, sustainable, and appropriate venue.

In 2003 the CDL implemented a content management system that supported improved management of digital objects by the CDL and allowed content contributors and data harvesters more direct access to the digital object repository. MOAC-only digital objects can be drawn from the repository, removed from their respective finding aids, and presented in multiple online venues. For example, BAM/PFA worked with CDL staff to implement a customized stylesheet that queries the repository for MOAC-only contributions, then makes MOAC images and EAD object-level descriptions (removed from the respective finding aids) available for searching and viewing in the MOAC project Web site hosted by BAM/PFA (www.bampfa.berkeley.edu/moac). Access to MOAC content is featured prominently on the MOAC Web site. The site also serves as a central location for project reports and technical specifications, which may be of limited interest (and perhaps confusing) to the public, but is of use to partners and cultural heritage professionals (including funders).

MOAC digital objects have been picked up as a collective product from the repository and made accessible in a CDL initiative testing image presentation software for classroom instruction and research use (Farley, 2004). The CDL also made MOAC and other OAC digital resources available for harvesting—using the Open Archives Initiative Protocol for Metadata Harvesting—by, for example, the OAIster project of University of Michigan Digital Library Production Services. MOAC content is now accessible from multiple online venues and each point of access has a distinct look, which may change asynchronously over time. Of the venues, the OAC provides access to all available digital objects from all OAC contributors (via an Image search), as well as access to the potentially greater context derived from aggregating the thousands-strong EAD finding aids (via a Finding Aids search). The MOAC Web site, on the other hand, aggregates and displays discrete digital objects (simple or complex METS digital objects, with accompanying EAD object-level records) contributed by MOAC partners to the OAC, an arrangement which may allow partners to examine more closely whether they use EAD data elements for objects in the same way and for the same purpose (it is assumed that sometimes they do and sometimes they do not) and address whether digital objects from different collections can in fact be pooled easily by users.

MOAC Classic partners collaborated in producing and delivering content to the extent that they agreed on best practice guidelines for implementing EAD for museum collections, provided feedback on OAC policies and initiatives (such as copyright, changes to the interface, adoption of standards like METS, updating OAC EAD best practice guidelines), and tested digital asset management tools for content preparation. MOAC partners did not devise an agreed-upon list of specific content to share in the OAC: for example, materials by or related to a particular maker or representing certain subject matter. (In the past, partners have discussed the possibility of identifying specific content of potential value to teachers to meet state curricular standards.) Nor did museum partners intend to package their entire holdings into EAD finding aids. Each partner decided which context and objects to share based on their institution's priorities and sensibilities, such as the perceived significance of a collection or objects to the institution and its audiences. The result is that MOAC finding aids are fundamentally stand-alone packages, which may be seen as a hodgepodge of available, even potentially related content, or seen as informative from a historical or institutional perspective.

Each partner determined the readiness of information about collections and objects to be pushed from local systems and published in the OAC as finding aids. Because partners share core information about objects in other public access projects, it is important (for purposes of consistency and trust) that the information not be changed for a given project, including MOAC.

The MOAC Community Toolbox

MOAC partners sought to implement cost-effective mechanisms by which to gather, package, and contribute images and metadata to content aggregation projects like the OAC because, for instance, implementing standards such as EAD may not be simple or routine for museums. To this end, BAM/PFA developed and distributed a digital asset management tool that some partners used to produce EAD and METS encoded files from data imported mainly from collections management systems and image files. BAM/PFA wrote scripts for the tool to facilitate export of files compliant with MOAC guidelines for EAD and METS, guidelines that are generally in sync with OAC best practice guidelines. The ease of conversion and export offered by the tool meant that MOAC partners did not need in-depth, hands-on experience with EAD. A general understanding of EAD (as well as METS) was sufficient for the purposes of the project because, for example, no museum partner expected to produce EAD finding aids for their entire holdings. The OAC has also made available to all content contributors an EAD Toolkit (www.cdlib.org/inside/projects/oac/toolkit) that includes resources and self-service tools to, among other things, help contributors create and validate finding aids against OAC EAD best practice guidelines. Use of self-service validation tools implies that a contributor must know enough about EAD to understand why a document is invalid if they are to fix the problem.

A simple, effective tool was desired by partners because the museum professionals that select and describe objects may or may not be responsible for or interested in managing the technical end of pushing content out of local systems and into packages for distribution to other online venues. The MOAC tool freed partners from learning to encode documents in EAD or METS, allowing partners to concentrate on building content. In its first iteration, the digital asset management tool exported data to standards used only in the MOAC project, was distributed with limited technical support, and was unforgiving in that a strict ingest methodology was required if the scripts were to work. Partners will improve, further develop, and share this tool in the MOAC Community Toolbox project.

MOAC II USER EVALUATION

MOAC partners are motivated, in part, to contribute to the OAC and other content aggregation projects because they want their content to be seen and used in education, research, and engagement. Decisions made by each partner about what content to share, how to describe and contextualize objects, and how to present that content and context online in venues like the OAC, provided a conceptual basis for a MOAC spin-off project. In MOAC II User Evaluation partners conducted a formal evaluation of the MOAC testbed to better understand the degrees to which partners' efforts to deliver resources—such as EAD finding aids and digital objects delivered independently of finding aids—on the Web are useful for education and research contexts in particular.[4]

For this project MOAC partners collaborated with researchers from the UCLA Graduate School of Education and Information Studies. The UCLA research team, led by Anne Gilliland,[5] conducted the evaluation using a triangulated approach similar to that used by Gilliland in a previous evaluation of the OAC (Gilliland-Swetland, 1998). UCLA researchers brought to MOAC II extensive experience in conceptualizing and executing user evaluations in addition to research expertise in EAD, digital content, and uses of primary sources in education and research contexts. The research team could generalize information emerging from MOAC II by connecting and comparing that information to past and contemporaneous evaluative work, thereby contributing to this area of research.

UCLA researchers and information technology staff established a MOAC II project Web site (www.gseis.ucla.edu/~moac) to disseminate information about the project and gather data from evaluation participants. The Web site included a private, shared workspace for MOAC II partners, featuring an online calendar and message board. Theoretically, a project with such geographically distant partners as MOAC (with partners located in Northern and Southern California) might benefit from an electronic workspace as a supplement to meetings; however, partners communicated largely through a MOAC email distribution list and engaged in a few face-to-face meetings.

As content producers and providers, museum and library partners offered considerable feedback on the MOAC II research methodology and assisted in recruiting participants and collecting data. MOAC partners brought an understanding of audiences and content to bear when vetting the research methodology—helping to ensure that the methodology was right for these institutions, this project, and the cultural heritage community. As practitioners, partners also had concerns about balancing what their institutions' perceived to be important to share (and how) with what targeted audiences wanted and expected in a content aggregation portal. Formal evaluation would provide a basis for MOAC II partners to consider and recommend strategies for improving the value and use of online resources like MOAC for education and research contexts.

While MOAC content may be of interest to people browsing online collections for entertainment or looking for images to use in commercial ventures, MOAC II partners limited the user groups investigated in the evaluation to key constituencies of the OAC and MOAC partner institutions: K–12 teachers, university students, academics in the humanities and social sciences, and information professionals working in museums, libraries, and archives. These constituencies may be motivated to find and use information about cultural objects in their work or learning activities and may be interested in accessing physical objects after discovering relevant images and descriptions online.

During the timeframe of the project, and notwithstanding exhaustive efforts, the MOAC II research team did not receive the number of potential evaluation participants anticipated and, of the individuals invited to participate from the resulting pool, few were willing to participate. Recruiting academics for the evaluation was a major challenge, with university students, in particular, being difficult to recruit because of human subjects constraints. The response rate was disappointing, but appeared to be in line with experiences of other researchers studying similar user constituencies.

The research team learned that participation in the study was the first exposure to MOAC as a resource for most MOAC II subjects. While recruitment might have been stymied in part by the typically hectic schedules of individuals in the targeted user constituencies, it may also have had to do with a potential participant's lack of awareness and investment in MOAC as a resource. Providing access may not be enough to have content exposed and used, especially if users are not aware of its availability or do not see how the content—as it is selected and presented—may relate to their needs or interests. Partners had steadily disseminated information about the MOAC collaboration and its projects to the museum, library, and archive communities. With few exceptions, however, it was clear that partners had not reached out sufficiently to intended users with news and demonstrations of MOAC's availability as a resource for learning and research. Partners need to broaden and/or refresh publicity about the availability and value of MOAC if they are to engage and collect feedback from users more routinely, just as institutions seek to engage and build relationships with onsite visitors.

MOAC II partners sought to discover why people use cultural objects and their metadata, what people need to understand about cultural objects and their metadata as sources of information, and how cultural objects (digital or physical) and their metadata might be used to enhance learning, teaching, and research activities. The research team planned initially to evaluate user needs and behaviors using the MOAC testbed positioned within the OAC. These plans changed, however, as the MOAC content and OAC evolved noticeably (but not very rapidly), with additions made to content (new images, new descriptions), changes made in packaging (adopting METS and opening a digital object repository), and redesigns in interfaces (OAC and MOAC Web sites).

Given the evolving nature of the MOAC testbed, the research team conducted a multifaceted evaluation, with the goal of obtaining a better understanding of what it means to provide access to and use information about cultural content and context in general, as well as examining aspects related specifically to MOAC. The research team gathered and triangulated quantitative and qualitative data from several sources, including pre-existing data, transaction logs, feedback forms, detailed questionnaires, and think-aloud sessions with selected participants. The research protocols were developed in phases, over two years, in an effort to keep the research instruments in step with MOAC content, interfaces, and points of access, and to incorporate any directions gleaned from preceding phases of the evaluation—for example, it was anticipated that data collected using the detailed questionnaires would affect the task-oriented think-aloud sessions.

The MOAC II research methodology is described elsewhere (Gilliland-Swetland, Chandler, & White, 2004); however, one aspect of data collection is particularly pertinent for this case study description. To get a picture of the usage of MOAC content, the research team asked partners to provide pre-existing statistical or qualitative data that suggested a potential correlation between accessing MOAC content via the Web and increased use of physical objects. Such data could help partners learn what, if any, impact the availability of the online resources in which they are investing is having upon other access services. Whereas library partners had this data at the ready, museum partners did not. This may be because museum professionals do not ordinarily track in a centralized manner how patrons learn about objects in their collections because many museum departments receive inquiries about objects. There is not a central point of inquiry, such as a reference desk, for museum collections. Responses to this request highlighted a difference between museums and libraries in their readiness to conduct ongoing evaluations of their digital content and services. Partners will publish a final report in 2006 for the collaborative three-year research project, including findings from the user evaluation. The following summarizes interim observations developed from exploring assumptions and experiences with all MOAC II partners, by listening to select subjects, and from complementary research (Duff, Craig, & Cherry, 2004; Harley et al., 2004).

MOAC partners' decisions about standards, selection, description, contextualization, and imaging are biased, in that decisions are often conditioned by curatorial and scholarly practices and informed by institutional resources and sensitivities. Partners chose to test EAD because, for one, EAD appeared to provide a means for partners to communicate some context to users about the contents of a collection—that is, a context identified or selected by partners at the time of contribution to the OAC. Museums recontextualize objects by moving them in and out of arrangements based on curatorial interests or scholarly reinterpretation. METS appeared to provide a means for users (including museum professionals) to gather and mix contributed objects with other content (such as personal collections), and begin to recontextualize objects into new, perhaps ephemeral, compilations based on the users' immediate interests, much more easily than digging for objects buried in an EAD finding aid.

Access to selected, aggregated digital content on one Web site can be attractive and useful for users if and when the content and its presentation fit the context(s) of user needs and behaviors. Users' interests in and understandings of cultural content can be biased as well, in that user reactions to partner decisions may be filtered by contexts such as a user's experience with or knowledge of cultural content (visual, textual, and contextual), the user's available resources (time, tools, and resources like primary source materials and peers), and motivations and information needs at that moment.

Share the Content

MOAC partners expected that people familiar with cultural content—such as information professionals and academics—would want direct access to authoritative, persistent information about cultural objects. MOAC partners deliver basic, unmediated data about objects through object-level descriptions and linked images. It may be assumed that each partner expresses information about objects in slightly or noticeably dissimilar ways, depending on partners' expectations and abilities for describing objects and producing images.

MOAC II subjects expressed general trust in descriptions and images delivered online by museums, libraries, and archives. While trust is not an absolute given for information professionals and academics, subjects trusted digital content from cultural institutions because, for example, these institutions have curatorial expertise and direct access to physical objects, access to other primary source materials, and often, access to artists/makers. Trust in online information may be similar to museum visitors trusting that the objects and wall text displayed in galleries and the information communicated by docents and audio guides is vetted and authoritative.

While users may trust information from cultural institutions and expect that information to be available in future, a more problematic issue is whether the digital content is actually useful for or of interest to intended audiences. It can be assumed that existing contributions to MOAC may not satisfy entirely every nuanced, particular inquiry of the user constituencies targeted in MOAC II. For example, an educator familiar with the collecting activities of MOAC partner institutions may expect to find photographs of a specific cultural group made by the same cultural group. The educator will not find representations of such photographs in MOAC unless, of course, partners have decided to contribute the desired objects to the OAC and have explicitly described or have implemented a system that can cull out data about objects in line with the user's understanding and choice of words.

Partners have been very selective of the content they wish to share, based on institutional priorities and sensibilities, such as the content curators believe best represents an institution's holdings or the readiness of content for publication. A selective approach may disappoint users of MOAC if partners have not shared the expected or most desirable content for the users' needs and interests. Some MOAC II subjects expected the online collections to represent significant portions of partner institutions' holdings and expected partners to share, through images and descriptions, what they knew about objects and makers: for instance, where the object was made and why or how the object or its maker fits within a larger cultural history.

Museum partners contribute to the OAC a small subset of the descriptive data and digital images that their institutions produce (or could produce) about objects. For example, a partner may choose to represent a complex object such as a multi-page pamphlet or album with a single digital image of the illustrated cover. Some people, however, will want direct online access to the interior of the object—such as a transcription of the pamphlet's text or a readable digital capture. More information is better for some users, as they try to determine the potential fitness of cultural objects (in digital or physical form) for their needs or interests, at the time.

Decisions about what content and which subset of data to contribute may reflect a partner's perception of the purpose of sharing. For instance, partners might contribute single, still images and accompanying descriptions to the OAC in order to expose content to known and potential users. This is exposure and discovery at the broadest level, giving users a general idea of the contents of a collection, rather than attempting to deliver digital reproductions and data as near to comprehensive as possible. That said, many MOAC II subjects did not, at the time of participation, expect content aggregation portals like the OAC to be the sole sources of information about an area of interest. Information professionals and academics, in particular, expected to search and collect information from a variety of sources (e.g., books, museum Web sites, articles) and viewed MOAC and the OAC as one stop on the way to conducting research or finding materials for teaching purposes.

Share the Context

MOAC II partners learned that many people—regardless of their named user constituency in MOAC II—will at times want objects to be delivered in some context: for instance, with explanations from content providers about the connections between objects presented and connections between objects, people or institutions, time, and events. MOAC partners attempted to use the hierarchical structure of an EAD finding aid to present some context about a collection of objects (e.g., context inherited from previous owners or imposed by a MOAC contributor). Similarly, at least one partner linked METS digital objects for complex works, such as albums and artists' books, to their associated finding aids in an attempt to provide users with more complete information about and experiences with particular works (i.e., through navigating multiple digital images representing a work).

MOAC partners and the OAC have demonstrated that content from museums, libraries, and archives can be aggregated insofar as users can search across the entire OAC collection of finding aids and retrieve lists of finding aids that may have related content—if, that is, all content contributors have accounted for potential relationships through descriptions and terminology. It remains to be tested whether the intended contexts or concepts that hold together collections of particular objects in some meaningful way can also be aggregated by users in a content aggregation portal like the OAC, especially when context is expressed differently by contributors with diverse professional and institutional practices, priorities, and sensibilities.

Opening the METS digital repository increased the ways in which users can find and view MOAC digital content: that is, simple and complex objects represented by images and associated object-level descriptions, removed from EAD finding aids or collection guides. METS may allow users to get straight to and aggregate digital objects, when there is related content to aggregate. This line of direct entry into discrete objects may be appropriate for people who know what they want and/or people who find finding aids to be of limited use or understanding and want relevant information to be surfaced quickly and easily. Making discrete digital objects available in content aggregation projects or union databases can also decontextualize objects, or remove objects from an inherited or purposefully imposed context. Should content like MOAC digital objects be self-explanatory, or is a reference or another intermediary service necessary for understanding cultural objects as sources of information? Several MOAC II subjects freely offered interpretations of images retrieved in MOAC based on their familiarity with the content and interest in subject matter, with or without commenting directly on the object-level descriptions or context presented by MOAC partners.

Lessons Learned for MOAC Partners

MOAC partners know that their decisions about what content to share, how to describe and contextualize objects, and how to present that content and context online, in venues like the OAC, can directly affect how people find, make sense of, and use their content. Results from the MOAC II study may suggest that changes to MOAC content and presentation are needed, especially if MOAC as a resource is to accommodate varying user needs and expectations for gathering, mixing, and integrating cultural content into their work or learning. As long as partners share content and express context in content aggregation portals, they will likely attempt to balance what users may want and expect with what the partners' institutions are prepared or motivated to share, often with curatorial and scholarly interests in mind.

From the MOAC II results and any subsequent evaluative work MOAC partners will consider what they can, should not, and cannot change about the content, packaging, and presentation of their contributions to the OAC, and will identify with whom responsibility lies for decision making (e.g., whether potential enhancements to object descriptions or online presentations should be made at the partner or OAC level).

MOAC partners will assess the feasibility of integrating what users want into partners' routine production and contribution activities. Museum partners, for instance, may be interested in contributing more digital objects and deeper metadata so that greater portions of their holdings are represented online, through readable images and fuller descriptions of what is known or interpreted about objects, makers, and subjects.

Work expended by partners on MOAC must complement institutional goals for access and dissemination, including sharing information about objects and collections in multiple, co-existing online venues. Partners will not want to change the descriptive data standards utilized in local systems (such as collections management systems or digital asset management tools) because partners may use those standards to promote consistency across descriptions and projects (albeit understanding the issue of idiosyncrasy noted above). Partners may not be willing to change the contextual expressions or descriptions packaged and delivered, in this case, via the OAC.

Museum and library partners participate in OAC working groups (along with other OAC contributors) charged with developing best practices for producing and delivering digital content. While MOAC is a significant component of the OAC and has provided a valuable testbed for demonstrating the capabilities of EAD and METS for aggregating varied content, MOAC is not the singular concern of the OAC. As the content aggregation portal, the OAC is responsible for managing the technological infrastructure, including content ingestion, search and retrieval capabilities, access files, and user interface(s) of the OAC. From the evaluation, MOAC II partners will recommend strategies to the CDL for improving OAC digital resources by, for instance, designing task- or community-specific navigation and retrieval mechanisms and presentation interfaces. However, museum and library partners alone do not have the requisite decision-making power to make changes to the technological infrastructure of the OAC. That said,

MOAC partners have some flexibility in developing the look of the MOAC-only content delivered in the MOAC project Web site hosted by BAM/PFA, which will make it easier for partners to implement some recommendations about interface design that emerge from the MOAC II study on the MOAC Web site.

CONCLUSION

MOAC is an evolving collaboration and resource. MOAC Classic, MOAC II User Evaluation, and MOAC Community Toolbox demonstrate partners' commitments to collaboration as a model for building and sustaining cultural information networks, especially for improving the ways in which digital content is produced, packaged, and presented. Readying content for delivery online and monitoring its relevance and accessibility over time requires significant resources and diverse expertise, as evidenced by the MOAC collaboration, in which partners brought wide-ranging experiences in content, digitization, standards, technology, and audiences (and which might be mirrored in cross-departmental collaborations within a single institution).

MOAC partners are motivated to share content because they want objects—in digital and physical form—to be seen and used in education, research, and engagement. For content to be seen and used though, it may not be enough to open access or disseminate content in a passive way. Just as museums, for example, engage seriously in planning, funding, and programming activities when presenting content and context in physical spaces, it is important and useful for museums, and other cultural institutions, to incorporate similar activities into sharing content and context online, which might include finding a balance between what institutions perceive to be important to share and what audiences want and expect to find, use, and experience online.

ENDNOTES

1. MOAC is a partnership between the California Digital Library, the UC Berkeley Bancroft Library, and several museums including UC Berkeley Art Museum and Pacific Film Archive, Japanese American National Museum, Oakland Museum of California, UC Berkeley Phoebe A. Hearst Museum of Anthropology, UCLA Grunwald Center for the Graphic Arts, UCLA Fowler Museum of Cultural History, UCR/California Museum of Photography, and the San Francisco Museum of Modern Art.
2. A data structure standard identifies and defines the data elements that comprise information records for a particular community.
3. Data content standards provide directions for entering data in elements. Data value standards are controlled vocabularies and thesauri that list or organize hierarchically terms or names used in elements.
4. A subset of MOAC partners collaborated formally in the MOAC II User Evaluation: California Digital Library, Online Archive of California, Bancroft

Library, Berkeley Art Museum and Pacific Film Archive, Grunwald Center for the Graphic Arts, Phoebe A. Hearst Museum of Anthropology, and the San Francisco Museum of Modern Art.

5. The MOAC II research team included Anne Gilliland, Kathleen Svetlik, Carina MacLeod, Sibyl Roud, and Elizabeth Spatz of UCLA; Layna White, Richard Rinehart, and Robin Chandler of MOAC.

Section 7

Conclusions

18 Information Professionals in Museums

Paul F. Marty

Florida State University

As museum professionals and visitors become more information-savvy, and their information needs and expectations become more complex, the role of information professionals working in museums worldwide has increased dramatically in importance. In some sense, of course, nearly all museum professionals can be considered information professionals, as most deal with some aspect of museum information resources on a daily basis (Orna & Pettitt, 1998). In addition, many traditional museum careers, such as registrar or librarian, require a great deal of experience and expertise working with information resources (Koot, 2001; Reed & Sledge, 1998). There is a clear sense, however, that a new role for the information professional is emerging in museums.

The evolution of this new role has coincided with the growth of museum informatics, and the role of the museum information professional is one that is expected to grow in importance in the future. Museum professionals are increasingly concerned with the ability of museums to function in the information society, to meet user needs, and to ensure that the right information resources are available at the right time and place, inside or outside the museum (Marty, 2004b). The success of museums in the future will depend largely on the work of information professionals specifically trained to deal with the problems of museum informatics and the museum's information needs (Marty, 2005). As Hamma (2004a) writes, "Adding information management as an integral part of a museum's routine activities will or should change the organization with the addition of at least some new staff, new skill sets and a new management effort" (p. 12).

The modern museum needs individuals on staff who are capable of setting information policy, managing information resources, administering content management systems, implementing metadata standards, evaluating information interfaces, etc. While some of these tasks may be performed by existing employees, a growing number of museums are seeking individuals with the skills to fill new positions specifically designed for these new responsibilities. Outside of small museums, it is rare that one individual would be responsible for performing all of these tasks. Depending on the size of the museum, these responsibilities might be met by several individuals or by an entire department. Those individuals might also possess many different titles, including Webmaster, database designer, technical support specialist, information resource manager, and chief information officer.

Despite their importance for today's museums (Coburn & Baca, 2004; Grant, 2001; Hamma, 2004b; Hermann, 1997; Roberts, 2001), little is known about the roles these individuals play; museum information professionals are among the least studied of all consumers and producers of museum resources (cf. Gilliland-Swetland & White, 2004). Research is needed that explains the roles information professionals play in museums, how their roles have evolved over time, what skills they need to fulfill their responsibilities, and how they have adapted to meet the challenges of museum informatics. Little is known about how museum information professionals characterize their own work, the challenges they face, and the strategies they employ.

To date, only a handful of studies have focused on the nature and behavior of information professionals in museums. Bernier and Bowen (2004) evaluated the information behaviors of museum professionals online, in particular their use of online discussion forums. Gilliland-Swetland and White (2004) studied the ability of museum information professionals to use metadata standards that provide access to museum information online. Marty (2004b) discussed the changing role of museum information professionals as technologies change, and how these individuals adapt their work practices to coincide with these new technologies (cf. Marty, 2006). Haley Goldman and Haley Goldman (2005) explored Web development as a profession in museums, interviewing museum webmasters and asking about their work, their sources of inspiration, and their ideas about the future of museum Web site design (cf. Marty, 2004a).

The main challenge in studying museum information professionals is that this position is one that can entail a variety of responsibilities and require a variety of skills. This challenge becomes obvious when one examines the descriptions of these individuals written over the past decade, as well as the job descriptions written by museum administrators seeking database managers, information architects, or similar positions.

For example, when describing this new position, Hermann (1997) wrote, "Perhaps there is a new role in museums for an 'information manager' who is charged with caring for the museum's information. The information manager should know standards, understand how they can be implemented for each museum system, and, most important, advocate their use within the museum. Few, if any, museums have a staff position with this title now, but the function will become increasingly important as we integrate information systems into our daily work" (p.75).

Similarly, when writing about possible museum careers, Glaser and Zenetou (1996) argued, "An information manager is a relatively new position for museums, but one of great importance for efficient operations and record-keeping. [. . .] The information manager is responsible for facilitating the flow of information within an institution and between the institution and the public. [. . .] Broadly interpreted, the activities include ongoing analysis and implementation of the ways in which the institution collects, stores, and disseminates information" (p.103).

In 2001, the Natural History Museum in London advertised a search for an "experienced, energetic, and innovative manager who can lead the dynamic development of the Museum's information infrastructure." This person's job would be to "coordinate the strategic development and positioning of the Museum so that external and internal customers derive optimum benefit from the information it generates and holds [. . .] by advocating, developing, and managing systems and technologies for collecting, sharing and disseminating information about the natural world gathered from both within and without the museum."

Similarly, the Field Museum in Chicago searched in 2002 for someone whose duties would include "developing a comprehensive information management architecture and database design for the scientific collections; acting as the liaison between technology and scientific staff; and designing, bidding, and implementing a coordinated collections management system." The same museum searched again in 2003 for another individual who would be "responsible for design, development, and support of museum databases and database interfaces, [. . .] for leading development projects, [. . . and] for defining the museum's strategic approach to information storage."

These job descriptions and position announcements illustrate the challenge of defining an evolving position that is still in the process of defining itself. The role of museum information professional is changing so rapidly that any attempt to define precisely the responsibilities associated with it likely would become quickly outdated, especially if one attempts to specify particular skills or technologies as requirements. In seeking individuals capable of filling such a position, then, museum administrators need employees who are just as capable of helping them define the very role they need to play in the museum

Despite these problems, the benefits that information professionals bring to the modern museum are very clear. Information and communication technologies in museums are changing so rapidly that most museums remain desperate for employees who can guide them through the basic technology hazards of planning digitization projects, purchasing collections information systems, or joining online data sharing consortia. Museum administrators know full well that if a museum is to be an active participant in the information society, someone at the museum needs to be well versed in metadata standards, controlled vocabularies, database design, and the latest Internet issues. While some technical jobs (including Web design) can be and will continue to be outsourced, museums that do not have at least some information technology skills in-house will likely find themselves paying increasingly expensive consultants if they wish to continue to meet the constantly evolving demands of their information-savvy audiences.

Even more important than the technical skills they possess, however, is how museum information professionals use their abilities to help or hinder the museum's users in accessing the museum's information resources. While there is no question that specific technical skills are crucial (someone needs to be able to build Web-enabled databases and museum Web sites), information professionals in the modern museum need to employ their skills while working with users of the museum's resources to ensure that new technologies are meeting the users' needs. They need to look beyond their specific technical abilities, and see that it is not what they can do that is important, but how they do it from the users' perspective. It is important to view information technology as a means to an end, one that allows museums to reach their users by incorporating new technologies into different aspects of museum activities.

If museums are to live up to their visitors' changing needs and expectations over time, someone in the museum needs to be responsible for tracking and evaluating the evolution of their information needs. The challenge of providing access to information in museums lies not with designing exhibits, building collections databases, or creating museum Web sites; it lies with ensuring that those information technologies are truly meeting the needs of the museum's visitors. To help museums keep up with changing expectations of museum users, the successful museum information professional needs to work closely with a broad spectrum of the museum's users, examining their information needs from multiple perspectives, and advocating those needs to appropriate individuals in the museum.

The users of museum information resources, be they online visitors searching the museum's online collection databases or employees looking for important documents on the museum's intranet, frequently have no voice in the design, development, and implementation of the museum's information systems and technologies. If museums are to meet the changing needs of their users, someone in the museum must work as the users' advocate, approaching problems from a user-centered perspective. When user needs are not being met, the museum information professional will likely be the museum staff member most attuned to the user's needs and best suited to argue the user's case. They are likely best positioned to serve as user-centered mediators between the museum and its users, advocating information needs for multiple users, and arguing for such issues as usability, open access, standardized data, and the availability of appropriate information resources to all users. They ensure, as much as humanly possible, that each user has a successful interaction with the museum's information resources.

To do so successfully, museum information professionals need to play an active role in guiding the future of information work in museums. In doing so, they will likely find themselves advocating, establishing, and administering information policies—making decisions that may influence entire museums as well as the activities of museum professionals and visitors. They will guide attempts to gather data about the needs of the museum's users, by running online surveys or interviewing museum visitors in the galleries. They will make decisions about the nature of information resources to be provided to users, deal with resulting questions about the quality or quantity of those resources, and ensure that the museum's information-related activities stay aligned with the museum's overall mission and goals.

Facing such an awesome responsibility may be difficult for many museum information professionals, requiring them to assert unfamiliar and uncomfortable roles, and placing them in the potentially awkward position of redefining the museum and their own place in it simultaneously. Despite these difficulties, museum information professionals are best qualified to evolve a basic understanding of the information resources a museum should provide and the role museums play in the information society. To do so, they must establish themselves as the best (and perhaps only) individuals qualified to make decisions about such issues as digitization policies, metadata standards, digital rights management, and so on. They need to take advantage of their unique position to understand changing capabilities, needs, and expectations at an organizational level.

In their new role, the ability to configure a server or complete a catalog record becomes less important for museum information professionals than the ability to identify and solve problems of information access, provision, and communication (cf. Marty, 2006). As new technologies emerge and needs and expectations evolve, the physical manifestations (e.g. Web sites, databases, etc.) of their skills may change, but the underlying importance of the skills they offer will not. The true value of museum information professionals lies not in their ability to solve individual technology problems; it lies in their ability to comprehend the future of information work in museums and the impact of information technologies on the job of the museum professional.

The real challenge of museum informatics, therefore, is not implementing the technology, but understanding the theories and principles behind the technology, and how they affect the people who access information in the museum environment. Solving the problems of museum informatics will require a number of individuals to bring a wide variety of skills—including basic information technology skills as well as an understanding of information organization, information architecture, and knowledge management—to the museum environment and the museum culture. For most museums, these individuals will come from the inside; current museum professionals will have to come to terms with these issues, develop the necessary skills, and present themselves as individuals with the proven skills and abilities to lead the museum into the future.

Doing this well will likely require forming strong collaborations with information professionals in other information organizations. These collaborations will help museums to stay abreast of the rapidly changing world of museum informatics; and they have already proven valuable for increasing the general understanding of how the principles and theories of information science apply to the museum, and how museums in particular are dealing with the new challenges and implications of information technology in the museum. It is in this way that museums will evolve the environments they need to address fundamental questions about the sociotechnical interactions of people, information, and technology in museums.

19 Curating Collections Knowledge
Museums on the Cyberinfrastructure

Jennifer Trant

Archives & Museum Informatics

INTRODUCTION

As museums move more of their programming into digital space, effective program delivery will be enabled by an integrated information management strategy that views collections documentation as an asset: a key, collection-related resource that forms an integral part of the value and appreciation of collections themselves and supports the ability of the museum to fulfill its mission. Investments in interpretation will be leveraged and re-used, as part of the institution's knowledge of its collection rather than created and cast off when the temporal context for their initial use passes. Museum professionals, in all parts of the institution, contribute to this developing corporate memory, building institutional knowledge as part of the cultural trust of the museum. Providing public access to this knowledge, through exhibition, publication, and reference, and conserving it for future generations, is emerging as a key ethical responsibility of museum work. Curating collections knowledge requires an acknowledgment of a multiplicity of information sources, inside and outside the institution, each of which contributes to a developing understanding.

Ensuring that collections are enlivened by context requires an active approach to museum documentation. Rather than passively recording information about a work of art, artifact, or specimen, museums are challenged to acknowledge information sources beyond the museum, and change their practices to incorporate new perspectives into both interpretation and documentation. Positioning museum objects in a networked information space as part of the "cyberinfrastructure" also positions them in a social space, and invites their creative and recombinant use. As communities respond to collections, so must museums respond to communities, ensuring that a diversity of voices provide context in the future as well as the present. The richer web of context provided by multiple interpretations adds to the value of cultural objects, enriching their position within collections and the position of collections within societies.

Museum collections are relevant when they are used. As museums take their place in the cyberinfrastructure as a trusted source of specific kinds of information and experience, they also assume additional responsibilities to maintain that position, and support those who use their collections. In a distributed knowledge society, use will be initiated from outside as well as inside the institution. The creations that result from that use, whether scholarly or personal, will be expressed in numerous, unanticipated forms. This new documentary heritage should itself become a part of the curatorial trust; recording it an active challenge.

CREATING COLLECTIONS INFORMATION

Museums have collections at their core. As defined by the International Council of Museums (2004), they "preserve, interpret, and promote aspects of the natural and culture inheritance of humanity" through an active stewardship of artifacts that represent cultural or scientific history and knowledge. That stewardship includes physical care of the collections, through proper storage, management, and conservation.

Managing information about collections has emerged in the past two decades as an activity as important as managing those collections. While initially somewhat controversial (Sledge, 1988), this responsibility is now reflected in the ICOM Code of Ethics, the document of the "museum professional" that "sets minimum standards of conduct and performance" (International Council of Museums (ICOM), 2004, p. Introduction). The growing awareness of the curatorial importance of collections documentation may reflect a more critical consciousness of the nature of collecting itself. The encyclopedic representative museum of Franz Boas (1907) has given way to a more situational understanding of the history of museums and their collections:

> Collecting, as a core museum practice, is complex and largely beyond scientific rationalism. It is an act of authorship and connoisseurship. It is a physical interpretation of a set of circumstances or body of potential data. The object is thus placed within a collection according to an individual's beliefs (Knell, 2003, p. 137).

This Post-Structuralist awareness of the importance of context in the assertion of meaning is embodied through conscious expressions of the significance of an object at the time it enters the collection in documents such as Acquisition Reports or an Acquisitions Justification (Buck & Gilmore, 1998) and through repeated re-documentation whenever a work is included in an exhibition, published in a book or article, hung in a gallery, or otherwise engaged in the service of the museum's educational or research mission. Comprehensive collections documentation practices record all these representations of the significance of an artifact.

Conceptualizing collections information in a central rather than supporting role may also have its roots in the developing understanding of the museum as a social institution, of significance in the maintenance of the "civil society." As museums reorient their institutional focus towards society and community, communication and engagement gain priority: "It is important for museums actively to use information to create understanding or to help their audiences exploit effectively the information resources in their self-directed quest for knowledge" (MacDonald & Alsford, 1991, p. 306).

A tension has emerged between the knowledge of collections that the museum creates, and knowledge that is created external to the institution, in an individually- or socially-defined context. MacDonald and Alsford (1991) posit that museums are involved in the "generation," "perpetuation," "organization," and "dissemination" of knowledge. Active interpretation is assumed, as museums are responsible for "both making information readily available and ensuring that its users have the ability to comprehend it" (MacDonald & Alsford, 1991, p. 307). But they don't see museums as receptors of knowledge that is created elsewhere. Works in the museum's collection are not seen as part of a broader story that involves others as well as museum professionals in its telling.

Authenticity and quality may set museums apart in the information landscape (Trant, 1998), but static assertions of value stand in conflict with the emerging conversational metaphors of information use in the museum context (Dietz, Besser, Borda, Geber, & Lévy, 2004). With individuals making their own meaning—for example, curating their own exhibitions—what happens to the role of the museum? It no longer has the luxury of being the sole arbiter of an object's interpretation. Rather than being challenged, and feeling a competitive need to assert authority in interpretation, museums could make a significant contribution through the longer-term management of the knowledge about objects in their collections, knowledge with a provenance as varied as the objects themselves. Just as museums acknowledge that theirs is not an "Unassailable Voice" (Walsh, 1997), they must accept responsibility for curating the context of their collections as told in many voices.

COLLECTING COLLECTIONS INFORMATION, HISTORICALLY

Collections Inventories

The role of basic collections documentation in responsible stewardship was established in large-scale inventory projects that provide the foundation for museum computing activity. With an emphasis on fiduciary responsibility, these projects emerged in a main-frame computing environment, and dominated most of the first and second generations of museum computing

(Jones-Garmil, 1997). Motivated by the challenge to be accountable to governing bodies and the sense that "if you don't know where it is, you might as well not have it," museum inventory projects laid the groundwork for technology in museums by producing databases that summarized holdings, and recorded their vital statistics (acquisition, location, physical description), as outlined in, for example, the early CHIN Humanities Data Dictionary, maintained to this day as the foundation for Artefacts Canada (see Canadian Heritage Information Network, 2005).

Collections Management Processes

The next generation of collections documentation systems was designed to support collections management processes. Information was conceived as deriving from or supporting aspects of museum operations. A desire for internal efficiencies led to calls for integrated systems that removed redundancy and assigned appropriate controls on authorship. Functionally-based museum data standards like SPECTRUM (the most recent version of which is McKenna & Patsatzi, 2005) illustrate the role different elements of museum information play in conducting the business of the museum. Conceptually, these systems reflect the view that most museum documentation is created for internal use. Hence, it is tied to the functions and offices that are responsible for its birth. As an example, Paul Marty's (1999) discussion of the Spurlock Integrated Museum Management Information System shows how museums began to see collections information systems as tools to support the functioning of the museum. Collections information was no longer just a thing to be kept, but a thing to be used by museums themselves.

Beyond Accountability and Process: Collections Knowledge Made Public

Functionally derived conceptual models formed the basis of collections documentation systems and provided the motivations for their use. But networked access to collections information enables many unanticipated uses, most of which take place outside the museum. External communities of users were originally "supported" through access to collections information in the form of on-line databases, tools museum professionals hoped would lessen the burden of "reference" by enabling users to see "all of the collection, all of the time."

A desire to open up museum collections and enable their use motivated participants in the Museum Educational Site Licensing Project (MESL), one of the first large-scale museum information sharing projects. MESL was designed to explore the possibilities and limitations of museum collection information as an educational resource. Content from seven collections was made available to seven universities to mount on their campus networks and use as they saw fit. (For a summary of findings, see McClung & Stephenson,

1998; Stephenson & McClung, 1998; Trant, 1995, 1996b, 1996c. For project rationale, see Trant, 1995, 1996a, 1996b, 1996c.) While our initial sense that museum content would be worthwhile was reaffirmed, participating in MESL taught us a great deal about the nature of museum collections information and its potential use. This was the first time that museum collections documentation had been exchanged in any volume, and seeing it outside the local systems within which it was created and managed had an impact. New application contexts surfaced many presuppositions about meaning in museum documentation. While the museums tended to think about objects, academics thought about themes; museums valued technical descriptions of the physicality of a work, teachers wanted more about the context of its creation and use. The language and perspectives of the two communities didn't mesh. For individual museum objects to be useful in an educational environment, it seemed that had to be wrapped with interpretation: included in a lecture, referenced on a course Web page, analyzed in an assignment. For the users on MESL campuses, individual museum works were a point of departure, not the end in themselves that museum professionals thought they were.

MESL surfaced a disjunction between accessibility and usability. We might have been able to facilitate discovery of works in museum collections, but we relied entirely on users to create meaning from them. The on-line catalog was like open storage; collections were visible, but not necessarily comprehensible. The metadata that was supplied with MESL objects was rudimentary, based on assumptions about what label copy should be provided in a gallery. The MESL project was conceived before the World Wide Web, and implemented in the early days of Web-accessible resources, when simply populating the Web with content and then enabling others to find it were worthy goals. We lost sight of the fact that discovery was the beginning of the journey, not the end; that by sharing museum resources, we wanted to facilitate research and learning, not just show people our stuff. Much of the learning in MESL was in the museums as they watched how works from the collections were used in teaching and research. Seeing that perspective was invaluable.

Little of the investment that museums make in interpretation or contextualization that comes from the curatorial care of the object was visible in the documentation shared with university users in the MESL project. While much digital content is created as a by-product of museum activities, including exhibition development, publication, and education program development, this isn't considered part of the core documentation of museum works, and is rarely recorded systematically or shared broadly. This is not just a problem of text: visual materials often created as part of conservation (such as views under alternative lights sources, or views of a work under restoration) provide more detail about an object, installation photographs of exhibitions (such as the amazing glass plate negatives at the Art Institute of Chicago showing the galleries when the museum opened) document its presentation and interpretation, audio files created for a tour of an exhibition provide curatorial insight and analysis. But this contextual detail is often dispersed throughout the museum, and is not accessible to a researcher enquiring about the work itself, either in person or on-line, a failure of museum process that doesn't reflect the interests of the public in the many facets of an object and its interpretation.

An Example: Documenting Collections Care (Conservation)

Documenting the physical state of an artifact is a *sine qua non* of any intervention or treatment, whether preservation or conservation. Whenever a conservator intervenes in the physical state of an artifact, its informational content is altered. Documenting that intervention is an ethical imperative. But often, that knowledge is kept inside, the province of the department, not even the museum as a whole. The scientific tools conservators have at their disposal can also aid in the interpretation of objects. Non-interventionist techniques, such as examination under ultraviolet light or with X-radiography illustrate characteristics not visible to the naked eye. These can assist in the authentication of an artifact and in the understanding of its physical makeup and process of creation.

As more and more conservation documentation becomes available in digital form, the museum bears a responsibility to make these additional analytical sources available to scholars and to share the process and excitement of discovery with a broader public. Conservation has proven to be one of the more fascinating of the "behind-the-scenes" museum processes. Allowing "views" on the treatment of painting, for example through its public cleaning or treatment, illustrates the active nature of collections care and helps demystify the role of the museum in the preservation of culture (Minneapolis Institute of Arts, 2004; Sayre, 2000).

Expanding Our Concept of Collections Documentation

Embracing the reality that museum collections are multifaceted and that our knowledge about them grows and changes over time challenges conventional approaches to museum documentation. Static standards and cataloging strategies that focus on recording information must give way to extensible information architectures that are flexible enough to accommodate incremental growth and change. Multiple perspectives and multiple—at times conflicting—opinions reflect the richness of culture and the development of knowledge. Museum information in a public space should provide a record of this conversation and encourage its continuance.

The AMICO Library™ developed as a way for members of the Art Museum Image Consortium (AMICO) to enable educational access to multimedia documentation of their collections. It was conceived with an understanding that what is known about museum objects grows and changes over time. Its model was a growing, changing knowledge base rather than a static permanent collection catalog. AMICO members added to the public view of their collections in The AMICO Library when resources became available internally, or as was the case with audio files from Antenna Audio, when the consortium negotiated access to a body of material. The AMICO Data Specification (Art Museum Image Consortium, 2004) outlined a modular record format that allowed for a core catalog record to be augmented by any number of related visual, textual, or multimedia sources (Figure 19.1).

User feedback about The AMICO Library reaffirmed the value of contextual information, often as a "point of departure" for future enquiry. Students presented their own interpretations and contextualizations of works, based on an active reading and their own research. (For an example of student work, see Alperstein, 2003.)

CULTURAL CONTEXTS FOR COLLECTIONS

Site-specific and process-based contemporary works of art highlight the failure of traditional documentation practices. These works, such as

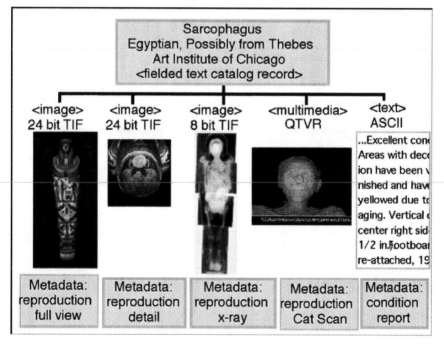

Figure 19.1 Each work in The AMICO Library is documented by a catalog record, an image file, and an image metadata record. Additional media files may also be present. Each of these has a metadata record. © Art Museum Image Consortium/Art Institute of Chicago

Antony Gormley's *Field,* may exist in both museum space and public space. Their creation may be collective, their perception varied and dependent upon where a viewer encounters the work. As an example, the American (http://www.antonygormley.com/walkthrough/239x2_american_field.htm) and European (http://www.antonygormley.com/full_list/283x5_european_field.htm) versions of the *Field* were installed in two different contexts, one speaking to the viewer in a neutral space, the other to the viewer in an art-historical context. The difference is palpable.

Installation photographs (such as http://www.antonygormley.com/photo_essays/field_pe03.htm) add an additional dimension. They document the creation of the work in a particular space, providing an essential perspective on variable works, particularly helpful when they are re-installed in a different gallery context.

But often a full understanding requires knowledge of a work's creation; knowing that the *Asian Field*—another iteration of this work comprising 190,000 figures—was made by a group of 350 people in the village of Xiangshan in a period of 5 days moves the figures from anonymous to personal. Hearing the artist speak adds an additional element of meaning. Gormley has said,

> *Field* is part of a global project in which the earth of a particular region is given form by a group of local people of all ages. It is made of clay, energised by fire, sensitised by touch and made conscious by being given eyes (British Council, 2003).

The work exists inside and outside the museum, in the mind of the artist and the eye of the viewer, in the participants who created it, and the teams that install it. In each context, different aspects of meaning resonate. Understanding the work requires appreciating those many perspectives. Documenting them requires new strategies and approaches (Rinehart, 1999, 2004).

Museum Objects Used in Many Different Ways

As *Field* shows, there are many approaches to collections documentation. What is recorded, however, is often colored by the museum's perception of the importance of the work:

> In the museum, the real object is capable of being an archival resource, a site of meaning making, a component in an educational programme, primary data in a research project and so on. But museums tend to select different objects for different jobs; a tattered item of costume might present a researcher with critical clues to an aspect of textile history but will never form an exhibit . . . (Knell, 2003, p. 139).

If what is valued is the assertion of the institution, then other aspects of interpretation are neglected, and therefore negated. But "Never" in a museum is likely to be "Now" in a research project.

An Example: Education and Interpretation

The provision of some context as scaffolding for self-directed learning about museum collections is a basis of technology-enabled interpretive strategy (Dierking & Falk, 1998). The STEM project at the Science Museum, London consciously "encourage[d] students and teachers to create their own perspectives, projects and educational resources related to the Science Museum, and to make them available on the Web by publishing them on their own servers" (Jackson, Bazley, Patten, & King, 1998). The resources of the Science Museum, London were presented as "raw material" to support a quest for understanding. But that understanding is not part of the Science Museum. The project architecture maintains the arms-length association between the museum and its third-party collaborators. The Science Museum has consciously chosen to point to resources rather than integrate the material into content for which the Science Museum takes "responsibility."

Information Architectures to Support Recombinant Use

Museums must take responsibility for the contextual construction of knowledge of and about collections objects that occurs as they are used in museum programs. Educational programs have led the way in using electronic sources to bridge the gap between work of art and viewer. For example, at the San Francisco Museum of Modern Art the *Making Sense of Modern Art* series (Samis, 1999; Samis & Wise, 2000) has provided a framework for the inclusion of artists' voices, one of a number of multiple points of view that are necessary to engage different types of audiences, and support different kinds of interpretation (MacDonald & Alsford, 1991, p. 309) and successful, "visitor-centered" experiences (Frey, 1998).

Some would argue that to keep everything that the museum produces would

> give permanence to acts which only exist temporarily in the museum. The more liberal interpretation of educators and exhibitors exists only for the period of engagement with an audience, while the objects themselves exist for the most part in the world of the specialist, where more pedantic forces of integrity and authenticity predominate and, ideally, associate an arcane dataset with the object (Knell, 2003, p. 139).

These words speak too harshly of the educator, but illustrate the hierarchy of value tacitly affirmed by many scholars. The retention of digital documentation made for many purposes may, to this point of view, be a compromise that the museum should not make for fear of sacrificing institutional goals of authority. But it is one that does a disservice to the active role of the museum in society and its long-term place as a cultural repository.

Extending knowledge of museum collections is one of the key aspects of museums' responsible stewardship. Traditionally, museums have focused on curatorial research on collections and the reporting of that research in exhibitions and publications. As part of a cyberinfrastructure, however, museums can become the site of research carried on by others, and the repository of results that relate to works in their collection. Re-use and re-interpretation are essential by-products of the public distribution of collections information. Museums can both encourage the creation of new knowledge about their collections and play an active role in recording it.

THE CHALLENGE OF RE-USE

If museum collections information is going to have a longer life and is going to become, in itself, a source for someone else's work, then the recording of the context of creation of that information becomes important. When that content is made explicit, for example through metadata about digital documents or the recording of authorship in on-line catalog entries, the institution is "released" from its burden of authority: the source becomes responsible, rather than the institution as a whole. The museum is playing a valuable role, but it is as conduit, not arbiter.

Could this be seen as an abrogation of responsibility? Not if the institution is enabled to record more information through this strategy. The challenge in providing on-line information about collections is often that of "getting it right," or overcoming the fear of "getting it wrong." If on-line provision of access to museum information is perceived as a traditional "publishing" function, then the quality of the data may not be up to "museum-standards" (one reason cited for not participating in CHIN's National Humanities Databases (CHIN, 2004, p. 6)). Explicit sourcing of data might free the institution from the self-imposed requirements that impede the flow of information.

A conscious expression of how much research has been conducted on an object helps internal or external readers of museum documentation. The seductive apparent veracity of a database record can be counteracted with an explicit indication of the "hardness" of a piece of data. At the Canadian Centre for Architecture (CCA), it was possible to overcome some curatorial reluctance to computerized documentation by including an explicit indication of "Catalog Status" (Trant, 1991, 1993). While this data element is not included in the records subsequently made available on-line, the existence of varying levels of research, and the changing nature of documentation is referenced in the preface to this searchable on-line database (Canadian Centre for Architecture (CCA), 2004). Explicitly or implicitly communicating functional context also helps the reader. For example, at the Fine Arts

Museums of San Francisco, presenting the volunteer-assigned keywords in the Thinker in alphabetical order re-enforced their role as access points rather than scholarly interpretation (Futernick, 2003).

If institutions are to manage content created for different audiences and delivered in different contexts, it needs to be sourced. There may be different standards of accountability in each of these areas that, if not explicitly referenced, could impede future interpretations. Without this, the perceived "equality" of a database might unintentionally privilege an interpretation, sliding subconsciously back into the "Unassailable Voice" of the audio guide (Walsh, 1997).

But in a context of multiple venues for user/collection interaction and multichannel communication, museums simply cannot control all that is said about their collections. Not trying to opens up the museum to the possibility that expertise exists elsewhere, and that the museum could benefit from the knowledge of many communities (Marty, 1999). It also positions the museum to be able to participate in dialogues that are situated in users' space as much as in museum-space.

Multiple Audiences with Different Needs

Museums have acknowledged that there are many ways to communicate their message. From the recognition that cultural heritage does not lie only in material objects, but also in intangibles—such as behaviors, beliefs, activities and processes—has come the appreciation that diverse media are required to capture and communicate the many facets of culture (Macdonald & Alsford, 1995, p. 132).

A diversity of media provides an opportunity to present multiple voices. But interactions with cultural heritage, both on-line and onsite, occur in different contexts. Models of those interactions can help museums structure the content that could support user engagement, and encourage the recording of insights that derive from user interactions with collections. In their exploration of interactive storytelling and user-driven narrative in Belgium, Pletinckx, Silberman, and Callebaut (2003) note the "different modalities by which people visit historical sites: coach tours with guide; small unguided groups of friends or family; and as individuals. Each visitor group type has different requirements" (p. 226), equally valid. To tailor information-based interactions in such diverse circumstances, museums must develop re-usable nuggets of information: "irreducible units of information from which the interactive stories are built" (Pletinckx, Silberman, & Callebaut, 2003, p. 226). This may require a disaggregation of traditional record-keeping strategies.

The ILEX system prototyped the delivery of customized content based on user interests, with dynamic hypertext (Hitzeman, Mellish, & Oberlander, 1997). Positing an interaction based on a personal tour through a gallery with a curator, the conversational model explored here included adaptations that were sensitive to the user's context, acknowledging levels of interest, expertise, and previous experience. Interpretive content was delivered in an adaptive manner, driven by the communicative goals of the curator. No provision was made for dialog or user contribution of content.

Further explorations of interactive storytelling in projects such as Telebuddy (Hoffmann & Goebel, 2003) attempt to personalize the museum experience and make it conversational, in this case by encouraging interaction between visitors onsite at the museums and those who might be online through the use of a fuzzy intermediary. A similar philosophy underlies the development of the Musée Transfer Suisse. Jaggi and Kraemer (2004) aimed

> to implement a new type of knowledge transfer from museums by digital storytelling, drawing on the often untold experience of life which museum collections hold. Virtual Transfer allows places and stories to be explored interactively in multimedia form, through a selection of objects and personalised modes of address (p. 5).

In all of these scenarios, the active construction of knowledge within a social group is enabled, but the museum itself is [willfully?] excluded from the conversation.

EXHIBITIONS AND PUBLIC ENGAGEMENT/ COLLECTIONS INFORMATION CONTEXT

Positioning museums as active players in an interconnected cyberinfrastructure requires a conscious awareness of where and how museum collections information (and contextual information about collections) is created and used. In scholarly terms, we need to recognize the role of the museum in the research process, and appreciate and support where and how users encounter collections (in surrogate or in situ) and what information needs they have in these contexts (Bearman & Trant, 1998). In public terms, we need to engage visitors on-line and onsite in the active interpretation and appreciation of the works in our care. Respect for our multiple audiences requires supporting the adaptive re-use of collections content in multiple contexts.

The maintenance of museum documentation for future use is a more complex question than preserving it for future access. For information about collections to have meaning, it must be available in the context of those collections, supporting their understanding and enabling their interpretation. So museums bring a different operational requirement to what has been called "digital preservation." The records of museum collections are rarely held in its archive; if they exist at all, museum archives are repositories for documents of institutional history and governance. Museum collections records remain "active records" as long as the object is in the collection.

Information about objects in the collection must be readily accessible and malleable, so that it can form the basis for a future re-interpretation.

So the strategies of preserving electronic records in museums will need to be different from those that focused on the stability of information. Yakel (2004) posits a model for the long-term management of digital assets in libraries and special collections (and, she says, museums though they are never discussed specifically). The "unit-based digital repository," the "institutional digital repository," and the "trusted digital repository" all assume a neutral "third-party" relationship between the custodian and the information being managed. The activity posited is to maintain the digital materials, not to use the information they contain. It is grounded in the theory that archives maintain evidentiary records of transactions in a content-neutral context.

Where museums differ from libraries and archives is in their active, programmatic use of the content in their collections, for interpretation, exhibition, and education. Museums are themselves one of the primary users of their collections documentation. Just as content is produced by museum business process, knowledge of collections informs those processes. This internal focus has shaped our understanding of museum documentation and our construction of systems to support it. It has mitigated against inter-institutional information-sharing; we're first trying to understand how to move information across museum departments before we move it outside the organization.

Our documentation strategies must become active, not passive. In order to make collections meaningful we need to record and represent the context created in the full range of institutional activities: programs, exhibitions, interpretation as well as collections care. We need to acknowledge that curation of knowledge adds meaning and value to collections. In the context of exploding volumes of information, when the value of filters and authoritative sources is increasing, museums are well positioned to help satisfy an increasing demand for knowledge (Hutter, 1998). We need to "go public" with the full range of resources at our disposal, rather than hoarding our knowledge for future internal use.

MUSEUMS AS PART OF THE CYBERINFRASTRUCTURE

Recent discussions about the nature of scholarship in the context of ubiquitous, networked information have drawn attention to the nature of the infrastructure required to support new forms of cultural creation and expression and new understandings and interpretations of the past, present, and future. John Unsworth, in the context of the American Council of Learned Societies, Commission on Cyberinfrastructure for the Humanities and Social Sciences (American Council of Learned Societies, 2004), has defined cyberinfrastructure as more than just hardware and software, more than bigger computer boxes and faster wires connecting them. The term describes new "research environments" in which disciplinary experts, in interdisciplinary teams, supported by specialized computational support staff, have global, instantaneous access to enormous computing resources (Unsworth, 2004).

This community of users assumes that the full content of museums will be one of those many resources, easily searchable, readily accessible, suitable for use and re-use, analysis, and representation.

What Role Might Museums Play There?

As part of the cyberinfrastructure, museums are custodians of information for future generations of humanities scholars. They are curators of knowledge as well as curators of collections. But to play this role they need to be connected, organized, available, engaged and of relevance: connected to each other and to the many communities that they serve; organized, so that the content in their care remains connected to related content in other institutions; available to a wide range of users in many different contexts; engaged with the active interpretation and documentation of their collections and with the users of those collections; and relevant because they are responsive to user needs and interests.

Authenticity in collections and in documentation is one of the key values that museums bring to an information environment. It is in providing the authentic re-presentation of artifacts that museum professionals of the future may find their most challenging and rewarding roles.

Much museum documentation is at best the raw material for scholarship; managing it is presenting primary sources in a more accessible way. Online cultural heritage documentation (in environments like the Web) offers the opportunity for museums to reach beyond their traditional local-service area, to provide service to a dispersed community of specialists and enthusiasts. Museums can move from only serving the "qualified researcher" (with an introduction) wishing to consult items from the collection in storage, to providing access to basic documentation to a worldwide audience.

What is gained in access, however, is sacrificed in mediation. It's not possible for museums to provide an interpretive wrapper around every work in every collection. Instead, other mediators are called upon to make meaning—teachers provide context for students, scholars provide new interpretations for their peers. The cyberinfrastructure provides a vehicle for others to make meaning about culture. Museums must be plugged into these grids, or they risk the well-being of their collections; they risk being bypassed by their core communities, simply out of ignorance.

Appreciating the interrelated role of collections and collections documentation reveals the importance of maintaining documentation of collections close to collections themselves. Only then is there a dialogue between the primary and secondary source, between the collections object and its growing, changing interpretation. To separate collections documentation from collections is to rob the artifacts of their meaning and their context, leaving them isolated, aesthetic objects or specimens, physically preserved and intellectually denuded.

Reconceptualizing the role of museum documentation as active curation of collections knowledge created inside and outside the institution enables museums to fulfill a broader role in society. Relevance is often defined in terms of "listening to our audiences"; responding to needs as "giving them what they want." But museum professionals need to hear what those outside the institution know, and record it, so that the knowledge is preserved to provide meaning to the collection. The museum information curator's selection, arrangement, and care[1] have as their object the cultural memory of the institution, a legacy to be guarded along with the physical preservation of objects themselves.

The vitality of collections derives from their use, interpretation, and re-interpretation. Museums can enable the creation of contemporary "sampled" culture through the re-use of the past (Federman, 2004, p. 11). The symbiotic relationship between scholar and collecting institution is strengthened when the scholar contributes to the institution's knowledge of its collections as well as benefiting from their use. By enabling easy access to the past, museums can support the creations of an ephemeral present, remixed and re-presented as cultural meaning is brought forward, re-interpreted, and re-invigorated.

The full implications of those assumptions are implicit in this scenario of use (a group of students is collecting information for an assignment), drawn from *Cyberinfrastructure for Education and Learning for the Future (CLEF): A Vision and Research Agenda*:

> When her digital agent tells Manuela that her friend Beth is near the town museum, she asks Beth to check out what's available. The students share the information they are collecting with each other and with other students back in school. The next day, the whole class interacts with these data to collate, represent, and analyse them. They compare their data to those collected by students in other schools and in previous years. Manuela sends some of her results of the geological survey back to the museum where it is incorporated into the exhibition. Several weeks later, a proud Manuela takes her parents back to the museum to see the results of her work (Computing Research Association, 2005).

What museum is ready to support Manuela? To have collections and the knowledge they represent readily available for re-use? To be open to the incorporation of that content into the work of others? To be able to incorporate the knowledge of others into their own interpretive [exhibition] and historical [documentary] space? To be a site of visible pride in the accomplishments of learning?

This scenario illustrates the unique cultural role museums play as a bridge: between formal and informal learning; between education and research; and between institutions and the public at large. Museums can be a positive, intergenerational public space that reflects community, on-line and off. But museums consistently sell themselves short when they don't share all they know. They ask users to take their assertion of the value of their collections on faith, rather than building on a moment of interest to transfer knowledge of the cultural context or significance of a work.

IMPACT ON MUSEUM PROFESSIONALS

Professionalization is often a process of setting apart, as the unique contributions of a group of people with identifiable skills are identified, acknowledged, and distinguished from the roles of others. But within the context of the cyberinfrastructure, museum professionals face a challenge of integration. Museum content has a vital role to play in the development of new social, educational, and research spaces. But museum professionals can only ensure that cultural institutions are relevant by changing their stance about the nature of their role: it is possible to contribute authenticity without demanding authority. Authenticity is a value; its maintenance an imperative in collections of lasting value. But demanding authority is an act, often of arrogance, that denies the contributions of others to the development of knowledge.

Within the rapidly developing environment of social computing, communities of practice are forming that could contribute significantly to the development of museums. Historically we have acknowledged that specialists (and awkwardly often enthusiasts) have a better understanding of aspects of museum collections than the professionals charged with their care. There is an opportunity for this knowledge to converge with that of the museum. Museums can build on their experiments in integrating information into physical spaces to integrate the information spaces of museums and their many communities of users.

By offering the content of their collections in both mediated and unmediated manners, museums can meet their educational missions, while proffering the potential for richer kinds of encounters. By ceding control over interpretation, museums can enable a richer kind of interaction, an authentic experience for some users that builds on their specific contexts and knowledge. The challenge for museums is to create the contexts within which these encounters can take place, and to ensure that that newly created context becomes part of the cultural legacy of the institution, a knowledge to be curated as it adds value to the collections. Curation is as much for the future as the past. It is a re-interpretation and recontextualization of knowledge as well as a preservation of artifacts. Curating collections knowledge ensures that the many facets of collections are carried forward and made available as the scaffolding on which new understandings are created. Curating knowledge of collections becomes a part of the cultural trust of museums.

ENDNOTES

1. See curate 2 v. (2004). *The Concise Oxford English Dictionary.* Oxford Reference Online.

20 The Future of Museums in the Information Age

Maxwell L. Anderson

Indianapolis Museum of Art

We are all beholden to technologists for their diligent efforts to improve museums. Notwithstanding extraordinary progress in this field over the course of a very few years, there are two incontrovertible facts about the proliferation of information technology since the 1990s: 1) developing proprietary solutions has been consistently shown to yield short-term pride but long-term problems, and 2) even jointly engineered solutions to any informatics challenge are destined to have a brief shelf life.

A CALL TO ACTION

The wealth of information in this book about information must accordingly end with a call to action: how can we cajole busy and techno-skeptical museum leaders into making collaborative investments in informatics? In the preceding pages we find indispensable accounts not only of our collective progress, but also of historical precedents, well-intentioned wrong turns, brilliant innovations swept aside by one-size-fits all commercial solutions, "inside baseball" conversations among specialists, and highly granular remedies for problems in standards and conventions.

Each of these accounts is important, but what this book must do above all is to lead museum professionals to distill its contents into a plan of action that will bridge not only potentially fractious departments, but also bridge museums, regions, nations, NGOs, and professionals the world over. The foibles of past and current technology solutions are of relative significance in this fast-paced world, where change is a constant. Far more important today is helping the non-profit community collectively pursue improved public service by challenging the quarter-to-quarter mentality of the technology sector. Public service begins with the traditional obligations of collecting, preserving, interpreting, and displaying, but through interactivity may now invite shared responsibility in shaping the institution from without.

The alternative to thoughtful engagement by museum leaders is too dismal to contemplate: the loss of copious born-digital information lacking a platform to migrate to, precious resources squandered on short-term or flawed solutions, and lost opportunities to serve digitally sophisticated audiences that may go elsewhere for stimulation.

MUSEUMS AND SOCIAL COMPUTING

For museum leaders open to learning, the accumulated expertise among specialists in information science and museology in this volume is both authoritative and providential. Close reading yields several key insights. Perhaps the most pressing of these is the wholesale change implied by social computing—by the transition from an input–output era to that of a porous and continuous authoring environment, open to anyone regardless of background, education, or location.

For many museum directors, an embrace of social tagging could engender the abdication of a leadership role and the possibility that a chorus of amateur enthusiasts might drown out the quiet authority of scholarship. Such fears are misplaced; there will always be a safe harbor for the institutional voice. But it must be acknowledged that the vast majority of potential museumgoers are only that—*potential* visitors. It is delusional to fear that inviting public commentary and engagement will be disruptive to anything but the fantasy that we are fulfilling our potential. Making the offerings of museums pertinent to as many people as possible is a primary obligation of our field. Within that experience museum leaders will of course provide that which they see fit to provide. But unless a museum's displays, interpretive insights, amenities, and services are manifestly open to all both onsite and online, the opportunity cost of everyone's effort will remain needlessly high.

The advent of folksonomy ushers in new ways of connecting our individual experiences through institutions, thereby allowing museums to enter into new relationships with their publics (Bearman & Trant, 2005). The interpretive responsibilities of museums are being cast in a new light, one far afield from the millennia-old model of master–pupil. Instead of simply dispensing knowledge, museums will soon be expected to offer a gateway to involvement that begins with scholarly offerings and begins afresh with community editing.

The resource implications of this change are potentially enormous. Today an online visitor already demands the same amount of institutional attention as an onsite visitor, and will probably demand more detailed information than a casual tourist. But the ideal point of departure in putting folksonomy to use demands agreement on how to assess different kinds of involvement and their value.

CALIBRATING LEARNING VERSUS ENTERTAINMENT

Museum administrators today measure institutional success very bluntly, based on bodies crossing their threshold instead of, for example, time spent in front of displays. Since U.S. museums are chartered with tax exemption as educational institutions, not as entertainment centers, the measurement of success should touch first on service to their core mission. Excessive reliance on raw attendance statistics makes no more sense than if universities were to assess their achievement primarily by the number of students enrolled. Such statistics are indicative of the girth of an institution, but offer little in an assessment of its overall performance.

For those enamored of girth for its own sake, the ratio of college applicants to places offered in the next freshman class is a legitimate measurement of a university's perceived standing among prospective students. Similarly, a museum's attendance with respect to the attendance at other cultural institutions locally, or with respect to a city's Metropolitan Statistical Area (MSA), can be revealing. But a blunt measurement without due consideration of the context in which a museum operates is simplistic, and has led museums to ape commercial indicators. This can have the effect of shortchanging visitors of an abiding educational focus in the choice of exhibitions presented, publications produced, and programs staged.

As a means to help a museum fulfill its mission, technology, too, should be introduced with an educational thrust. While there is debate about how best to evaluate the number and usage patterns of end users, both attendance and end user counts should take advantage of more sophisticated means of measurement. If the average viewing of a museum object lasts only a few seconds, and our goal is to prolong each encounter and its rewards, we should find a way through motion detection to measure the amount of time that people linger in front of displays that are considered particularly deserving. Just as people are curious about which movies prove popular with others, we should be in a position to know which displays end up being the most magnetic and analyze why, with the goal being close looking at other works as well.

Similarly, if the number of end users visiting a museum Web site is actually secondary to the amount of time that end users delve into rich multimedia assets, then we should devote more energy to the provision and quantification of such experiences than to driving and celebrating traffic for its own sake.

Measuring with insight is all the more necessary since the online museumgoer promises to become more transactional than a traditional visitor. He or she will expect that queries will be answered. If such queries require even rudimentary staff research, he or she will not be patient with a delay or a generic auto-reply. As public institutions, museums will have to develop protocols and mechanisms to cope with increasing expectations on the part of end users worldwide. In the last century we added telephone operators and, briefly, telex operators. The new model of collaborative authoring presages much more change to come in staff engagement with the public.

PAYING FOR INVESTMENTS IN TECHNOLOGY

Much of the adaptation required by technology's promise will demand new resources. The provision of experience is the primary calling of museums. However we end up agreeing on how to measure engagement most effectively, if millions of potential online visitors are incentivized to take advantage of a museum's improved online offerings, we will surely be led to question the current cost–benefit assumptions of museums. By way of example, U.S. art museums today have an extremely high ratio of cost per visitor, averaging over $40 a person, calculated by the size of operating budget divided by the number of onsite visitors. If museums chose to think of participation differently, a marginal investment in an online visitor could repay the museum handsomely—not in immediate cash return, but in demonstrating the value of the museum to a greater number of people. This in turn can have the effect of persuading funders—be they governments, foundations, corporations, or private individuals—to direct support to those museums that reach the most visitors in the most rewarding ways.

We know the following to be true: the promise of top-level funding from without is always greater than the potential for incremental gains from per-visitor earned income. By contrast, today's economic model is an inefficient and skewed one—as mentioned above, the more people that cross a threshold, the more important a museum is deemed to be. A generation ago, the primary basis for evaluation had to do with the depth of a museum's collection. Since the 1980s, museums have increasingly relied on entertainment-focused amenities, such as predictable blockbusters, larger shops and restaurants, and special events, to draw in visitors. But anyone responsible for supporting museums, whether government officials or ticket-buyers, should be asking the following: what are visitors experiencing? How much of their time is devoted to the consideration of objects in the museum's care? The investment required is in the provision of rich online experiences, from detailed encounters with illustrations of objects or other features of museum collections and offerings, to live chat sessions or Webcast events. All of which can be mission-focused and educational, and far less costly than brick-and-mortar-related expenses.

The myopia of imitating for-profit attractions will lead museums inexorably towards a highly speculative reliance on fickle markets. A far more rewarding approach to serving a mission would make a virtue of the necessity that museums care for collections, by offering in-depth encounters with the core of each museum's offerings. Part of the success of Dan Brown's *The Da Vinci Code* is explained by a public appetite for insight into the mysteries behind creativity, belief systems, and history. If museums can shed their instinct to imitate theme parks and malls, and focus instead on providing privileged access to authentic objects and experiences, the online encounter with museums can be no less rewarding than whatever fare popular culture has to dole out.

VALUING CREATIVE AUTHENTICITY
OVER PASSIVE ENCOUNTERS

Which brings us to another trend for museums in the information age: the incremental growth of media-based experiences is directly proportionate to the incremental rarity of handmade artifacts. Because of their increasing preciousness, it is reasonable to presume growth in the allure of authentic objects as virtual experiences multiply. A young suburban reared on mall culture, video games, iTunes, and IM has fewer opportunities to encounter authenticity as older generations construe it. Yet while an increasing number of artists, designers, and others are employing digital media and eschewing the traditional market-driven commodification of unique creative acts, there will always be objects that have to be cared for. Thus museums will increasingly find themselves both promoting illustrated or simulated encounters with real things, while caring for the real things themselves. How we make a virtue of this necessity is an interpretive challenge no smaller than the technical challenge of reducing the intrusiveness of hardware. But it will require agility in matching digital museum offerings with new learning habits.

With each half-decade of change in technology, we see young people increasingly reared in electronic cocoons that they are shaping. Their impatience with passive encounters is already transforming entertainment options, as movie theater attendance declines and video games and downloadable films with alternative endings increase. Museums have a plethora of challenges in keeping pace with this new and demanding audience. Ceding authority is never easy, but particularly so in the case of professional fetishists. Our resistance to community editing is natural enough for scholars trained in an academic environment. But opening museum content to contributions from the public will be essential as we seek to be relevant in a world that finds satisfaction in collaborative communication.

As keyboards give way to voice recognition, finite screens will give way to projections on any kind of flat surface, and Web pages will give way to more fluid learning environments through lightweight wearable gear including mobile phones. Museums' battle lines of just a few years ago—around how to write a good label—have swiftly given way to the myriad choices presented by podcasting, blogging, and other subversions of institutional insularity. Our collective embrace of bit streams created by and with others will be seen by some as threatening the very essence of museum culture. And, indeed, the wholesale abandonment of scholarly authority would spell the end of museums and their substitution by warehouses. But the goal should be to reward the fruits of disciplined research by making these eminently accessible to and annotated by a broad and diverse public.

The way forward will surely not be in the printed exhibition catalogue selling to 5% of audiences, or the Web site resembling a kinetic brochure, but in live, streaming, downloadable, and open-ended resources. We will have to learn to eschew certain traditional forms of gate-keeping and create new invitations to learn and collaborate in tandem with our audiences. The choice of museum programming itself will be under continuing pressure as resources shrink for the emphatically non-digital world of packing crates, jet fuel, dry wall, and technical expertise associated with museum displays.

How we cope with the fickle tastes of the marketplace will reflect the extent of our reliance on earned income. Those institutions fortunate enough to have sizable endowments can better insulate themselves from the vagaries of presumed public appetite, and can offer up museum experiences that they believe are especially deserving. The alternative is to be increasingly drawn into compromises akin to those in the for-profit world. If museums are to be educational preserves, they need to lead by example, rather than to be reflexively populist. If higher education were to cede decisions about what to teach to a plebiscite of students, the curricula at universities would see much of what is precious about human knowledge vanish under the waves of fashion. And ultimately, this is the preeminent challenge served up by the prevalence of digital media.

What may be obvious to a curator is far from obvious to a funder, trustee, government official, or visitor. Since the provision of a personal compass for understanding history, culture, and nature are among our key goals, we will have to establish and collectively advocate new ways of assessing the contributions of museums that imaginatively promote the value of serendipitous encounters. Old models, whether based on the late 19th century vision of museums as beneficent treasure houses ministering to the working class, or the late 20th century vision of museums as shopping malls with galleries attached, can mercifully become obsolete.

A new model, privileging public engagement with accessible scholarly resources, offers a promise of relevance in a world awash with consumer response driving corporate planning. Instead of using technology to make museums more efficient imitators of for-profit attractions, we should devote our energies to making museums more responsive to the perspectives of others, while arguing forcefully for the legitimacy of scholarly innovation. This approach will both underscore museums' educational benefits and encourage their vitality as a public resource, and can help wean us of a misguided pursuit of ticket sales to the exclusion of mission-focused goals.

CONCLUSION

A decade ago, in the introduction to *The Wired Museum*, I imagined a series of imminent changes to the museum landscape, while guardedly noting that "today's prognostications will fuel tomorrow's ironic reminiscences." Along those lines, I then assumed that voice recognition technology would be more universal than it has yet become, along with personally tailored content, three-dimensional imaging, the convergence of the computer and the television, and site licensing as a means of coping with copyright dilemmas. I also supposed that intranets would allow visitors to "eavesdrop on our activities"—but failed to predict that folksonomy would go so far as to allow them to be unmediated content creators. Mercifully I did promote more open sharing of the mechanics of museum operations, on the assumption that visitors allowed to bear witness to museum activities would be more engaged in the life of museums overall.

Notwithstanding the deliberate pace of technology's march in some respects, in others we have much promise and many choices ahead. The "wireless" museum is a radical variant of Malraux's museum without walls. The return of art to daily life, in the form of affordable high design, presages our return to a time during which the few will dictate less to the many. Artistic expression is no longer seen as the preserve of an eccentric in a garret, but as the birthright of any boy or girl with authoring software, who can instantly publish the results to a ubiquitous platform. The calibration of achievement in creativity was once made by the French Academy, then by museums, now by the marketplace, and soon enough by the appetites of millions online. While those who toiled in graduate school will bemoan the decline of standards, others with a longer view will have faith that truth and reason will prevail in an unfathomably vast sea of bit streams.

For museums to become relevant in societies offering instant access to digital equivalents of anything we can create or imagine, a healthy attitude would privilege involvement over silence, questioning over placid acceptance, and a commitment to providing challenging artistic experiences over pandering fare.

I continue to believe, a decade on, that deepened exposure to the latest thinking of experts will inspire people to visit museums and sidle up to original objects. The success of the "CSI" television series is a function of removing the veil from the mechanics of detective work, just as popular hospital dramas offer crash courses in diagnosing and treating ailments. Museums must be no less determined to provide behind-the-scenes access to how we evaluate and foster creativity, history, and science.

Like all of the assumptions we have collectively made about museum informatics since the 1990s, the safest ground in making predictions is there: that greater exposure to daily life in and around museums, such as that provided by this book, can only promote more interest and involvement by those who might be inclined to become regular visitors, both onsite and online.

References

Abell-Seddon, B. (1988). *Museum catalogues: A foundation for computer processing*. London: Clive Bingley.

Abell-Seddon, B. (1989). Reforming collection documentation: A new approach. *International Journal of Museum Management and Curatorship, 8*, 63–67.

Adams, C., Cole, T., DePaolo, C., & Edwards, S. (2001). Bringing the curatorial process to the web. In D. Bearman & J. Trant (Eds.), *Museums and the Web 2001: Selected papers from an international conference* (pp. 11–22). Pittsburgh, PA: Archives & Museum Informatics. Available from http://www.archimuse.com/mw2001/papers/depaolo/depaola.html

Adams, M., & Moussouri, T. (2002, May 17). *What is the nature of the interactive experience? Linking research and practice*. Paper presented at the 2002 International Conference on Interactive Learning in Museums of Art, Victoria and Albert Museum, London, England. Available from http://www.vam.ac.uk/res_cons/research/learning/

Allen, N. (2000). Collaboration through the Colorado Digitization Project, *First Monday, 5*(6). Available from http://firstmonday.org/issues/issue5_6/allen/

Allkin, R., White, R. J., & Winfield, P. J. (1992). Handling the taxonomic structure of biological data. *Mathematical and Computer Modelling, 16*(6–7), 1–9.

Alperstein, N. (2003). The Museum of Modern Art's Visual Thinking Curriculum: Getting ready for a visit to the Picasso-Matisse Exhibit at MoMA-QNS working with AMICO (Art Museum Images Consortium). Available from http://neme.alperstein.home.att.net/art2003/moma_amico.htm

American Association of Museums. (1984). *Museums for a new century*. Washington, DC: American Association of Museums.

American Association of Museums. (1992). *Excellence and equity: Education and the public dimension of museums*. Washington, DC: American Association of Museums.

American Association of Museums. (1994). *Careers in museums: A variety of vocations*. Washington, DC: American Association of Museums.

American Association of Museums. (n.d.). *ABCs of museums*. Available from http://www.aam-us.org/aboutmuseums/abc.cfm#how_many

American Council of Learned Societies. (2004). *Commission on cyberinfrastructure for the humanities & social sciences*. Available from http://www.acls.org/cyberinfrastructure/cyber.htm

American Library Association. (2005, September). *ALA library fact sheet 1*. Available from http://www.ala.org/Template.cfm?Section=Library_Fact_Sheets&Template=/ContentManagement/ContentDisplay.cfm&ContentID=22687

Anderson, M. (1999). Museums of the future: The impact of technology on museum practices. *Daedalus, 128*(3), 129–162.

Anderson, P., & Roe, B. C. (1993). *The museum impact evaluation study: Roles of affect in the museum visit and ways of assessing them. Volume one: Summary.* Chicago: Museum of Science and Industry.

Aoki, P., & Woodruff, A. (2000). Improving electronic guidebook interfaces using a task-oriented design approach. In *Conference proceedings on designing interactive systems: Processes, procedures, methods, and techniques* (pp. 319–325). New York: ACM Press.

Aoki, P. M., Grinter, R. E., Hurst, A., Szymanski, M. H., Thornton, J. D., & Woodruff, A. (2002). Sotto Voce: Exploring the interplay of conversation and mobile audio spaces. In *Proceedings of the SIGCHI conference on human factors in computing systems: Changing our world, changing ourselves* (pp. 431–438). New York: ACM Press.

Art Museum Image Consortium. (2004). *AMICO Data Specification.* Available from http://www.amico.org/docs/dataspec.html

Arthur, K., Byrne, S., Long, E., Montori, C. Q., & Nadler, J. (2004). *Recognizing digitization as a preservation reformatting method: Prepared for the Preservation of Research Library Materials Committee of the Association of Research Libraries.* Available from http://www.arl.org/preserv/digit_final.html

Arts, M., & Schoonhoven, S. (2005). Culture around the corner and its location-based application. In D. Bearman & J. Trant (Eds.), *Museums and the Web 2005: Selected papers from an international conference* (n.p.). Toronto, Canada: Archives & Museum Informatics. Available from http://www.archimuse.com/mw2005/papers/arts/arts.html

Augst, T., & Wiegand, W. (2003). *Libraries as agencies of culture.* Madison, WI: University of Wisconsin Press.

Baca, M. (1998a). A crosswalk of metadata standards. In M. Baca (Ed.), *Introduction to metadata: Pathways to digital information* (pp. 23–33). Los Angeles: Getty Information Institute. Online Edition, Version 2.1 available from the J. Paul Getty Trust Website: http://www.getty.edu/research/conducting_research/standards/intrometadata/3_crosswalks/index.html

Baca, M. (Ed.). (1998b). *Introduction to metadata: Pathways to digital information.* Los Angeles: Getty Information Institute. Online Edition, Version 2.1 available from the J. Paul Getty Trust Website: http://www.getty.edu/research/conducting_research/standards/intrometadata/index.html

Baca, M., & Harpring, P. (1996). Art information task force: Categories for the description of works of art. *Visual Resources, 11*(3–4), 241–436.

Baillie, A. (1996, November). *Empowering the visitor: The family experience of museums. A pilot study of ten family group visits to the Queensland Museum.* Paper presented at the 1996 Museums Australia Conference: Power & Empowerment, Preparing for the New Millennium, Sydney, Australia. Available from http://amol.org.au/evrsig/pdf/baillie96.pdf

Barbieri, T., & Paolini, P. (2000). Cooperative visits for museum WWW sites a year later: Evaluating the effect. In *Museums and the Web 2000: Selected papers from an international conference* (pp. 173–178). Pittsburgh, PA: Archives & Museum Informatics. Available from http://www.archimuse.com/mw2000/papers/barbieri/barbieri.html

Battles, M. (2003). *Library: An unquiet history.* New York: W. W. Norton.

Battro, A. M. (1999, September 13–18). *From Malraux's imaginary museum to the virtual museum.* Paper presented at the 10th World Federation of Friends of Museums Congress, Sydney, Australia. Available from http://www.byd.com.ar/vm99sep.htm

Baxandall, M. (1987). *Patterns of intention: On the historical explanation of pictures.* New Haven, CT: Yale University Press.

Bearman, D. (1987). *Functional requirements for collections management systems (No. 3)*. Pittsburgh, PA: Archives & Museum Informatics.

Bearman, D. (1988). Considerations in the design of art scholarly databases. *Library Trends, 37*(2), 206–219.

Bearman, D. (1990a). *Archives & museum data models and dictionaries (No. 10)*. Pittsburgh, PA: Archives & Museum Informatics.

Bearman, D. (1990b). *Functional requirements for membership, development & participation systems (no. 11)*. Pittsburgh, PA: Archives & Museum Informatics.

Bearman, D. (Ed.). (1991). *Hypermedia & Interactivity in Museums: Proceedings from ichim91*. Pittsburgh, PA: Archives & Museum Informatics.

Bearman, D. (1992a). *Archives and museum information systems (AMIS) specification. A report to the Research Libraries Group*. Pittsburgh, PA: Archives & Museum Informatics.

Bearman, D. (1992b). A user community discovers its standards. *Journal of the American Society for Information Science, 43*(8), 576–578.

Bearman, D. (1994a). Strategies for cultural heritage information standards in a networked world. *Archives and Museum Informatics, 8*(2), 93–106.

Bearman, D. (1994b, October 13). *Useful electronic space or virtual junkheap?* Paper presented at the 1994 Conference on Digital Cataloging, Washington, DC. Available from http://www.loc.gov/catdir/semdigdocs/bearman.html

Bearman, D. (1995a). Data relationships in the documentation of cultural objects. *Visual Resources, 11*, 295–306.

Bearman, D. (1995b). Information strategies and structures for electronic museums. In A. Fahy & W. Sudbury (Eds.), *Information: The hidden resource, museums and the Internet* (pp. 5–22). Cambridge, England: Museum Documentation Association.

Bearman, D. (Ed.). (1995c). *Hands on: Hypermedia & interactivity in museums: Selected papers from the third International Conference on Hypermedia and Interactivity in Museums*. Pittsburgh, PA: Archives & Museum Informatics.

Bearman, D. (1997). New economic models for administering cultural intellectual property. In K. Jones-Garmil (Ed.), *The wired museum: Emerging technology and changing paradigms* (pp. 231–266). Washington, DC: American Association of Museums.

Bearman, D., & Cox, L. (1990). *Directory of software for archives and museums (No. 12)*. Pittsburgh, PA: Archives & Museum Informatics.

Bearman, D., & Garzotto, F. (Eds.). (2001). *International Cultural Heritage Informatics Meeting: Proceedings from ichim01*. Pittsburgh, PA: Archives & Museum Informatics.

Bearman, D., Miller, E., Rust, G., Trant, J., & Weibel, S. (1999). A common model to support interoperable metadata: Progress report on reconciling metadata requirements from the Dublin Core and INDECS/DOI communities. *D-Lib Magazine, 5*(1). Available from http://www.dlib.org.proxy.lib.fsu.edu/dlib/january99/bearman/01bearman.html

Bearman, D., & Perkins, J. (1993). The standards framework for computer interchange of museum information. *Spectra, 20*(2 & 3), 1–61.

Bearman, D., & Peterson, T. (1991). Retrieval requirements of faceted thesauri in interactive information systems. In S. M. Humphrey & B. H. Kwasnik (Eds.), *Advances in Classification Research: Proceedings of the 1st ASIS SIG/CR Classification Research Workshop* (pp. 9–24). Medford, NJ: Learned Information.

Bearman, D., & Szary, R. (1986). Beyond authority control: Authorities as reference files in a multi-disciplinary setting. In K. E. Markey (Ed.), *Authority control symposium* (pp. 69–78). Tuscon, AZ: ARLIS/NA.

Bearman, D., & Trant, J. (Eds.). (1997). *Museum interactive multimedia 1997: Cultural heritage systems design and interfaces: Selected Papers from the fourth*

International Conference on Hypermedia and Interactivity in Museums. Pittsburgh, PA: Archives & Museum Informatics.

Bearman, D., & Trant, J. (1998a). Economic, social, technical models for digital libraries of primary resources. *New Review of Information Networking, 4,* 71–91.

Bearman, D., & Trant, J. (1998b). Unifying our cultural memory: Could electronic environments bridge the historical accidents that fragment cultural collections? In L. Dempsey, S. Criddle & R. Heseltine (Eds.), *Information Landscapes for a Learning Society, Networking and the Future of Libraries* (Vol. 3).

Bearman, D., & Trant, J. (Eds.). (1999). *Cultural heritage informatics: Selected papers form ichim99.* Pittsburgh, PA: Archives & Museum Informatics.

Bearman, D., & Trant, J. (Eds.). (2003). *International Cultural Heritage Informatics Meeting: Proceedings from ichim03* [CD-ROM]. Toronto, Canada: Archives & Museum Informatics.

Bearman, D., & Trant, J. (Eds.). (2004). *International Cultural Heritage Informatics Meeting: Proceedings from ichim04* [CD-ROM]. Toronto, Canada: Archives & Museum Informatics.

Bearman, D., & Trant, J. (Eds.). (2005). *International Cultural Heritage Informatics Meeting: Proceedings from iChim05* [CD-ROM]. Toronto, Canada: Archives & Museum Informatics.

Bearman, D., Trant, J., Chun, S., Jenkins, M., Smith, K., Cherry, R., et al. (2005). Social terminology enhancement through vernacular engagement: Exploring collaborative annotation to encourage interaction with museum collections. *D-Lib Magazine, 9*(11). Available from http://www.dlib.org/dlib/september05/bearman/09bearman.html

Bearman, D., & Vulpe, M. (1985). *CMASS (Collections Management Action Support System): A statement of problem.* Washington, DC: Smithsonian Institution, Office of Information Resource Management.

Bearman, D., & Wright, B. (1992). *Directory of software for archives and museums.* Pittsburgh, PA: Archives & Museum Informatics.

Bearman, D., & Wright, B. (1994). *Directory of software for archives and museums.* Pittsburgh, PA: Archives & Museum Informatics.

Bennett, N., & Sandore, B. (2001). The IMLS digital cultural heritage community project: A case study of tools for effective project management and collaboration. *First Monday, 6* (7). Available from http://firstmonday.org/issues/issue6_7/bennett

Bennett, N., & Trofanenko, B. (2002). Digital primary source materials in the classroom. In D. Bearman & J. Trant (Eds.), *Museums and the Web 2002: Selected papers from an international conference* (pp. 149–156). Pittsburgh, PA: Archives & Museum Informatics. Available from http://www.archimuse.com/mw2002/papers/bennett/bennett.html

Bergengren, G. (1979). Towards a total information-system + museum computerization. *Museum, 30*(3–4), 213–217.

Bernier, R., & Bowen, J. (2004). Web-based discussion groups at stake: The profile of museum professionals online. *Program: Electronic Library & Information Systems, 38,* 120–137.

Besser, H. (1997a). The changing role of photographic collections with the advent of digitization. In K. Jones-Garmil (Ed.), *The wired museum: Emerging technology and changing paradigms* (pp. 115–128). Washington, DC: American Association of Museums.

Besser, H. (1997b). The transformation of the museum and the way it's perceived. In K. Jones-Garmil (Ed.), *The wired museum: Emerging technology and changing paradigms* (pp. 153–170). Washington, DC: American Association of Museums.

Besser, H., & Stephenson, C. (1996). The Museum Educational Site Licensing Project: Technical issues in the distribution of Museum images and textual data to universities. In J. Hemsley (Ed.), *E.V.A. '96 London (Electronic Imaging and the Visual Arts)*. Hampshire, England: Vasari Ltd.

Beynondavies, P., Tudhope, D., Taylor, C., & Jones, C. (1994). A semantic database approach to knowledge-based hypermedia systems. *Information and Software Technology, 36*(6), 323–329.

Bishoff, L., & Allen, N. (2004). *Business planning for cultural heritage institutions: A framework and resource guide to assist cultural heritage institutions with business planning for sustainability of digital asset management programs*. Washington, DC: Council on Library and Information Resources.

Bissel, T., Bogen, M., Hadamschek, V., & Riemann, C. (2000). Protecting a museum's digital stock through watermarks. In D. Bearman & J. Trant (Eds.), *Museums and the Web 2000: Selected papers from an international conference* (pp. 73–82). Pittsburgh, PA: Archives & Museum Informatics. Available from http://www.archimuse.com/mw2000/papers/bissel/bissel.html

Blackaby, J. R. (1997). Integrated information systems. In K. Jones-Garmil (Ed.), *The wired museum: Emerging technology and changing paradigms* (pp. 203–230). Washington, DC: American Association of Museums.

Blackaby, J., & Greeno, P. (1988). *The revised nomenclature for museum cataloging: A revised and expanded version of Robert G. Chenhall's system for classifying man-made objects*. Walnut Creek: AltaMira Press.

Boas, F. (1907). Some principles of museum administration. *Science, 25*(650), 921–933.

Boehner, K., Gay, G., & Larking, C. (2005). Drawing evaluation into design for mobile computing: A case study of the Renwick Gallery's handheld education project. *Journal of Digital Libraries, Special Issue on Digital Museum, 5*(3), 219–230.

Boiko, B. (2002). *Content management bible*. New York: Wiley.

Booth, B. (1998). Information for visitors to cultural attractions. *Journal of Information Science, 24*(5), 291–303.

Borysewicz, S. (1998). Networked media: The experience is closer than you think. In S. Thomas & A. Mintz (Eds.), *The virtual and the real: Media in the museum* (pp. 103–117). Washington, DC: American Association of Museums.

Bowen, J. (1999). Time for renovations: a survey of museum websites. In D. Bearman & J. Trant (Eds.), *Museums and the Web 1999: Selected papers from an international conference* (pp. 163–174). Pittsburgh, PA: Archives & Museum Informatics. Available from http://www.archimuse.com/mw99/papers/bowen/bowen.html

Bowen, J. P., & Filippini-Fantoni, S. (2004). Personalization and the web from a museum perspective. In D. Bearman & J. Trant (Eds.), *Museums and the Web 2004: Selected papers from an international conference* (pp. 63–78). Toronto, Canada: Archives & Museum Informatics. Available from http://www.archimuse.com/mw2004/papers/bowen/bowen.html

Bower, J. M. (1993). Vocabulary control and the virtual database. *Knowledge Organization, 20*(1), 4–7.

British Council. (2003). *Asian field*. Available from http://www.britishcouncil.org/arts-art-sculpture-antony-gormley-asian-field.htm

Brown, B., & Chalmers, M. (2003). Tourism and mobile technology. In K. Kuutti, E. H. Karsten, G. Fitzpatrick, P. Dourish & K. Schmidt (Eds.), *ECSCW 2003: Proceedings of the 8th European conference on computer supported cooperative work, 14–18 September 2003, Helsinki, Finland* (pp. 335–354). Dordrecht, The Netherlands: Kluwer Academic Publishers.

Brown, B., MacColl, I., Chalmers, M., Galani, A., Randell, C., & Steed, A. (2003). Lessons from The Lighthouse: Collaboration in a shared mixed reality system. In *Proceedings of the SIGCHI conference on human factors in computing systems* (pp. 577–584). New York: ACM Press.

Brown, J. S., & Duguid, P. (1991). Organizational learning and communities of practice: Toward a unified view of working, learning and innovation. *Organization Science, 2*, 40–57.

Brown, J. S., & Duguid, P. (2000). *The social life of information*. Boston: Harvard Business School Press.

Bryman, A. (2001). *Social research methods*. Oxford, England: Oxford University Press.

Buck, R. A., & Gilmore, J. A. (Eds.). (1998). *The new museum registration methods*. Washington, DC: American Association of Museums.

Buckland, M. K. (1997). What is a "document?" *Journal of the American Society for Information Science, 48*(9), 804–809.

Burkaw, G. E. (1997). *Introduction to museum work*. Walnut Creek, CA: AltaMira Press.

California State University Monterey Bay. (n.d.). *How to write a policy*. Available from http://policy.csumb.edu/develop/write.html

Cameron, F. (2003). Digital futures I: Museum collections, digital technologies, and the cultural construction of knowledge. *Curator, 46*, 325–340.

Cameron, F., & Kenderdine, S. (Eds.). (2007). *Theorizing digital cultural heritage: A critical discourse*. Cambridge, MA: MIT Press.

Canadian Centre for Architecture. (2004). *Collections online*. Available from http://www.cca.qc.ca/pages/Niveau3.asp?page=catalogue_collection &lang=eng

Canadian Heritage Information Network. (1999). *Information technology in Canadian museums: A survey by the Canadian Heritage Information Network*. Available from http://www.chin.gc.ca/English/Reference_Library/Information_Technology/index.html

Canadian Heritage Information Network. (2004). *CHIN 2004 National Membership Study*. Available from http://www.chin.gc.ca/English/Members/Reports/Membership_Survey

Canadian Heritage Information Network. (2005). *CHIN Data Dictionary: Humanities*. Available from http://daryl.chin.gc.ca:8000/BASIS/chindd/user/wwwhe/SF

Cannon-Brookes, P. (1992). The nature of museum collections. In J. Thompson (Ed.), *Manual of curatorship* (pp. 500–512). London: Butterworths.

The Can-Spam Act of 2003, Pub. L. No. 108-187, S.877. (2003). Available from http://www.spamlaws.com/federal/can-spam.shtml

Carliner, S. (2003). Modeling information for three-dimensional space: lessons learned from museum exhibit design. *Technical Communication, 50*(4), 554–570.

Case, M. (Ed.). (1995). *Registrars on record: Essays on museum collections management*. Washington, DC: Registrars Committee of the American Association of Museums.

Chadwick, J., (1998). *Public Utilization of Museum-based World Wide Web Sites*. Unpublished doctoral dissertation, University of New Mexico.

Chadwick, J., & Boverie, P. (1999). A survey of characteristics and patterns of behavior in visitors to a museum web site. In D. Bearman & J. Trant (Eds.), *Museums and the Web 1999: Selected papers from an international conference* (pp. 154–162). Pittsburgh, PA: Archives & Museum Informatics. Available from http://www.archimuse.com/mw99/papers/chadwick/chadwick.html

Chadwick, J., Falk, J. H. & O'Ryan, B. (2000). *Assessing Institutional Web Sites*. Council on Library and Information Resources. Available from http://www.clir.org/pubs/reports/pub88/appendix2.html

Chalmers, M., & Galani, A. (2004). Seamful interweaving: Heterogeneity in the theory and design of interactive systems. In *Proceedings of the 2004 conference on designing interactive systems: Processes, practices, methods, and techniques* (pp. 243–252). New York: ACM Press.

Chandler, B., Andrews, W. J. H., & Rossi-Wilcox, S. M. (1994). *Treasures of art and science at Harvard University.* Cambridge, MA: Harvard University Museums Council.

Chandler, R. L. (2002). Museums in the Online Archive of California (MOAC): Building digital collections across libraries and museums, *First Monday, 7*(5). Available from http://firstmonday.org/issues/issue7_5/chandler/index.html

Chavan, V., & Krishnan, S. (2003). Natural history collections: A call for national information infrastructure. *Current Science, 84*(1), 34–42.

Checkland, P., & Howell, S. (1998). *Information, systems and information systems: Making sense of the field.* Chichester, NY: Wiley.

Chenhall, R. G. (1975). *Museum cataloging in the computer age.* Nashville, TN: American Association for State and Local History.

Chenhall, R. G. (1978). *Nomenclature for museum cataloging: A system for classifying man-made objects.* Nashville, TN: American Association for State and Local History.

Chenhall, R. G., & Vance, D. (1988). *Museum collections and today's computers.* Westport, CT: Greenwood Press.

Cheverst, K., Davies, N., Mitchell, K., Friday, A., & Efstratiou, C. (2000). Developing a context-aware electronic tourist guide: Some issues and experiences. In *Proceedings of the SIGCHI conference on human factors in computing systems* (pp. 17–24). New York: ACM Press.

Children's Online Privacy Protection Act of 1998, 15 U.S.C. §§ 91. (2003). Available from http://www.ftc.gov/ogc/coppa1.htm

Coburn, E., & Baca, M. (2004). Beyond the gallery walls: Tools and methods for leading end-users to collections information. *Bulletin of the American Society for Information Science and Technology, 30*(5), 14–19. Available from http://www.asis.org/Bulletin/Jun-04/coburn_baca.html

Computing Research Association. (2005). *Cyberinfrastructure for education and learning for the future: A vision and research agenda.* Available from http://www.cra.org/reports/cyberinfrastructure.pdf

Consortium for Interchange of Museum Information. (2000). *Guide to best practice: Dublin Core.* Available from http://www.cimi.org/public_docs/meta_bestprac_v1_1_210400.pdf

Corcoran, F., Demaine, J., Picard, M., Dicaire, L.-G., & Taylor, J. (2002). INUIT3D: An interactive virtual 3D web exhibition. In D. Bearman & J. Trant (Eds.), *Museums and the Web 2002: Selected papers from an international conference* (n.p.). Pittsburgh, PA: Archives & Museum Informatics. Available from http://www.archimuse.com/mw2002/papers/corcoran/corcoran.html

Cowton, J. (Ed.). (1997). *SPECTRUM: The UK museum documentation standard.* Cambridge: Museum Documentation Association.

Creighton, R. A., & Crockett, J. J. (1971). SELGEM: A system for collection management. *Smithsonian Institute Information Systems Innovations, 2*(3), 1–26.

Cultural Applications: Local Institutions Mediating Electronic Resources. (2005, February). *Guidelines on social inclusion.* Available from http://www.calimera.org/Lists/Guidelines/Social_inclusion.htm

Cunliffe, D., Kritou, E., & Tudhope, D. (2001). Usability evaluation for museum web sites. *Museum Management and Curatorship, 19*(3), 229–252.

Danilov, V. J. (1994). *Museum careers and training: A professional guide.* Westport, CT: Greenwood Press.

Data Protection Act 1998. United Kingdom: Queen's Printer of Acts of Parliament. (1998). Available from http://www.hmso.gov.uk/acts/acts1998/19980029.htm#aofs

Davenport, T. H., & Prusak, L. (1997). *Information ecology: Mastering the information and knowledge environment.* New York: Oxford University Press.

Deakin University, Cultural Heritage Centre for Asia and the Pacific. (2002). *A study into the key needs of collecting institutions in the heritage sector.* Melbourne, Australia: Deakin University Press. Available from http://sector.amol.org.au/ publications_archive/national_policies/key_needs

DeAngelis, I. (2002). Collections management: Hypothetical cases, acquisitions, deaccessions and loans. In T. A. Lipinksi (Ed.), *Libraries, museums and archives: Legal issues and ethical challenges in the new information era* (pp. 83–94). Lanham, MD: Scarecrow Press.

Devine, J., Gibson, E., & Kane, M. (2004). *What clicks? Electronic access to museum resources in Scotland and e-learning opportunities using museum resources.* Available from http://www.hunterian.gla.ac.uk/what_clicks/index.shtml

Devine, J., & Hansen, C. (2001). Light from SHADE: Educational access to museum collections through digital technologies. *Spectra, 28*(2), 38–40.

Devine, J., & Hansen, C. C. (2003). SHADE Smithsonian-Hunterian Advanced Digital Experiments. In D. Bearman & J. Trant (Eds.), *Museums and the Web 2003: Selected papers from an international conference* (n.p.). Pittsburgh, PA: Archives & Museum Informatics. Available from http://www.archimuse.com/mw2003/papers/devine/devine.html

Devine, J., & Welland, R. (2000). Cultural computing: Exploiting interactive digital media. *Museum International, 52*(1), 32–35.

Diamond, J. (1986). The behavior of family groups in science museums. *Curator, 29*(2), 139–154.

Di Blas, N., Gobbo, E., & Paolini, P. (2005). 3D worlds and cultural heritage: Realism vs. virtual presence. In D. Bearman & J. Trant (Eds.), *Museums and the Web 2005: Selected papers from an international conference* (pp. 183–194). Toronto, Canada: Archives & Museum Informatics. Available from http://www.archimuse.com/mw2005/papers/diBlas/diBlas.html

Di Blas, N., Guermandi, M. P., Orsini, C., & Paolini, P. (2002). Evaluating the features of museum websites: (The Bologna report). In D. Bearman & J. Trant (Eds.), *Museums and the Web 2002: Selected papers from an international conference* (pp. 179–188). Pittsburgh, PA: Archives & Museum Informatics. Available from http://www.archimuse.com/mw2002/papers/diblas/diblas.html

Dierking, L. D., & Falk, J. H. (1994). Family behavior and learning in informal science settings: A review of the research. *Science Education, 78*(1), 57–72.

Dierking, L. D., & Falk, J. H. (1998). Understanding the museum experience: A review of visitor research and its applicability to museum web sites. In D. Bearman & J. Trant (Eds.), *Museums and the Web 1998: Selected papers from an international conference* [CD-ROM] (n.p.). Pittsburgh, PA: Archives & Museum Informatics. Available from http://www.archimuse.com/mw98/papers/dierking/dierking_paper.html

Dierking, L. D., Luke, J. J., Foat, K. A., & Adelman, L. (2001, November/December). The family and free choice learning. *Museum News, 80*(6), 38–43, 67–69.

Dietz, S. (1998). Curating (on) the web. In D. Bearman & J. Trant (Eds.), *Museums and the Web 1998: Selected papers from an international conference* [CD-ROM] (n.p.). Pittsburgh, PA: Archives & Museum Informatics. Available from http://www.archimuse.com/mw98/papers/dietz/dietz_curatingtheweb.html

Dietz, S., Besser, H., Borda, A., & Gerber, K. (2004). *Virtual Museum (of Canada): The next generation.* Available from http://www.chin.gc.ca/English/Members/Rethinking_Group/

The Digital Millennium Copyright Act, Pub. L. No. 105-304, 112 Stat 2860. (1998). Available from the U.S. Copyright Office Website at www.copyright.gov/legislation/dmca.pdf

Doerr, M. (1997). Reference information acquisition and coordination. *Proceedings of the American Society for Information Science Annual Meeting, 34,* 295–312.

Doerr, M., Hunter, J., & Lagoze, C. (2003). Towards a core ontology for information integration, *Journal of Digital Information*, 4(1). Available from http://jodi.tamu.edu/Articles/v04/i01/Doerr/

Dooley, J. M. (Ed.). (1998). *Encoded archival description: Context, theory, and case studies*. Chicago: Society of American Archivists.

Doty, P. (1990). Automating the documentation of museum collections. *Museum Management and Curatorship*, 9(1), 73–83.

Douma, M., & Henchman, M. (2000). Bringing the object to the viewer: Multimedia techniques for the scientific study of art. In D. Bearman & J. Trant (Eds.), *Museums and the Web 2000: Selected papers from an international conference* (pp. 59–64). Pittsburgh, PA: Archives & Museum Informatics. Available from http://www.archimuse.com/mw2000/papers/doumahenchman/doumahenchman.html

Drucker, P. (1994). The age of social information. *The Atlantic Monthly*, 274(5), 53–70.

Duff, W. M., Craig, B., & Cherry, J. (2004). Historians' use of archival sources: Promises and pitfalls of the digital age. *The Public Historian*, 26(2), 7–22.

Dworman, G., Kimbrough, S., & Patch, C. (2000). On pattern-directed search of archives and collections. *Journal of the American Society for Information Science*, 51(1), 14–23.

Dyson, M., & Moran, K. (2000). Informing the design of Web interfaces to museum collections. *Museum Management and Curatorship*, 18, 391–406.

Economou, M. (1996). Museum collections and the information superhighway. *Spectra*, 23(1), 7–10.

Economou, M. (1998). The evaluation of museum multimedia applications: Lessons from research. *Museum Management and Curatorship*, 17(2), 173–187.

Economou, M. (2003). New media for interpreting archaeology in museums: Issues and challenges. In M. Doerr & A. Sarris (Eds.), *Proceedings from CAA2002 (Computer Applications in Archaeology International Conference)* (pp. 371–375). Athens, Greece: Hellenic Ministry of Culture.

Electronic Communications Privacy Act, 18 U.S.C. §§ 2510. Available from http://floridalawfirm.com/privacy.html

Electronic Privacy Information Center (EPIC). (2003, March 6). *The Supreme Court set to review Alaska's Megan's Law*. Available from http://www.epic.org/privacy/meganslaw/

Ellenbogen, K. M. (2002). Museums in family life: An ethnographic case study. In G. Leinhardt & K. Crowley (Eds.), *Learning conversations: Explanation and identity in museums* (pp. 81–101). Mahwah, NJ: Erlbaum.

Ellenbogen, K.M. (2003). *From dioramas to the dinner table: An ethnographic case study of the role of science museums in family life*. Dissertation Abstracts International, 64(03), 846A. (University Microfilms No. AAT30-85758)

Ellenbogen, K. M., Luke, J. J., & Dierking, L. D. (2004). Family learning research in museums: An emerging disciplinary matrix? *Science Education*, 88(Supplement 1), S48–S58.

Ellin, E. (1969). Museums and the computer, an appraisal of new potential. *Computers and the Humanities*, 4(1), 25–30.

Evans, J., & Sterry, P. (1999). Portable computers and interactive multimedia: A new paradigm for interpreting museum collections. *Archives and Museum Informatics*, 13, 113–126.

The Exploratorium. (2001, October). *Electronic guidebook forum report*. Available from http://www.exploratorium.edu/guidebook/forum_report.pdf

Fahy, A. (Ed.). (1995). *Collections management*. London: Routledge.

Falk, J. H. (1993). *Leisure decisions influencing African American use of museums*. Washington, DC: American Association of Museums.

Falk, J. H. (1998, March/April). Visitors: Who does, who doesn't and why. *Museum News, 77*(2), 38–43.

Falk, J. H. (2006). The impact of visit motivation on learning: Using identity as a construct to understand the visitor experience. *Curator, 49*(2), 151–166.

Falk, J. H., & Dierking, L. D. (1992). *The museum experience*. Washington, DC: Whalesback Books.

Falk, J. H., & Dierking, L. D. (2000). *Learning from museums: Visitor experiences and the making of meaning*. Walnut Creek, CA: AltaMira Press.

Falk, J. H., Moussouri, T., & Coulson, D. (1998). The effect of visitors' agendas on museum learning. *Curator, 41*(2), 106–120.

Falk, J.H. & Storksdieck, M. (2005). Using the Contextual Model of Learning to understand visitor learning from a science center exhibition. *Science Education* 89, 744–778.

Fallows, D. (2005). *Search engine users: Internet searchers are confident, satisfied and trusting—but they are also unaware and naïve*. Washington, DC: Pew Foundation. Available from http://www.pewinternet.org/pdfs/PIP%5FSearchengine%5Fusers.pdf

Farley, L. (2004, May). Digital images come of age, *Syllabus*. Available from http://www.campus-technology.com/article.asp?id=9363

Federman, M. (2004). *The ephemeral artefact: Visions of cultural experience*. Paper presented at eCulture Horizons: From Digitisation to Creating Cultural Experience(s), Salzburg, Austria. Available from http://www.mcluhan.utoronto.ca/EphemeralArtefact.pdf

Fink, E. (1999). The Getty Information Institute: A retrospective. *D-Lib Magazine, 5*(3). Available from http://dlib.anu.edu.au/dlib/march99/fink/03fink.html

Fleck, M., Frid, M., Kindberg, T., O'Brien-Strain, E., Rajani, R., & Spasojevic, M. (2002, April-June). From informing to remembering: Ubiquitous systems in interactive museums. *Pervasive Computing, 1*(2), 13–21.

Flintham, M., Benford, S., Anastasi, R., Hemmings, T., Crabtree, A., Greenhalgh, C., et al. (2003). Where on-line meets on the streets: Experiences with mobile mixed reality games. In *Proceedings of the SIGCHI conference on human factors in computing systems* (pp. 569–576). New York: ACM Press.

Freedom of Information Act, 5 U.S.C. § 552, As Amended by Pub. L. No. 104-231, 110 Stat. 3048. (1996 & West Supp. 2004).

Frey, B. S. (1998). Superstar museums: An economic analysis. *Journal of Cultural Economics, 22*, 113–125.

Frost, C. O. (1999). Cultural heritage outreach and museum/school partnerships: Initiatives at the School of Information, University of Michigan. In D. Bearman & J. Trant (Eds.), *Museums and the Web 1999: Selected papers from an international conference* (pp. 223–229). Pittsburgh, PA: Archives & Museum Informatics. Available from http://www.archimuse.com/mw99/papers/frost/frost.html

Frost, C. O. (2001). Engaging museums, content specialists, educators, and information specialists: A model and examples. In D. Bearman & J. Trant (Eds.), *Museums and the Web 2001: Selected papers from an international conference* (pp. 177–188). Pittsburgh, PA: Archives & Museum Informatics. Available from http://www.archimuse.com/mw2001/papers/frost/frost.html

Futernick, B. (2003). Personal communication on access points in The Thinker, http://www.thinker.org.

Gaitatzes, A., Papaioannou, G., & Christopoulos, D. (2004). The ancient Olympic games: Being part of the experience. In K. Cain, Y. Chrysanthou, F. Niccolucci & N. Silberman (Eds.), *VAST 2004: The 5th international symposium on virtual reality, archaeology and cultural heritage* (pp. 19–28). Aire-la-Ville, Switzerland: Eurographics. Available from http://www.eg.org/EG/DL/WS/VAST/VAST04/019028.pdf.abstract.pdf;internal&action=paperabstract.action

Galani, A. (2003). Mixed reality museum visits: Using new technologies to support co-visiting for local and remote visitors. *Museological Review Extra, 10*, 1–15.

Galani, A., & Chalmers, M. (2002). Can you see me? Exploring co-visiting between physical and virtual visitors. In D. Bearman & J. Trant (Eds.), *Museums and the Web 2002: Selected papers from an international conference* (pp. 31–40). Pittsburgh, PA: Archives & Museum Informatics. Available from http://www. archimuse.com/mw2002/papers/galani/galani.html

Galani, A., & Chalmers, M. (2003, September 10–12). *Far away is close at hand: Shared mixed reality museum experience for local and remote museum companions* [CD-ROM]. Paper presented at the 2003 International Cultural Heritage Informatics Meeting (ICHIM), Paris, France. Available from the Archives & Museum Informatics Website, http://www.archimuse.com

Galani, A., & Chalmers, M. (2004). Production of pace as a collaborative activity. In *CHI '04 extended abstracts on human factors in computing systems* (pp. 1417–1420). New York: ACM Press.

Garfinkel, H. (1967). *Studies in ethnomethodology*. Cambridge, England: Polity Press.

Garret, J. (2005, June 27). *ISO archiving standards—Overview*. Available from http://ssdoo.gsfc.nasa.gov/nost/isoas

Gautier, T. G. (1979). Automated collection documentation system at the National Museum of Natural History, Smithsonian Institution. *Museum, 30*(3–4), 160–168.

Geser, G. (2004). Introduction and overview. In G. Geser & J. Pereira (Eds.), *DigiCULT Thematic Issue 5: Virtual Communities and Collaboration in the Cultural Sector* (pp. 5–6). Austria: DigiCULT. Available from http://www.digicult. info/downloads/digicult_thematicissue5_january_2004.pdf

Geser, G., & Pereira, J. (Eds.). (2004). *DigiCULT Thematic Issue 5: Virtual Communities and Collaboration in the Cultural Sector*. Austria: DigiCULT. Available from http://www.digicult.info/downloads/digicult_thematicissue5_january_2004. pdf

Gillard, P. (2002). Cruising through History Wired. In D. Bearman & J. Trant (Eds.), *Museums and the Web 2002: Selected papers from an international conference* (n.p.). Pittsburgh, PA: Archives & Museum Informatics. Available from http:// www.archimuse.com/mw2002/papers/gillard/gillard.html

Gillard, P., & Cranny-Francis, A. (2002). Evaluation for effective Web communication: An Australian example. *Curator, 45*, 35–49.

Gilliland-Swetland, A. J. (1998). Evaluation design for large-scale, collaborative online archives: Interim report of the Online Archive of California Evaluation Project. *Archives and Museum Informatics, 12*(3/4), 177–203.

Gilliland-Swetland, A. J. (2000). *Setting the stage*. Available from http://www.getty. edu/research/conducting_research/standards/intrometadata/2_articles/index.html

Gilliland-Swetland, A. J. (2001). Popularizing the finding aid: Exploiting EAD to enhance online browsing and retrieval in archival information systems by diverse user groups. *Journal of Internet Cataloging, 4*(3/4), 199–225.

Gilliland-Swetland, A. J., Chandler, R. L., & White, L. (2004). We're building it, will they use it? the MOAC II evaluation project. In D. Bearman & J. Trant (Eds.), *Museums and the Web 2004: Selected papers from an international conference* (n.p.). Toronto, Canada: Archives & Museum Informatics. Available from http:// www.archimuse.com/mw2004/papers/g-swetland/g-swetland.html

Gilliland-Swetland, A. J., & White, L. (2004). Museum information professionals as providers and users of online resources. *Bulletin of the American Society for Information Science and Technology, 30*(5), 23–27.

Gladney, H. M., Mintzer, F., & Schiattarella, F. (1997, July/August). Safeguarding digital library contents and users: Digital images of treasured antiquities, *D-Lib*

Magazine. Available from http://www.dlib.org/dlib/july97/vatican/07gladney. html

Gladney, H., Mintzer, F., Schiattarella, F., Bescos, J., & Treu, M. (1998). Digital access to antiquities. *Communications of the ACM, 41*(4), 49–57.

Glaser, J. R., & Zenetou, W. A. (1996). *Museums: A place to work.* New York: Routledge.

Goldman, K. H., & Schaller, D. (2004). *Exploring motivational factors and visitor satisfaction in on-line museum visits.* Arlington, VA: Archives & Museum Informatics.

Gorman, M. (Ed.). (2004). *The concise AACR2* (4th ed.). Chicago: American Library Association.

Graham, C. H., Ferrier, S., Huettman, F., Moritz, C., & Peterson, A. T. (2004). New developments in museum-based informatics and applications in biodiversity analysis. *Trends in Ecology & Evolution, 19*(9), 497–503.

Grant, A. (2001). 'Cataloguing is dead: Long live the cataloguers!' The changing role of museum information professionals in mediating museum knowledge. *mda Information, 5*(3), 19–22.

Green, D. (1999). Best practices in networking cultural heritage resources: Where to start? *Spectra, 26*(1), 10–11.

Grinter, R. E., Aoki, P. M., Hurst, A., Szymanski, M. H., Thornton, J. D., & Woodruff, A. (2002). Revisiting the visit: Understanding how technology can shape the museum visit. In *Proceedings of the 2002 ACM conference on computer supported cooperative work* (pp. 146–155). New York: ACM Press.

Grose, T. O. (1996). Reading the bones: Information content, value, and ownership issues raised by the Native American Graves Protection and Repatriation Act. *Journal of the American Society for Information Science, 47*(8), 624–631.

Guralnick, R. P. (1995). Weaving towards the web. *Spectra, 22*(4), 10–12.

Haley Goldman, K. & Dierking, L.D. (2005). Free-Choice Learning Research and the Virtual Science Center: Establishing a Research Agenda. In L. Tan and R. Subramaniam (Eds.), *E-Learning and Virtual Science Centers* (pp. 28–50). Hershey, PA: Information Science Publishing.

Haley Goldman, K., & Schaller, D. (2004). Exploring Motivational Factors and Visitor Satisfaction in On-line Museum Visits. In D. Bearman & J. Trant (Eds.), *Museums and the Web 2004: Selected papers from an international conference* (n.p.). Toronto, Canada: Archives & Museum Informatics. Available from http:// www.archimuse.com/mw2004/papers/haleyGoldman/haleyGoldman.html

Haley Goldman, K., & Wadman, M. (2002). There's Something Happening Here, What It Is Ain't Exactly Clear. In D. Bearman & J. Trant (Eds.), *Museums and the Web 2002: Selected papers from an international conference* (n.p.). Pittsburgh, PA: Archives & Museum Informatics. Available from http://www.archimuse. com/mw2002/papers/haleyGoldman/haleygoldman.html

Haley Goldman, M., & Haley Goldman, K. (2005). Whither the web: Professionalism and practices for the changing museum. In D. Bearman & J. Trant (Eds.), *Museums and the Web 2005: Selected papers from an international conference* (n.p.). Toronto, Canada: Archives & Museum Informatics. Available from http:// www.archimuse.com/mw2005/papers/haleyGoldman/haleyGoldman.html

Hamma, K. (2004a, June/July). Becoming Digital. *Bulletin of the American Society for Information Science and Technology, 30*(5). Available from http://www.asis. org/Bulletin/Jun-04/hamma.html

Hamma, K. (2004b). The role of museums in online teaching, learning, and research, *First Monday, 9*(5). Available from http://firstmonday.org/issues/issue9_5/ hamma

Harley, D., Henke, J., Head, A., Miller, I., Nasatir, D., & Sheng, X. (2004). *The use of digital resources in humanities and social science undergraduate education, first year report*. Center for Studies in Higher Education, UC Berkeley. Available from http://digitalresourcestudy.berkeley.edu

Harms, I., & Schweibenz, W. (2001). Evaluating the usability of a museum web site. In D. Bearman & J. Trant (Eds.), *Museums and the Web 2001: Selected papers from an international conference* (pp. 43–54). Pittsburgh, PA: Archives & Museum Informatics. Available from http://www.archimuse.com/mw2001/papers/schweibenz/schweibenz.html

Hazan, S. (2004). Weaving community webs: A position paper. In G. Geser & J. Pereira (Eds.), *DigiCULT Thematic Issue 5: Virtual Communities and Collaboration in the Cultural Sector* (pp. 7–10). Austria: DigiCULT. Available from http://www.digicult.info/downloads/digicult_thematicissue5_january_2004.pdf

Heath, C., & vom Lehn, D. (2002). *Misconstruing interactivity*. Paper presented at the 2002 Interactive Learning in Museums of Art and Design conference, Victoria and Albert Museum, London, England. Available from http://www.vam.ac.uk/res_cons/research/learning/index.html

Heimlich, J.E. (Ed.) (2005). Special Issue: Free-choice learning and the environment. *Environmental Education Research* 11(3).

Hein, G. (1998). *Learning in the museum*. London: Routledge.

Hein, H. S. (2000). *The museum in transition: A philosophical perspective*. Washington, DC: Smithsonian Institution Press.

Heller, J. (1973, November). *GRIPHOS: General retrieval and information processor for humanities oriented studies (Tech. Rep.)*: State University of New York at Stony Brook, Department of Computer Science.

Hensen, S. L. (1989). *Archives, personal papers, and manuscripts: A cataloging manual for archival repositories, historical societies, and manuscript libraries* (2nd ed.). Chicago: American Library Association.

Herman, D. L., Johnson, K., & Ockuly, J. (2004). What clicked? An interim report on audience research and media resources. In D. Bearman & J. Trant (Eds.), *Museums and the Web 2004: Selected papers from an international conference* (n.p.). Toronto, Canada: Archives & Museum Informatics. Available from http://www.archimuse.com/mw2004/papers/ockuly/ockuly.html

Hermann, G. (1995). Sights on the web. *Spectra*, 22(4), 13–14.

Hermann, G. (1997). Shortcuts to Oz: Strategies and tactics for getting museums to the emerald city. In K. Jones-Garmil (Ed.), *The wired museum: Emerging technology and changing paradigms* (pp. 65–92). Washington, DC: American Association of Museums.

Hershman, L. (2001, August 8). *The difference engine #3, 1995–1998 A.D.* Available from http://www.lynnhershman.com/investigations/surveillance/de3/de3.html

Hertzum, M. (1998). A review of museum web sites: In search of user-centered design. *Archives and Museum Informatics*, 12(2), 127–138.

Hilke, D. D. (1987). Museums as resources for family learning: Turning the question around. *The Museologist*, 50(175), 14–15.

Hindmarsh, J., Heath, C., vom Lehn, D., & Cleverly, J. (2002). Creating assemblies: Aboard the Ghost Ship. In *Proceedings of the 2002 ACM conference on computer supported cooperative work* (pp. 156–165). New York: ACM Press.

Hitzeman, J., Mellish, C., & Oberlander, J. (1997). Dynamic generation of museum web pages: The intelligent labelling explorer. *Archives and Museum Informatics*, 11(2), 107–115.

Hoffmann, A., & Goebel, S. (2003). Designing collaborative group experience for museums with Telebuddy. In D. Bearman & J. Trant (Eds.), *Museums and the Web 2003: Selected papers from an international conference* (n.p.). Toronto, Canada: Archives & Museum Informatics. Available from http://www.archimuse.com/mw2003/papers/hoffmann/hoffmann.html

Hood, M. G. (1983, April). Staying away: Why people choose not to visit museums. *Museum News, 61*(4), 50–57.

Hood, M. G. (1989). Leisure criteria of family participation and non-participation in museums. In B. Butler & M. Sussman (Eds.), *Museum visits and activities for family life enrichment* (pp. 151–169). New York: Haworth Press.

Hooper-Greenhill, E. (1992). *Museums and the shaping of knowledge.* London: Routledge.

Hooper-Greenhill, E. (1999). *The educational role of the museum.* London: Routledge.

Hsi, S. (2003). A study of user experiences as mediated by nomadic web content in a museum. *Journal of Computer Assisted Learning, 19*, 308–319.

Hsi, S., & Fait, H. (2005). RFID enhances visitors' museum experience at the Exploratorium. *Communications of the ACM, 48*(9), 60–65.

Hunter, J. (2002). Combining the CIDOC CRM and MPEG-7 to describe multimedia in museums. In D. Bearman & J. Trant (Eds.), *Museums and the Web 2002: Selected papers from an international conference* (pp. 73–84). Pittsburgh, PA: Archives & Museum Informatics. Available from http://www.archimuse.com/mw2002/papers/hunter/hunter.html

Hutter, M. (1998). Communication productivity: A major cause for the changing output of art museums. *Journal of Cultural Economics, 22*(2–3), 99–112.

Hyvönen, E., Junnila, M., Kettula, S., Saarela, S., Salminen, M., Syreeni, A., et al. (2004). Finnish museums on the semantic web: The user's perspective on MuseumFinland. In D. Bearman & J. Trant (Eds.), *Museums and the Web 2004: Selected papers from an international conference* (pp. 21–32). Toronto, Canada: Archives & Museum Informatics. Available from http://www.archimuse.com/mw2004/papers/hyvonen/hyvonen.html

Institute of Museum and Library Services. (2002). *Status of technology and digitization in the nation's museums and libraries 2002 report.* Available from http://www.imls.gov/publications/TechDig02/

International Council of Museums. (2002). *The CIDOC relational data model.* Available from http://www.willpowerinfo.myby.co.uk/cidoc/model/relational.model/

International Council of Museums. (2004). *ICOM code of ethics for museums — 2004 edition.* Available from http://www.icom.museum/ethics.html

Jackson, R., Bazley, M., Patten, D., & King, M. (1998). Using the web to change the relation between a museum and its users. In D. Bearman & J. Trant (Eds.), *Museums and the Web 1998: Selected papers from an international conference* [CD-ROM] (n.p.). Pittsburgh, PA: Archives & Museum Informatics. Available from http://www.archimuse.com/mw98/papers/jackson/jackson_paper.html

Jaén, J., Bosch, V., Esteve, J. M., & Mocholí, J. A. (2005). MoMo: A hybrid museum infrastructure. In D. Bearman & J. Trant (Eds.), *Museums and the Web 2005: Selected papers from an international conference* (pp. 141–150). Toronto, Canada: Archives & Museum Informatics. Available from http://www.archimuse.com/mw2005/papers/jaen/jaen.html

Jaggi, K., & Kraemer, H. (2003). *The Virtual Transfer Musée Suisse: Digital technology reshaping the strategy of the Swiss National Museum.* Paper presented at the International Cultural Heritage Informatics Meeting (ichim03) [CD-ROM], Paris, France. Available from the Archives & Museum Informatics Website, http://www.archimuse.com

Jaggi, K., & Kraemer, H. (2004). The Virtual Transfer Musée Suisse. *ICOM News, 3*, 5. Available from http://icom.museum/pdf/E_news2004/p5_2004-3.pdf

Johnson, D. P., Sr. (1998, December). Scholars help Bosnia rebuild destroyed libraries. *Northeast News.* Available from http://www.wrmea.com/backissues/1298/9812064.html

Johnston, L., & Jones-Garmil, K. (1996). So you want to build a web page. *Spectra,* *23*(1), 25–27.

Jones-Garmil, K. (1995). AAM Reports: Bringing the Internet to AAM. *Spectra,* *22*(4), 6–7.

Jones-Garmil, K. (1997a). Laying the foundation: Three decades of computer technology in the museum. In K. Jones-Garmil (Ed.), *The wired museum: Emerging technology and changing paradigms* (pp. 35–62). Washington, DC: American Association of Museums.

Jones-Garmil, K. (Ed.). (1997b). *The wired museum: Emerging technology and changing paradigms.* Washington, DC: American Association of Museums.

Keene, S. (1998). *Digital collections, museums and the information age.* Oxford: Butterworth-Heinemann.

Kenderine, S. (1999). Inside the meta-center: a cabinet of wonder. In D. Bearman & J. Trant (Eds.), *Museums and the web 1999: Selected papers from an international conference* (pp. 175–186). Pittsburgh, PA: Archives & Museum Informatics. Available from http://www.archimuse.com/mw99/papers/kenderine/kenderine. html

Knell, S. J. (2003). The shape of things to come: Museums in the technological landscape. *Museums and Society, 1*(3), 132–146.

Koot, G.-J. (2001). Museum librarians as information strategists. *INSPEL, 35,* 248–258.

Kraus, A. (2000, September/October). Sold! To the curator with the fastest modem. *Museum News, 79*(5), 12–16.

Kravchyna, V., & Hastings, S. (2002). Informational value of museum Web sites. *First Monday, 7*(2). Available from http://firstmonday.org/issues/issue7_2/kravchyna

Lagoze, C., & Hunter, J. (2001). The ABC ontology and model. *Journal of Digital Information, 2*(2). Available from http://jodi.tamu.edu/Articles/v02/i02/Lagoze/

Lagoze, C., & Van de Sompel, H. (2003). The making of the open archives initiative protocol for metadata harvesting. *Library Hi-Tech, 21*(2), 118–128.

Lanzi, E. (1998). *Introduction to vocabularies: Enhancing access to cultural heritage information.* Los Angeles: Getty Trust Publications.

Lee, E. (2004). Building interoperability for United Kingdom historic environment information resources. *Lecture Notes in Computer Science, 3232,* 179–185.

Lees, D. (Ed.). (1993). *Museums and interactive multimedia: Selected papers from the second International Conference on Hypermedia and Interactivity in Museums.* Pittsburgh, PA: Archives & Museum Informatics.

Lévy, P. (1995). *Qu'est ce que le virtuel?* Paris: La Découverte.

Light, R. B., Roberts, D. A., & Stewart, J. D. (1986). *Museum documentation systems.* London: Butterworths.

Lloyd, R. (1999). *Metric mishap caused loss of NASA orbiter.* Available from http://www.cnn.com/TECH/space/9909/30/mars.metric.02/index.html

Loomis, R. J., Elias, S. M., & Wells, M. (2003). *Website availability and visitor motivation: An evaluation study for the Colorado Digitization project.* Available from http://www.cdpheritage.org/resource/reports/loomis_report.pdf

Loran, M. (1999). *Survey of information technology training needs and provision in the U.S. museum community.* Cambridge, MA: Harvard University, Peabody Museum of Archaeology and Ethnology.

Lord, B., & Lord, G. D. (1997). *The manual of museum management.* Walnut Creek, CA: AltaMira Press.

Luhila, M. (2001). Networked communication: Are museums in Sub-Saharan Africa ready? *mda Information, 5*(3) 71–73.

Lusaka, J., & Strand, J. (1998, November/December). The boom—And what to do about it: Strategies for dealing with an expanding field. *Museum News, 77*(6), 54–60.

Lynch, C. A. (2002). Digital collections, digital libraries and the digitization of cultural heritage information. *First Monday, 7*(5). Available from http://firstmonday. org/issues/issue7_5/lynch/index.html

MacDonald, G. (1988, September/October). The future of museums in the global village. *Museum News, 66*(7), 69–71.

MacDonald, G., & Alsford, S. (1991). The museum as information utility. *Museum Management and Curatorship, 10*(3), 305–311.

MacDonald, G. F., & Alsford, S. (1995). Museums and theme parks: Worlds in collision? *Museum Management and Curatorship, 14*(2), 129–147.

MacDonald, S. (1993). The enigma of the visitor sphinx. In S. Bicknell & G. Farmelo (Eds.), *Museum visitor studies in the 90s* (pp. 77–81). London: Science Museum.

Malaro, M. C. (2002). Legal and ethical foundations of museum collecting policies. In T. A. Lipinksi (Ed.), *Libraries, museums and archives: Legal issues and ethical challenges in the new information era* (pp. 69–82). Lanham, MD: Scarecrow Press.

Malraux, A. (1951). Le musée imaginaire [The imaginary museum]. In A. Malraux (Ed.), *Les voix du silence*. Paris: Nouvelle Revue Française, Gallimard.

Manning, A., & Sims, G. (2004). The Blanton iTour: An interactive handheld museum guide. In D. Bearman & J. Trant (Eds.), *Museums and the Web 2004: Selected papers from an international conference* (n.p.). Toronto, Canada: Archives & Museum Informatics. Available from http://www.archimuse.com/ mw2004/papers/manning/manning.html

Martin, D. (2000). Audio guides. *Museum Practice, 5*(1), 71–81.

Martin, W., Rieger, R., & Gay, G. (1999). Designing across disciplines: Negotiating collaborator interests in a digital museum project. In D. Bearman & J. Trant (Eds.), *Cultural heritage informatics 1999: Selected papers from ichim99* (pp. 83–90). Pittsburgh, PA: Archives & Museum Informatics.

Marty, P. F. (1999a). Museum informatics and collaborative technologies: The emerging socio-technological dimension of information science in museum environments. *Journal of the American Society for Information Science, 50*(12), 1083–1091.

Marty, P. F. (1999b). Online exhibit design: Building a museum over the world wide web. In D. Bearman & J. Trant (Eds.), *Museums and the Web 1999: Selected papers from an international conference* (pp. 207–216). Pittsburgh, PA: Archives & Museum Informatics. Available from http://www.archimuse.com/mw99/ papers/marty/marty.html

Marty, P. F. (2000). Museum informatics: sociotechnical infrastructures in museums. *Bulletin of the American Society for Information Science and Technology, 26*(3), 22–24. Available from http://www.asis.org/Bulletin/Mar-00/marty.html

Marty, P. F. (2002). *Museum informatics and the evolution of an information infrastructure in a university museum*. Unpublished doctoral dissertation, University of Illinois, Urbana-Champaign.

Marty, P. F. (2004a). The changing role of the museum webmaster: Past, present, and future. In D. Bearman & J. Trant (Eds.), *Museums and the Web 2004: Selected papers from an international conference* (n.p.). Toronto, Canada: Archives & Museum Informatics. Available from http://www.archimuse.com/mw2004/ papers/marty/marty.html

Marty, P. F. (2004b). The evolving roles of information professionals in museums. *Bulletin of the American Society for Information Science and Technology, 30*(5), 20–23. Available from http://www.asis.org/Bulletin/Jun-04/marty.html

Marty, P. F. (2005). So you want to work in a museum? Guiding the careers of future information professionals in museums. *Journal of Education for Library and Information Science, 46*, 115–133.

Marty, P. F. (2006). Meeting user needs in the modern museum: Profiles of the new museum information professional. *Library & Information Science Research*, *28*(1), 128–144.

Marty, P. F., Rayward, W. B., & Twidale, M. B. (2003). Museum informatics. *Annual Review of Information Science and Technology, 37*, 259–294.

Mason, I. (2002). The social webs that must be woven: Information management, museums and the knowledge industry. *mda Information, 5*(5), 15–24.

Mattes, J. (2001). Access for all? Consideration of multimedia content and design to ensure access for users with disabilities. *mda Information, 5*(3), 43–46.

McClung, P., & Stephenson, C. (Eds.). (1998). *Images online: Perspectives on the museum educational site licensing project.* Los Angeles: Getty Research Institute.

McKenna, G., & Patsatzi, E. (2005). *SPECTRUM: The UK museum documentation standard.* Cambridge, England: Museum Documentation Association.

McManus, P. M. (1987a). *Communication with and between visitors to a science museum.* Unpublished doctoral dissertation, Kings College, London.

McManus, P. M. (1987b). It's the company you keep . . . The social determinants of learning-related behaviour in a science museum. *International Journal of Museum Management and Curatorship, 6*, 263–270.

McManus, P. M. (1988). Good companions . . . More on the social determination of learning-related behaviour in a science museum. *International Journal of Museum Management and Curatorship, 7*, 37–44.

Minneapolis Institute of Arts. (2004). *Restoration online: Restoring a masterwork.* Available from http://www.artsmia.org/restoration-online/

Mintz, A. (1998). Media and museums: A museum perspective. In S. Thomas & A. Mintz (Eds.), *The virtual and the real: Media in the museum* (pp. 19–35). Washington, DC: American Association of Museums.

Mintzberg, H. (1989). *Mintzberg on management: Inside our strange world of organizations.* New York: The Free Press.

Mintzer, F., Braudaway, G., Giordano, F., Lee, J., Magerlein, K., D'Auria, S., et al. (2001). Populating the Hermitage Museum's new web site. *Communications of the ACM, 44*(8), 52–60.

Mintzer, F. C., Lirani, A. C., Magerlein, K. A., Pavani, A. M. B., Schiattarella, F., Boyle, L. E., et al. (1996). Toward online, worldwide access to Vatican Library materials. *IBM Journal of Research and Development, 40*(2), 139–162.

Moen, W. (1998). Accessing distributed cultural heritage information. *Communications of the ACM, 41*(4), 45–48.

Morrissey, K., & Worts, D. (1988). A place for the muses? Negotiating the role of technology in museums. In A. Mintz & S. Thomas (Eds.), *The virtual and the real: Media in the museum* (pp. 147–171). Washington, DC: American Association of Museums.

Moussouri, T. (1997). *Family agendas and family learning in hands-on museums.* Unpublished doctoral dissertation, University of Leicester, England.

Moussouri, T., Nikiforidou, A., & Gazi, A. (2003). Front-end and formative evaluation of an exhibition on Greek mathematics. In *Current trends in audience research and evaluation* (Vol. 16) (pp. 42–47). San Francisco: American Association of Museums.

Muller, K. (2002). Museums and virtuality. *Curator, 45*, 21–33.

Mulrenin, A. (Ed.). (2002). *The DigiCULT Report: Technological landscapes for tomorrow's cultural economy: Unlocking the value of cultural heritage.* Luxembourg: European Commission, Directorate-General Information Society. Available from http://www.digicult.info/pages/report.php

Museum Documentation Association. (2005). *SPECTRUM.* Cambridge, England: MDA.

Museums Libraries and Archives Council. (2005). *Accessibility of museum, libraries, and archive websites: The MLA audit.* London: City University.

National Information Standards Organization. (2004a). *Framework of guidance for building good digital collections (2nd ed.).* Available from http://www.niso.org/framework/Framework2.html

National Information Standards Organization. (2004b). *Understanding metadata.* Available from http://www.niso.org/standards/resources/Understanding Metadata.pdf

National Library of Australia. (1999). *Preservation metadata for digital collections.* Available from http://www.nla.gov.au/preserve/pmeta.html

National Library of Australia. (n.d.). *Government policy and the information superhighway.* Available from http://www.nla.gov.au/lis/govnii.html

Native American Graves Protection and Repatriation Act, 25 U.S.C. 3001 et seq. (1990). Available from http://www.cr.nps.gov/nagpra/MANDATES/25USC3001 etseq.htm

Oberlander, J., Mellish, C., O'Donnell, M., & Knott, A. (1997). Exploring a gallery with intelligent labels. In *Museum interactive multimedia 1997: Cultural heritage systems design and interfaces: Selected papers from ICHIM 97, the Fourth International Conference on Hypermedia and Interactivity in Museums, Paris, France, 3–5 September 1997* (pp. 153–161). Pittsburgh, PA: Archives & Museum Informatics.

Ochoa, T. (2001). Bridgeman Art Library v. Corel Corporation: Three possible responses. *Spectra, 28*(3), 32–35.

Ockuly, J. (2003). What clicks? An interim report on audience research. In D. Bearman & J. Trant (Eds.), *Museums and the Web 2003: Selected papers from an international conference* (n.p.). Toronto, Canada: Archives & Museum Informatics. Available from http://www.archimuse.com/mw2003/papers/ockuly/ockuly.html

Oldenburg, Ray. (1999) *The Great Good Place: Cafes, Coffee Shops, Bookstores, Bars, Hair Salons, and Other Hangouts at the Heart of a Community.* New York: Marlowe and Company.

Opperman, R., Specht, M., & Jaceniak, I. (1999). Hippie: A nomadic information system. In H.-W. Gellersen (Ed.), *Handheld and ubiquitous computing: First international symposium, HUC'99, Karlsruhe, Germany, September 27–29, 1999, proceedings* (pp. 330–333). Berlin: Springer.

Orna, E. (1999). *Practical information policies.* Brookfield, VT: Gower.

Orna, E. (2001). *The knowing museum.* Paper presented at the 2001 Annual Conference of the Museum Documentation Association, Cambridge, England. Available from http://www.mda.org.uk/404.htm?404;http://www.mda.org.uk/conference2001/papers.htm

Orna, E., & Pettitt, C. (1980). *Information handling in museums* (1st ed.). London: Clive Bingley.

Orna, E., & Pettitt, C. (1987). *Information policies for museums.* London: Museum Documentation Association.

Orna, E., & Pettitt, C. (1998). *Information management in museums* (2nd ed.). Aldershot, Hampshire, England: Gower.

Packer, J. & Ballantyne, R. (2002). Motivational factors and the visitor experience: A comparison of three sites. *Curator 45,* 183–198.

Pantalony, R. E. (2001). The carrot vs. the stick: Can copyright enhance access to cultural heritage resources in the networked environment? In D. Bearman & J. Trant (Eds.), *Cultural heritage informatics 1999: Selected papers from ichim99* (pp. 225–232). Pittsburgh, PA: Archives & Museum Informatics.

Paolini, P., Barbieri, T., Loiudice, P., Alonzo, F., Zanti, M., & Gaia, G. (2000). Visiting a museum together? How to share a visit to a virtual world. *Journal of the American Society for Information Science, 51*(1), 33–38.

Parker, E. B. (1987). *LC thesaurus for graphic materials: Topical terms for subject access*. Washington, DC: Cataloging Distribution Service, Library of Congress.

Parker, S., Waterson, K., Michaluk, G., & Rickard, L. (2002). *Neighbourhood renewal and social inclusion: The role of museums, archives and libraries*. Available from http://www.mla.gov.uk/documents/neighbourhood.pdf

Parry, R. (2001). Including technology. In J. Dodd & R. Sandell (Eds.), *Including museums: Perspectives on museums, galleries and social inclusion* (pp. 110–114). Leicester, England: Research Centre for Museums and Galleries, University of Leicester.

Parry, R. (2005). Digital heritage and the rise of theory in museum computing. *Museum Management and Curatorship, 20,* 333–348.

Parry, R., & Arbach, N. (2005). The localized learner: Acknowledging distance and situatedness in online museum learning. In D. Bearman & J. Trant (Eds.), *Museums and the Web 2005: Selected papers from an international conference* (pp. 67–76). Toronto, Canada: Archives & Museum Informatics. Available from http://www.archimuse.com/mw2005/papers/parry/parry.html

Paterno, F., & Mancini, C. (2000). Effective levels of adaptation to different types of users in interactive museum systems. *Journal of the American Society for Information Science, 51*(1), 5–13.

Peacock, D. (2002). Statistics, structures, and satisfied customers: Using web log data to improve site performance. In D. Bearman & J. Trant (Eds.), *Museums and the Web 2002: Selected papers from an international conference* (pp. 157–166). Pittsburgh, PA: Archives & Museum Informatics. Available from http://www.archimuse.com/mw2002/papers/peacock/peacock.html

Peacock, D., Ellis, D., & Doolan, J. (2004). Searching for meaning: Not just records. In D. Bearman & J. Trant (Eds.), *Museums and the Web 2004: Selected papers from an international conference* (pp. 11–20). Toronto, Canada: Archives & Museum Informatics. Available from http://www.archimuse.com/mw2004/papers/peacock/peacock.html

Pearce, S. (1986, March). Thinking about things: Approaches to the study of artefacts. *Museum Journal,* 198–201.

Perkins, J. (2001b). A new way of making cultural information resources visible on the web: Museums and the open archive initiative. In D. Bearman & J. Trant (Eds.), *Museums and the web 2001: Selected papers from an international conference* (pp. 87–92). Pittsburgh, PA: Archives & Museum Informatics. Available from http://www.archimuse.com/mw2001/papers/perkins/perkins.html

Peterson, T. (1990). *Art and architecture thesaurus*. New York: Oxford University Press.

Pew Internet and the American Life Project, Pew Trust. (2006). *Demographics of Internet Users*. Available from http://www.pewinternet.org/trends.asp

Pham, A. (2004, October 4). Art that goes on the blink. *Los Angeles Times,* p. A1.

Pitti, D. V., & Duff, W. M. (Eds.). (2001). *Encoded archival description on the Internet*. New York: Haworth Information Press.

Pletinckx, D., Silberman, N., & Callebaut, D. (2003). Heritage presentation through interactive storytelling: A new multimedia database approach. *Journal of Visualization and Computer Animation, 14*(4), 225–231.

Porter, M. F. (1979). Establishing a museum documentation system in the United Kingdom. *Museum, 30*(3–4), 169–178.

Prensky, M. (2001). Digital natives, digital immigrants. *On the Horizon, 9*(5), 1–6.

Prochaska, A. (2001). Librarians as curators. *Record, 103*(9). Available from http://www.la-hq.org.uk/directory/record/r200109/article3.html

Rabinovitch, V., & Alsford, S. (2002, May). *Museums and the Internet: Reflections on eight years of Canadian experience*. Paper presented at the 6th World

Colloquium of the International Association of Museums of History, Lahti, Finland. Available from http://www.civilization.ca/academ/articles/rabi_01e.html

Rand, J. (2001). The 227-mile museum, or a visitors' bill of rights. *Curator, 44*(1), 7–14.

Ranganathan, S. R. (1933). *Colon classification.* Madras, India: Madras Library Association.

Rayward, W. B. (1998). Electronic information and the functional integration of libraries, museums and archives. In E. Higgs (Ed.), *History and Electronic Artefacts* (pp. 207–224). Oxford, England: Oxford University Press.

Rayward, W. B., & Twidale, M. B. (2000). From docent to cyberdocent: Education and guidance in the virtual museum. *Archives and Museum Informatics, 13*, 23–53.

Reed, P. A., & Sledge, J. (1998). Thinking about museum information. *Library Trends, 37*, 220–231.

Research Libraries Group. (1994). AMIS: Archives and museum information service demonstration [Videotape]: (Available from the Research Libraries Group, 2029 Stierlin Court, Suite 100, Mountain View, CA).

Research Libraries Group. (2004). *Automatic exposure: Capturing technical metadata for digital still images.* Available from http://www.rlg.org/longterm/ae_whitepaper_2003.pdf

Rinehart, R. (1999). Conceptual and intermedia arts online: The challenge of documenting and presenting non-traditional art collections. In D. Bearman & J. Trant (Eds.), *Museums and the Web 1999: Selected papers from an international conference* (n.p.). Pittsburgh, PA: Archives & Museum Informatics. Available from http://www.archimuse.com/mw99/papers/rinehart/rinehart.html

Rinehart, R. (2001). Museums and the Online Archive of California. *Spectra, 28*(1), 20–27.

Rinehart, R. (2003). MOAC: A report on integrating museum and archive access in the Online Archive of California. *D-Lib Magazine, 9*(1). Available from http://www.dlib.org/dlib/january03/rinehart/01rinehart.html

Rinehart, R. (2004). Archiving the Avant-Garde, Documenting and Preserving Digital/Variable Media Art. Available from http://www.bampfa.berkeley.edu/about_bampfa/avantgarde.html

Rinehart, R., Elings, M. W., & Garcelon, E. (1998). The Robert Honeyman Jr. Collection digital archive: EAD and the use of library and museum descriptive standards. *Archives and Museum Informatics, 12*(3), 205–219.

Roberts, A. (2001). The changing role of information professionals in museums. *mda Information, 5*(3), 15–18.

Roberts, D. A. (1985). *Planning the documentation of museum collections.* Cambridge, England: Museum Documentation Association.

Roberts, D. A. (1988). *Collections management for museums.* Cambridge, England: Museum Documentation Association.

Roberts, D. A. (Ed.). (1990). *Terminology for museums: Proceedings of an international conference held in Cambridge, England, 21–24 September 1988: The Second Conference of the Museum Documentation Association.* Cambridge, England: Museum Documentation Association.

Roberts, D. A. (Ed.). (1993). *European museum documentation strategies and standards: Proceedings of an international conference held in Canterbury, England, 2–6 September 1991: The fifth international conference of the Museum Documentation Association.* Cambridge, England: Museum Documentation Association.

Roberts, L. C. (1997). *From knowledge to narrative: Educators and the changing museum.* Washington, DC: Smithsonian Institution Press.

Rodger, E. J., Jorgensen, C., & D'Elia, G. (2005). Partnerships and collaboration among public libraries, public broadcast media, and museums: Current context and future potential. *Library Quarterly, 75*(1), 42–67.

Roller, H. U. (1976). Museum didactics and application of documentation: Typology of exhibits in cultural history museums. *Zeitschrift Fur Volkskunde, 72*(1), 90–92.

Rosenfeld, S. B. (1980). Informal learning in zoos: Naturalistic studies of family groups. *Dissertation Abstracts International, 41*(07). (University Microfilms No. AAT80-29566).

Roussou, M. (2002). Virtual heritage: From the research lab to the broad public. In F. Niccolucci (Ed.), *Virtual archaeology: Proceedings of the VAST 2000 Euroconference* (pp. 93–100). Oxford: Archaeopress. Available from http://makebelieve. gr/mr/research/papers/VAST/VAST_00/mroussou_VAST00_press.pdf

Roussou, M. (2004). Learning by doing and learning through play: An exploration of interactivity in virtual environments for children. *ACM Computers in Entertainment, 2*(1), 1–23.

Roussou, M., Johnson, A. E., Moher, T. G., Leigh, J., Vasilakis, C., & Barnes, C. (1999). Learning and building together in an immersive virtual world. *PRESENCE: Teleoperators and Virtual Environments, 8*(3), 247–263.

Rush, C., & Chenhall, R. (1979). Computer and registration: Principles of information management. In D. Dudley & I. Wilkenson (Eds.), *Museum registration methods* (pp. 319–339). Washington, DC: American Association of Museums.

Ryder, M., & Wilson, B. (1996). Affordances and constraints of the Internet for learning and instruction. In M. R. Simonson, M. Hays & S. Hall (Eds.), *18th annual proceedings of selected research and development presentations at the 1996 National Convention of the Association for Educational Communications Technology* (pp. 642–654). Ames, IA: Iowa State University.

Sacks, H. (1998). *Lectures on conversation* (Vol. I & II). Cambridge, MA: Blackwell.

Samis, P. (1999). Artwork as interface. *Archives and Museum Informatics, 13*(2), 191–198.

Samis, P. (2001, September 7). *Points of departure: Integrating technology into the galleries of tomorrow.* Paper presented at the 2001 International Cultural Heritage Informatics Meeting (ichim01), Milan, Italy. CD-ROM available from the Archives & Museum Informatics Website: http://www.archimuse.com

Samis, P., & Wise, S. (2000). Making the punishment fit the crime: Content-driven multimedia development. In D. Bearman & J. Trant (Eds.), *Museums and the Web 2000: Selected papers from an international conference* (n.p.). Pittsburgh, PA: Archives & Museum Informatics. Available from http://www.archimuse. com/mw2000/papers/samis/samis.html

Sanders, M. B., & Perkins, J. A. (1999). *Model for museum information management (Version 0.11).* Available from http://www.cimi.org/old_site/documents/ IIM_model.doc

Sarasan, L., & Neuner, A. M. (1983). *Museum collections and computers: Report of an ASC survey.* Lawrence, KS: Association of Systematics Collections.

Sarraf, S. (1999). A survey of museums on the web: Who uses museum websites? *Curator, 42,* 231–2443.

Sayre, S. (2000). Sharing the experience: The building of a successful online/on-site exhibition. In D. Bearman & J. Trant (Eds.), *Museums and the Web 2000: Selected papers from an international conference* (pp. 13–20). Pittsburgh, PA: Archives & Museum Informatics. Available from http://www.archimuse.com/ mw2000/papers/sayre/sayre.html

Sayre, S., & Wetterlund, K. (2003). Addressing multiple audiences with multiple interfaces to the AMICO library. In D. Bearman & J. Trant (Eds.), *Museums and the Web 2003: Selected papers from an international conference* (n.p.). Pittsburgh, PA: Archives & Museum Informatics. Available from http://www.archimuse.com/mw2003/papers/sayre/sayre.html

Schaller, D. T., & Allison-Bunnell, S. (2005). Learning styles and online interactives. In D. Bearman & J. Trant (Eds.), *Museums and the Web 2005: Selected papers from an international conference* (n.p.). Toronto, Canada: Archives & Museum Informatics. Available from http://www.archimuse.com/mw2005/papers/schaller/schaller.html

Schwarzer, M. (2001a). *Graduate training in museum studies: What students need to know*. Washington, DC: American Association of Museums.

Schwarzer, M. (2001b, July/August). Art and gadgetry: The future of the museum visit. *Museum News, 80*(4), 36–41, 68, 73.

Schweibenz, W. (1998). The virtual museum: New perspective for museums to present objects and information using the Internet as a knowledge base and communication system. In H. Zimmerman & V. Schramm (Eds.), *Knowledge management und kommunikationssysteme* (pp. 185–200). Konstanz, Germany: UKV.

Scott, C. (1995). *To market, to market: The inbound tourist and museums*. Paper presented at the Museums Australia annual conference, Brisbane, Australia.

Scottish Museums Council. (2004). *A national ICT strategy for Scotland's museums*. Available from http://www.scottishmuseums.org.uk/information_services/publications_intro.asp

Semper, R., Wanner, N., Jackson, R., & Bazley, M. (2000). Who's out there? A pilot user study of educational web resources by the Science Learning Network. In D. Bearman & J. Trant (Eds.), *Museums and the Web 2000: Selected papers from an international conference* (pp. 179–200). Pittsburgh, PA: Archives & Museum Informatics. Available from http://www.archimuse.com/mw2000/papers/semper/semper.html

Shane, J. (1997). The virtual visit—virtual benefits? In G. Farmelo & J. Carding (Eds.), *Here and now: Contemporary science and technology in museums and science centres* (pp. 189–196). London: Science Museum.

Shapiro, M. (1999). Developing museum information policies. In D. Bearman & J. Trant (Eds.), *Museums and the web 1999: Selected papers from an international conference* (n.p.). Pittsburgh, PA: Archives & Museum Informatics. Available from http://www.amico.org/docs/papers/1999/shapiro.infopolicy.99.pdf

Shapiro, M. (2000a). Developing information policies for the web (Workshop). In D. Bearman & J. Trant (Eds.), *Museums and the Web 2000: Selected papers from an international conference* (n.p.). Pittsburgh, PA: Archives & Museum Informatics. Abstract available from http://www.archimuse.com/mw2000/abstracts/prg_80000175.html

Shapiro, M. (2000b). *Electronic information policies*. Paper presented at the American Association of Museums' "Current Issues in Intellectual Property" seminar, Philadelphia, PA.

Shapiro, M., & Miller, B. (1999). *A museum guide to copyright and trademark*. Washington, DC: American Association of Museums.

Sherman, A. (2002). An assessment of the effect of handheld application on the enhancement of the field museum visitor experience. *Spectra, 29*(2/3).

Sherwood, L. E. (1997). Moving from experiment to reality: Choices for cultural heritage institutions and their governments. In K. Jones-Garmil (Ed.), *The wired museum: Emerging technology and changing paradigms* (pp. 129–152). Washington, DC: American Association of Museums.

Shreeves, S. L., Kaczmarek, J., & Cole, T. W. (2003). Harvesting cultural heritage metadata using the OAI protocol. *Library Hi-Tech, 21*(2), 159–169.

Siegfried, S. L., & Bernstein, J. (1991). Synoname: The Getty's new approach to pattern-matching for personal names. *Computers and the Humanities, 25*(4), 211–226.

Silveira, M., Pinho, M., Gonella, A., Herrmann, M., & Calvetti, P. (2005). Using mobile devices to help teachers and students during a visit to a museum. In D. Bearman & J. Trant (Eds.), *Museums and the Web 2005: Selected papers from an international conference* (n.p.). Toronto, Canada: Archives & Museum Informatics. Available from http://www.archimuse.com/mw2005/papers/silveira/silveira. html

Silverman, L. (1990). *Of us and other "things": the content and functions of talk by adult visitor pairs in an art and a history museum.* Unpublished doctoral dissertation, University of Pennsylvania.

Sims, R. (1997, January 27). *Interactivity: A forgotten art?* Available from http://www2.gsu.edu~wwwitr/docs/interact/

Sledge, J. (1995). Points of view. In D. Bearman (Ed.), *Multimedia computing and museums: Selected papers from the Third International Conference on Hypermedia and Interactivity in Museums* (pp. 335–346). Pittsburgh, PA: Archives & Museum Informatics.

Sledge, J. E. (1988). Survey of North American collections management systems and practices. In D.A. Roberts (Ed.), *Collections Management for Museums* (pp. 9–17). Cambridge, UK: Museum Documentation Association.

Smith, K. (2004). Tour museums with your mobile phone—Thanks to the Scottish ICT strategy. *Museums Journal, 104,* 8.

Smith, L. (Ed.). (2000). *Building the digital museum: A national resource for the learning age.* London: Museums, Libraries, and Archives Council.

Squire, K. D. & Steinkuehler, C. A. (2006). Generating CyberCulture/s: The case of Star Wars Galaxies. In D. Gibbs & K. L. Krause (Eds.), *Cyberlines 2.0 Languages and cultures of the Internet.* Albert Park, Australia: James Nicholas Publishers.

Steinkuehler, C. A. (2005). The new third place: Massively multiplayer online gaming in American youth culture. *Tidskrift Journal of Research in Teacher Education, 3,* 17–32.

Stephenson, C., & McClung, P. (Eds.). (1998). *Delivering digital images: Cultural heritage resources for education.* Los Angeles: Getty Research Institute.

Steuer, J. (1995). Defining virtual reality: Dimensions determining telepresence. In F. Biocca & M. R. Levy (Eds.), *Communication in the age of virtual reality* (pp. 33–55). Hillsdale, NJ: Erlbaum.

Stone-Miller, R. (n.d.). Curator's choice: Ancient bat flute. *Antiquities and Fine Art.* Available from http://www.antiquesandfineart.com/articles/article.cfm?request =338

Streten, K. (2000). Honored guests: Towards a visitor centered Web experience. In D. Bearman & J. Trant (Eds.), *Museums and the Web 2000: Selected papers from an international conference* (n.p.). Pittsburgh, PA: Archives & Museum Informatics. Available from http://www.archimuse.com/mw2000/papers/streten/streten.html

Sumption, K. (1999). *Virtual Museums—primitive rebirth or synthetic recovery? An examination of the role and value of "real" museum experience in the development of on-line education programs at the Powerhouse Museum.* Unpublished Master's Thesis, University of Sydney.

Sumption, K. (2000). Meta-centers: Do they work and what might the future hold? A case study of Australian Museums On-line. In D. Bearman & J. Trant (Eds.), *Museums and the web 2000: Selected papers from an international conference* (pp. 105–112). Pittsburgh, PA: Archives & Museum Informatics. Available from http://www.archimuse.com/mw2000/papers/sumption/sumption.html

Szirtes, B. (1998). Virtual members: CHIN and the guide to Canadian museums and galleries. In D. Bearman & J. Trant (Eds.), *Museums and the web 1998: Selected papers from an international conference* (n.p.). Pittsburgh, PA: Archives & Museum Informatics. Available from http://www.archimuse.com/mw98/papers/szirtes/szirtes_paper.html

Taxén, G., & Frécon, E. (2005). The extended museum visit: Documenting and exhibiting post-visit experiences. In D. Bearman & J. Trant (Eds.), *Museums and the Web 2005: Selected papers from an international conference* (n.p.). Toronto, Canada: Archives & Museum Informatics. Available from http://www.archimuse.com/mw2005/papers/taxen/taxen.html

Taylor, T. (2004). *Mummy: The inside story*. London: The British Museum Press.

Teather, L. (1998). A museum is a museum is a museum . . . or is it? Exploring museology and the web. In D. Bearman & J. Trant (Eds.), *Museums and the Web 1998: Selected papers from an international conference* [CD-ROM] (n.p.). Pittsburgh, PA: Archives & Museum Informatics.

Teather, L., & Wilhelm, K. (1999). Web musing: Evaluating museums on the web from learning theory to museology. In D. Bearman & J. Trant (Eds.), *Museums and the Web 1999: Selected papers from an international conference* (pp. 131–143). Pittsburgh, PA: Archives & Museum Informatics. Available from http://www.archimuse.com/mw98/papers/teather/teather_paper.html

Thomas, S., & Mintz, A. (Eds.). (1998). *The virtual and the real: Media in the museum*. Washington, DC: American Association of Museums.

Thomas, W. A., & Carey, S. (2005). Actual/virtual visits: What are the links? In D. Bearman & J. Trant (Eds.), *Museums and the Web 2005: Selected papers from an international conference* (n.p.). Toronto, Canada: Archives & Museum Informatics. Available from http://www.archimuse.com/mw2005/papers/thomas/thomas.html

Tolva, J. (2005). Recontextualizing the collection: Virtual reconstruction, replacement, and repatriation. In D. Bearman & J. Trant (Eds.), *Museums and the Web 2005: Selected papers from an international conference* (pp. 173–182). Toronto, Canada: Archives & Museum Informatics. Available from http://www.archimuse.com/mw2005/papers/tolva/tolva.html

Trahanias, P. E., Burgard, W., Haehnel, D., Moors, M., Schulz, D., Baltzakis, H., et al. (2003). Interactive tele-presence in exhibitions through web-operated robots. In *Proceedings of the 11th international conference on advanced robotics (ICAR)* (pp. 1253–1258). New York: IEEE.

Trant, J. (Ed.). (1991). *Collections documentation guide*. Montreal, Canada: Canadian Centre for Architecture.

Trant, J. (1993). On speaking terms: Towards virtual integration of art information. *Knowledge Organization, 20*(1), 8–11.

Trant, J. (1995). The Museum Educational Site Licensing Project. *Spectra, 22*(3), 19–21.

Trant, J. (1996a). *Enabling educational use of museum digital materials: The Museum Educational Site Licensing (MESL) project*. Paper presented at the Electronic Imaging and Visual Arts Conference, Florence, Italy.

Trant, J. (1996b). The Museum Educational Site Licensing (MESL) project: An update. *Spectra, 23*(3), 32–34.

Trant, J. (1996c). New models for distributing digital content: The Museum Educational Site Licensing project. In P. B. Heydorn & B. Sandore (Eds.), *Digital imaging access and retrieval* (pp. 29–41). Urbana-Champaign: University of Illinois.

Trant, J. (1998). When all you've got is "the real thing": Museums and authenticity in the networked world. *Archives and Museum Informatics, 12*(2), 107–125.

Trant, J., Bearman, D., & Richmond, K. (2000). Collaborative cultural resource creation: The example of the Art Museum Image Consortium. In D. Bearman & J. Trant (Eds.), *Museums and the Web 2000: Selected papers from an international conference* (pp. 39–52). Pittsburgh, PA: Archives & Museum Informatics. Available from http://www.archimuse.com/mw2000/papers/trant/trant.html

Uniting and Strengthening America by Providing Appropriate Tools Required to Intercept and Obstruct Terrorism (USA PATRIOT ACT) Act of 2001, Pub. L. No. 107-56, 115 Stat. 272. (2001). Available from http://frwebgate.access.gpo.gov/cgibin/getdoc.cgi?dbname=107_cong_public_laws&docid=f:publ056.107

University of Minnesota. (n.d.). *Guide to writing university policy.* Available from http://www.fpd.finop.umn.edu/groups/ppd/documents/information/guide_to_writing.cfm

Unsworth, J. M. (2004, April 26). *Cyberinfrastructure for the humanities and social sciences.* Paper presented at the 2004 Research Libraries Group annual meeting, Washington, DC. Available from http://www3.isrl.uiuc.edu/~unsworth/Cyberinfrastructure.RLG.html

Vance, D. (1975). Museum computer network: Progress report. *Museologist, 135,* 3–10.

Vance, D. (1986). The Museum Computer Network in context. In R. B. Light, D. A. Roberts & J. D. Stewart (Eds.), *Museum documentation systems: Developments and applications* (pp. 37–47). London: Butterworths.

Vance, D., & Chenall, R. (1988). *Museum collections and today's computers.* Westport, CN: Greenwood Press.

Van Someren Cok, N. (1981). *All in order: Information systems for the arts, including the National Standards for Arts Information Exchange.* Washington, DC: National Assembly of State Arts Agencies.

Visual Resources Association. (2006). *Cataloguing Cultural Objects: A Guide to Describing Cultural Works and Their Images.* Chicago: American Library Association. Available from http://www.vraweb.org/ccoweb/

Vlahakis, V., Karigiannis, J., Tsotros, M., Gounaris, M., Almeida, L., Stricker, D., et al. (2001). Archeoguide: First results of an augmented reality, mobile computing system in cultural heritage sites. In *Proceedings of the 2001 conference on virtual reality, archaeology, and cultural heritage* (pp. 131–140). New York: ACM Press.

vom Lehn, D. (2002). *Exhibiting interaction: Conduct and participation in museums and galleries.* Unpublished doctoral dissertation, Kings College, University of London, London.

vom Lehn, D., & Heath, C. (2003). *Displacing the object: Mobile technologies and interpretive resources.* Paper presented at the Paper presented at the 2003 International Cultural Heritage Informatics Meeting, Paris, France. CD-ROM available from the Archives & Museum Informatics Website: http://www.archimuse.com

Vulpe, M. (1986). Collections Management Action Support System – CMASS: A statement of problem. *International Journal of Museum Management and Curatorship, 5,* 349–356.

Wakkary, R., & Evernden, D. (2005). Museum as ecology: A case study analysis of an ambient intelligent museum guide. In D. Bearman & J. Trant (Eds.), *Museums and the Web 2005: Selected papers from an international conference* (pp. 151–164). Toronto, Canada: Archives & Museum Informatics. Available from http://www.archimuse.com/mw2005/papers/wakkary/wakkary.html

Wallace, B., & Jones-Garmil, K. (1994, July/August). Museums and the Internet: A guide for the intrepid traveler. *Museum News, 73*(4), 32–36, 57–62.

Walsh, P. (1997). The web and the unassailable voice. *Archives and Museum Informatics, 11*(2), 77–85.

Walter, T. (1996). From museum to morgue? Electronic guides in the Roman Bath. *Tourism Management, 17*(4), 241–245.

Washburn, W. E. (1984, February). Collecting information, not objects. *Museum News, 62*(3), 5–15.

Weber, C. (2002). Designing, drafting and implementing new policies. In T. A. Lipinksi (Ed.), *Libraries, museums and archives: Legal issues and ethical challenges in the new information era* (pp. 303–319). Lanham, MD: Scarecrow Press.

Weil, S. E. (1999). From being about something to being for somebody: The ongoing transformation of the American museum. *Daedalus, 128*(3), 229–258.

Weil, S. E. (2002). From being about something to being for somebody: The ongoing transformation of the American museum. In S. E. Weil (Ed.), *Making museums matter* (pp. 28–52). Washington, DC: Smithsonian Institution Press.

Wheeler, T., & Gleason, T. (1994). *Digital photography and the ethics of photofiction: Four tests for assessing the reader's qualified expectation of reality.* Available from http://www.journalism.bsu.edu/classes/pfarmen/ethics

White, L. (2004). Museum informatics: Collections, people, access, and use. *Bulletin of the American Society for Information Science and Technology, 30*(5), 9–10. Available from http://www.asis.org/Bulletin/Jun-04/white.html

Williams, D. W. (1987). *A guide to museum computing.* Nashville, TN: American Association for State and Local History.

Wisser, K. M. (2004). Museum metadata in a collaborative environment: North Carolina ECHO and the North Carolina Museums Council Metadata Working Group. In D. Bearman & J. Trant (Eds.), *Museums and the Web 2004: Selected papers from an international conference* (n.p.). Toronto, Canada: Archives & Museum Informatics. Available from http://www.archimuse.com/mw2004/papers/wisser/wisser.html

Woodruff, A., Aoki, P. M., Grinter, R. E., Hurst, A., Szymanski, M. H., & Thornton, J. D. (2002). Eavesdropping on electronic guidebooks: Observing learning resources in shared listening environments. In D. Bearman & J. Trant (Eds.), *Museums and the Web 2002: Selected papers from an international conference* (pp. 21–30). Pittsburgh, PA: Archives & Museum Informatics. Available from http://www.archimuse.com/mw2002/papers/woodruff/woodruff.html

Woodruff, A., Szymanski, M. H., Aoki, P. M., & Hurst, A. (2001). The conversational role of electronic guidebooks. In G. D. Abowd, B. Brumitt & S. Shafer (Eds.), *Ubicomp 2001: Ubiquitous computing: Third international conference, Atlanta, Georgia, USA, September 30–October 2, 2001, proceedings* (pp. 187–208). New York: Springer.

Yakel, E. (2004). Digital assets for the next millennium. *OCLC Systems and Services: International Digital Library Perspectives, 20*(3), 102–105.

Zorich, D. M. (1997). Beyond bitslag: Integrating museum resources on the internet. In K. Jones-Garmil (Ed.), *The wired museum: Emerging technology and changing paradigms* (pp. 171–202). Washington, DC: American Association of Museums.

Zorich, D. M. (1999). *Introduction to managing digital assets: Options for cultural and educational organizations.* Los Angeles: Getty Trust Publications.

Zorich, D. M. (2000, January/February). Copyright in the digital age. *Museum News, 79*(1), 36–45, 66–67.

Zorich, D. M. (2003). *Developing intellectual property policies: A how-to guide for museums.* Ottawa, Canada: Candian Heritage Information Network. Available from http://www.chin.gc.ca/English/Intellectual_Property/Developing_Policies/index.html

Zweizig, D., & Dervin, B. (1977). Public library use, users, uses: Advances in knowledge of the characteristics and needs of the adult clientele of American public libraries. *Advances in Librarianship, 7*, 231–255.

Contributors

Maxwell L. Anderson became director and CEO of the Indianapolis Museum of Art in June 2006. Prior to that he served as a principal with AEA Consulting LLC of New York and London, specializing in strategic planning for the cultural sector, from 2004–2006, having previously served as a museum director since 1987 and a curator of Greek and Roman Art at the Metropolitan Museum of Art from 1981–87. Anderson received an AB from Dartmouth (1977) with highest distinction in Art History, and AM (1978) and PhD (1981) degrees from Harvard. Long committed to the use of new technology on art museums, he wrote the introduction to *The Wired Museum* (Washington, D.C., American Association of Museums, 1997).

Murtha Baca holds a PhD in art history and Italian language and literature from the University of California, Los Angeles (UCLA). She is head of the Getty Vocabulary Program and the Digital Resource Management Department at the Getty Research Institute in Los Angeles. Her publications include *Introduction to Art Image Access* (Getty Research Institute, 2002) and *Introduction to Metadata* (third revised edition forthcoming from Getty Research Institute publications), and she is a member of the Visual Resources Association editorial team that produced *Cataloging Cultural Objects: A Guide to Describing Cultural Works and Their Images* (ALA Editions, 2006). Murtha has taught workshops and seminars on metadata, visual resources cataloging, and thesaurus construction at museums, universities, and other organizations in North and South America and in Europe; she teaches a graduate seminar on metadata in the School of Information Studies at UCLA.

David Bearman is president of Archives & Museum Informatics. He consults on issues relating to electronic records and archives, integrating multi-format cultural information and museum information systems and was founding editor of the quarterly journal *Archives and Museum Informatics,* published by Kluwer Academic Publishers, in The Netherlands through 2000. Since 1991, he has organized and chaired the biennial International Cultural Heritage Informatics Meetings (ICHIM), and more recently the annual Museums and the Web Conferences (1997–) as well as directing numerous educational seminars and workshops on related topics. Bearman is the author of over 165 books and articles on museum and archives information management issues.

Matthew Chalmers is a reader in Computing Science at the University of Glasgow, and a co-investigator in the EPSRC Equator IRC. He has published widely at an international level, in topics including the use of philosophical hermeneutics to discuss and design computer systems, photorealistic computer graphics techniques, the nature of the museum visit experience, and fast techniques for non-linear multidimensional scaling. His industrial research experience started at Rank Xerox EuroPARC, working on the earliest pervasive or ubiquitous computing technology. He led the information access and visualization research group at UBS Ubilab, in Zurich, and was co-director of research at the Kelvin Institute, a government-funded technology transfer center. He held a visiting research fellowship at the University of Hokkaido before joining the University of Glasgow. He is on the editorial board for the *Information Visualization* book series from Springer Verlag, is an associate editor for the *Journal of Information Visualization,* and is an associate chair for the premier conference in human computer interaction, ACM CHI.

Erin Coburn is manager of Collections Information at the J. Paul Getty Museum. Formerly, she was data standards administrator in the same department where her work focused on the use of data standards and the creation and use of controlled vocabularies for creating and accessing information on the Getty Museum's collection, and providing metadata for the Museum's collection online. As manager of Collections Information, she oversees all aspects of data creation, data management, and data dissemination relating to the collection of the Getty Museum. Erin is a member of the board of directors of the Museum Computer Network (MCN), and of the advisory committee for the Visual Resources Association's project and publication *Cataloging Cultural Objects: A Guide to Describing Cultural Objects and Their Images* (Chicago: ALA Editions, 2006). Her publications include "Creating a User-Friendly Kiosk System for Museum Visitors: The Getty Art Access Project" (*VRA Bulletin,* vol. 27, no. 2, Summer 2000) and (with Murtha Baca) "Beyond the Gallery Walls: Tools and Methods for Leading End-Users to Collections Information" in *ASIST Bulletin* (June/July 2004).

Jim Devine is head of multimedia at the Hunterian Museum and Art Gallery, and honorary lecturer in Computing Science in the University of Glasgow, Scotland. He is a graduate of the University of Glasgow where he took joint honors in Archaeology and Classical Civilisation. For over ten years he has been combining his interests in cultural heritage with his interest in the potential for new technology to present e-learning opportunities, in an interactive way, to a global audience. In 1994, he initiated a collaborative program between the Hunterian and the University's Department of Computing Science, which has led to a wide variety of innovative multimedia projects focused around the Hunterian's collections, and those of other cultural heritage organizations. He has supervised numerous multimedia academic student projects from the Computing Science Honours undergraduate, MSc. IT, and Ph.D. levels, as part of an ongoing collaborative effort in developing new applications for leading edge technologies.

Basil Dewhurst was a key architect of both Australian Museums Online (AMOL) and the Collections Australia Network (CAN). He has a strong interest in metadata and interoperability standards particularly for the discovery of cultural and natural heritage resources. He represented the Australian museum sector on two Standards Australia committees and was a key Australian member of the Committee for the Interchange of Museum Information (CIMI). Following his work on AMOL and CAN he managed the Powerhouse Museum Image Services department which creates, manages, preserves, and provides access to the Museum's digital assets. He currently works at the National Library of Australia managing the development of People Australia, an online service providing access to information about people and organizations as well as associated physical and digital resources.

Maria Economou is assistant professor in Museology and New Technologies at the Department of Cultural Technology and Communication, University of the Aegean. She studied Archaeology at the Aristotle University of Thessaloniki, Greece, holds an MA in Museum Studies from the University of Leicester and a DPhil from the University of Oxford. She worked as assistant curator for IT at the Pitt Rivers Museum of the University of Oxford, as an independent consultant in the cultural sector, and as Lecturer at the Universities of Glasgow and Manchester in the UK. She specializes in the use of information and communication technology in the cultural sector on which she has written extensively. Her book *Museum: Warehouse or Live Organisation? Museological Issues and Questions* was published in Greek by Kritike in 2003. Other research interests include visitor studies in cultural organizations, the evaluation of exhibitions and communication in museums, and the use of ICT for presentation and interpretation.

Kirsten Ellenbogen is the director of Evaluation and Research in Learning at the Science Museum of Minnesota. She focuses her research on the role of museums in family life, designing exhibits to encourage science conversations, and the ways in which scientific visualization technology can be used to engage the public in exploring scientific data and understanding complex phenomenon. She began her work in museums as a demonstrator at the Detroit Science Center in 1987. Her award-winning exhibition development work has focused on inquiry experiences, multimedia interactives, and dual-purpose spaces appropriate for both school and family groups. She was the project director of the Center for Informal Learning & Schools at King's College London at its inception. She was also an affiliated researcher of the Museum Learning Collaborative. Most recently, she was a Senior Associate at the Institute for Learning Innovation where she was the principle investigator of an NSF-funded initiative designed to coalesce the last decade of research on learning in museums into frameworks for practitioners. Recent publications have included chapters in *Learning Conversations in Museums: Perspectives on Object-Centered Learning in Museums*; *Research in Science Education: Past, Present, and Future;* the *Encyclopedia of Education;* and *Cognition in a Digital World.* She received her PhD in Science Education from Vanderbilt University and her BA from the University of Chicago.

John H. Falk is founder and president of the Institute for Learning Innovation, an Annapolis, Maryland-based non-profit learning research and development organization and professor, Free-Choice Science Learning, Oregon State University. He is known internationally for his research on free-choice learning. Dr. Falk has authored over one hundred scholarly articles and chapters in the areas of learning, biology, and education, more than a dozen books as well as helped to create several nationally important out-of-school educational curricula. Some notable books include *The Museum Experience* (1992), *Learning from Museums: Visitor Experiences and the Making of Meaning* (2000), *Free-Choice Science Education: How People Learn Outside of School* (2001); *Lessons without Limit: How Free-Choice Learning Is Transforming Education* (2002) and *Thriving in the Knowledge Age: New Business Models for Museums and Other Cultural Institutions* (2006). Dr. Falk serves on the editorial boards of the journals *Curator, Journal of Museum Education,* and *Science Education,* and such national advisory boards as the Institute for Museum and Library Services' 21st Century Learner Initiative, National Postal Museum, National Association for Research in Science Teaching taskforce on Informal Learning, and National Ecological Observatory Network. Before founding and directing the Institute for Learning Innovation, Dr. Falk worked at the Smithsonian Institution for fourteen years where he held a number of senior positions including special assistant for Education to the Assistant Secretary for Science and Director, Smithsonian Office of Educational Research. Dr. Falk received a joint doctorate in Biology and Education from the University of California at Berkeley. He also earned MA and BA degrees in Zoology and a secondary teaching credential in Biology and Chemistry from the same institution.

Areti Galani is a lecturer in Museum and Heritage Studies in the International Centre for Cultural and Heritage Studies at the University of Newcastle, UK. She teaches and researches the use of digital media in museums and the heritage sector, and especially the design, study, and understanding of digital applications for the purpose of interpretation, learning, and exhibition design. She is interested in the ways heterogeneous media, digital and traditional, are used in the course of a visiting activity as well as the way social interaction may shape people's engagement with cultural content both in museum environments and beyond. Areti has been involved in research at the Universities of Leicester, Nottingham, and Glasgow (Equator IRC) and in curatorial projects in Greece and the UK. She has reviewed papers for conferences and journals such as the *Virtual Reality Journal* and the , Her work has been published in conferences such as Museums and the Web, Design of Interactive Systems (DIS) and Human Factors in Computing Systems (CHI).

Kate Haley Goldman is a senior research associate at the Institute for Learning Innovation. Her work concentrates on furthering theory and practice of the use of technology for personal learning. Kate's research reflects the Institute's focus on changing the world of education by understanding, facilitating, advocating, and communicating about free-choice learning across the life span, working with a variety of free-choice learning institutions such as museums, other cultural institutions, public television stations, libraries, scientific societies, and humanities councils as well as schools and universities. Kate has worked on dozens of learning-based research and evaluation projects including projects at Networks Financial Institute, Disney's Animal Kingdom, Mystic Seaport, the Astronomy Society of the Pacific, the Cleveland Museum of Art, and the CONNECT project, a European Union-based educational technology collaboration. She has written (with Lynn Dierking) a chapter on establishing a research agenda for virtual museum research and has published several other papers including ones on motivation in online museum visits and Web site evaluation methodologies. Her current research priorities include investigation into the long-term impact of museum visits, the impact of changing technology on personal learning, and the nature and context of learning in online environments.

Sally Hubbard is digital projects manager for the Digital Resource Management Department at the Getty Research Institute in Los Angeles, working on the digitization, persistence, and continued accessibility of virtual collections created by the Research Institute. She was previously new media coordinator at the UCLA Film and Television Archive, where she developed educational/historical CDs and a prototype online encyclopedia for the newly developed Moving Image Archiving Studies graduate degree program. She has a master's degree in African Studies from the University of California, Los Angeles and a bachelor's degree in Anthropology from the University of London, UK. She served as co-chair of the Preservation Domain of the InterPARES 2 research project, and was editor of *Introduction to Imaging* (Getty Research Institute, 2003).

Katherine Burton Jones is the assistant dean for Information Technology and Media Services at the Harvard Divinity School. She is the research advisor for the Masters in Liberal Arts in Museum Studies and teaches at the Harvard Extension School in the Museum Studies Program. She holds a graduate degree in Anthropology from Florida State University. She is a member of the Harvard University Technology Architecture Group and sits on several Harvard-wide technology working groups. She is a former president of the Board of Directors of the Museum Computer Network. She is a member of the Museum Committee of the International Tennis Hall of Fame and on the Advisory Committee of the Mildred Morse Allen Center, Massachusetts Audubon Society. As an assistant director of the Harvard Peabody Museum, Katherine developed the online exhibit entitled "Against the Winds: Traditions of American Indian Running" as well as other information products. For many years she was the head of the Office of Information Services and Technology at the Harvard Peabody Museum. She is the Editor of and a contributor to *The Wired Museum: Emerging Technology and Changing Paradigms,* a book available from the American Association of Museums. She has consulted in the museum and government communities since 1985 primarily in the areas of strategic technology planning, project management, and database development.

Paul F. Marty is assistant professor in the College of Information at Florida State University. He has a background in ancient history and computer science engineering, and his PhD is from the Graduate School of Library and Information Science at the University of Illinois at Urbana-Champaign. Before arriving at FSU, he was director of Information Technology at the University of Illinois' Spurlock Museum. Dr. Marty's research and teaching interests include museum informatics, computer-supported cooperative work, information behavior, and usability engineering. He specializes in the study of museums as sociotechnical systems, and is particularly interested in the social implications of introducing new technologies into the museum environment. His current research focuses on the evolution of sociotechnical systems and collaborative work practices in museums, the usability of museum Web sites, the evolving roles of information professionals in museums, and the digital museum in the life of the user.

Darren Peacock is a researcher and consultant in museum information management and technology. He has worked for art, history, and science museums in Australia, Europe, and the US. He is currently a PhD candidate at the University of South Australia, where he is researching organizational change and the World Wide Web within the collections sector as well as teaching in the Arts and Cultural Management Program. He was formerly director of Information and Communications Technology at the National Museum of Australia.

Richard Rinehart is a digital media artist and director of Digital Media at the UC Berkeley Art Museum/Pacific Film Archive. Richard has taught digital media studio and theory in the UC Berkeley Department of Art Practice and has also been visiting faculty at the San Francisco Art Institute, UC Santa Cruz, San Francisco State University, Sonoma State University, and JFK University. Richard sits on the Executive Committee of the UC Berkeley Center for New Media and on the Board of Directors for New Langton Arts in San Francisco, where he also curates net.art. Richard manages research projects in the area of digital culture, including the NEA-funded project "Archiving the Avant Garde," a national consortium of museums and artists distilling the essence of digital art in order to document and preserve it.

Kevin Sumption has worked for over ten years as both a science and social history curator at the Powerhouse Museum and Australian National Maritime Museum. Prior to taking up his current role as associate director of Exhibitions and Planning, he was the National Project Manager of Australian Museums Online (AMOL) and curator of Information Technology at the Powerhouse Museum. Kevin is also a lecturer in Design History and Theory at the University of Technology, Sydney and for many years has focused his research on the creation of computer-based education programs. He is also a frequent invited speaker at conferences in France, Japan, New Zealand, Taiwan, the UK, and the USA.

Jennifer Trant is a partner in Archives & Museum Informatics, where she consults on digital cultural strategy and collaboration and is co-chair of the international conferences Museums and the Web and ICHIM (International Cultural Heritage Informatics Meeting). Trant was the executive director of the Art Museum Image Consortium (AMICO) from 1997–2005. She was editor-in-chief of *Archives and Museum Informatics*, the cultural heritage informatics quarterly from Kluwer Academic Publishers from 1997–2000, and has served on the program committees of the Joint Digital Libraries (JDL) and Digital Libraries (DL) conferences, the Culture Program Committee of the International World Wide Web Conference, and the Board of the Media and Technology Committee of the American Association of Museums. A specialist in arts information management, Trant has worked with automated documentation systems in major Canadian museums, including the National Gallery of Canada and the Canadian Centre for Architecture. She has been actively involved in the definition of museum data standards, participating in numerous committees and regularly publishing articles and presenting papers about issues of access and intellectual integration of collections. Her current interests center on the use information technology and communications networks to improve access to cultural heritage information, and to integrate culture fully into digital libraries for research, learning, and enjoyment. In addition to her consulting practice, Trant is enrolled in the

PhD program of the Faculty of Information Studies at the University of Toronto, where she is researching the role of folksonomy in museums.

Layna White is head of Collections Information and Access at the San Francisco Museum of Modern Art. Her department is responsible for advancing a collections management system, creating and managing visual documentation related to works of art, managing issues related to intellectual property, and supporting pluralistic and changing needs for access to and use of digital content related to works of art. Prior to SFMOMA, Layna was collections information manager at the UCLA Grunwald Center for the Graphic Arts, where she integrated cataloguing and collections management needs with onsite and online public access. She is a member of the Museums and Online Archives Collaboration (MOAC), a group investigating cross-community standards, digitization tools and workflows, and use of digital content by different audiences.

Diane M. Zorich is a cultural heritage consultant who specializes in the delivery of cultural information over digital networks. Before establishing her consultancy, Ms. Zorich was data manager at the Association of Systematics Collections in Washington, D.C., and documentation manager at the Peabody Museum of Archaeology and Ethnology at Harvard University. She is the author of *Introduction to Managing Digital Assets: Options for Cultural and Educational Organizations* (1999, The J. Paul Getty Trust), *Developing Intellectual Property Policies: A "How-To" Guide for Museums* (2003, Canadian Heritage Information Network), and *A Survey of Digital Cultural Heritage Initiatives and Their Sustainability Concerns* (2003, Council on Library and Information Resources). Ms. Zorich has graduate degrees in anthropology and museum studies, and is based in Princeton, NJ.

Index